W9-CSE-966

Contemporary Optics

OPTICAL PHYSICS AND ENGINEERING

Series Editor: **William L. Wolfe**
Optical Sciences Center, University of Arizona, Tucson, Arizona

M. A. Bramson
Infrared Radiation: A Handbook for Applications

Sol Nudelman and S. S. Mitra, Editors
Optical Properties of Solids

S. S. Mitra and Sol Nudelman, Editors
Far-Infrared Properties of Solids

Lucien M. Biberman and Sol Nudelman, Editors
Photoelectronic Imaging Devices
Volume 1: Physical Processes and Methods of Analysis
Volume 2: Devices and Their Evaluation

A. M. Ratner
Spectral, Spatial, and Temporal Properties of Lasers

Lucien M. Biberman, Editor
Perception of Displayed Information

W. B. Allan
Fibre Optics: Theory and Practice

Albert Rose
Vision: Human and Electronic

J. M. Lloyd
Thermal Imaging Systems

Winston E. Kock
Engineering Applications of Lasers and Holography

Shashanka S. Mitra and Bernard Bendow, Editors
Optical Properties of Highly Transparent Solids

M. S. Sodha and A. K. Ghatak
Inhomogeneous Optical Waveguides

A. K. Ghatak and K. Thyagarajan
Contemporary Optics

Kenneth Smith and R. M. Thomson
Computer Modeling of Gas Lasers

A Continuation Order Plan is available for this series. A continuation order will bring delivery of each new volume immediately upon publication. Volumes are billed only upon actual shipment. For further information please contact the publisher.

Contemporary Optics

A. K. Ghatak
and K. Thyagarajan

Indian Institute of Technology at New Delhi

PLENUM PRESS • NEW YORK AND LONDON

Library of Congress Cataloging in Publication Data

Ghatak, Ajoy K 1939-
Contemporary optics.

(Optical physics and engineering)
Bibliography: p.
Includes index.
1. Optics. 2. Thyagarajan, K., joint author. II. Title.
QC372.G48 535 77-21571
ISBN 0-306-31029-5

A Division of Plenum Publishing Corporation
227 West 17th Street, New York, N.Y. 10011

Printed in the United States of America

Preface

With the advent of lasers, numerous applications of it such as optical information processing, holography, and optical communication have evolved. These applications have made the study of optics essential for scientists and engineers. The present volume, intended for senior undergraduate and first-year graduate students, introduces basic concepts necessary for an understanding of many of these applications. The book has grown out of lectures given at the Master's level to students of applied optics at the Indian Institute of Technology, New Delhi.

Chapters 1–3 deal with geometrical optics, where we develop the theory behind the tracing of rays and calculation of aberrations. The formulas for aberrations are derived from first principles. We use the method involving Luneburg's treatment starting from Hamilton's equations since we believe that this method is easy to understand.

Chapters 4–8 discuss the more important aspects of contemporary physical optics, namely, diffraction, coherence, Fourier optics, and holography. The basis for discussion is the scalar wave equation. A number of applications of spatial frequency filtering and holography are also discussed.

With the availability of high-power laser beams, a large number of nonlinear optical phenomena have been studied. Of the various nonlinear phenomena, the self-focusing (or defocusing) of light beams due to the nonlinear dependence of the dielectric constant on intensity has received considerable attention. In Chapter 9 we discuss in detail the steady-state self-focusing of light beams.

In Chapter 10, we discuss, in reasonable detail, graded-index optical waveguides. This subject is of particular interest because of the use of laser beams in optical communication systems. Although the emphasis is on waveguides characterized by a parabolic variation of the refractive index, the analysis brings out many salient features of waveguides.

Evanescent waves, which are of great importance in electromagnetic theory, are discussed in Chapter 11; particular emphasis is given to the Goos–Hänchen shift.

There are many other interesting topics which could have been included, but then the size of the book would have been unmanageable.

The solved and unsolved problems form an important part of the book. Some of the phenomena are put in the form of problems (rather than treated in separate sections) to enable the reader to skip them (if desired) without any loss of continuity.

We have not made an attempt to refer to original works; instead, we have tried to refer to recent review articles, monographs, and research papers, which are usually available in most libraries and will enable the reader to learn more of this subject.

We would like to show our gratitude to Professor M. S. Sodha for his interest and encouragement in this endeavor and for his many valuable suggestions. We would also like to thank Dr. Kehar Singh, Dr. I. C. Goyal, Dr. Arun Kumar, Mr. Anurag Sharma, Ms. E. Khular, Ms. Aruna Rohra, Dr. Anjana Gupta, and Mr. B. D. Gupta for many stimulating discussions. Thanks are also due to Ms. D. Radhika, Ms. K. K. Shankari, Mr. T. N. Gupta, and Mr. N. S. Gupta for their help in the preparation of the manuscript.

We are grateful to Professor E. Wolf, Dr. H. Kogelnik, Dr. R. S. Sirohi, Mr. K. K. Gupta, and Mr. Anurag Sharma for providing some of the photographs appearing in the book and to various authors and publishers for their kind permission to use material from their publications. Finally we would like to thank Professor N. M. Swani, Director, I. I. T. Delhi, for his interest and support of this work.

New Delhi

Ajoy Ghatak
K. Thyagarajan

Contents

1

Paraxial Ray Optics

1.1. Introduction

Light is an electromagnetic wave and since electromagnetic waves are completely described by Maxwell's equations, it seems possible, in principle, to obtain all the laws of propagation of light as solutions of Maxwell's equations. This problem is, in general, difficult to solve and rigorous solutions may be obtained only for some simple systems.* Hence one is led to consider approximations that might give easily understood solutions and also describe the phenomenon well. One such approximation makes use of the fact that when the wavelength of light is extremely small compared to the dimensions of the system with which it interacts, one can, to a good approximation, neglect the finiteness of the wavelength. Indeed, the zero-wavelength approximation of wave optics is known as geometrical optics.

Geometrical optics employs the concept of rays, which are defined as the direction of propagation of energy in the limit $\lambda \to 0$. As will be shown in Chapter 4, the spreading of a light beam due to diffraction is entirely due to the finiteness of the wavelength. But when the wavelength is assumed to go to zero these diffraction effects also go to zero, so that one can form the infinitesimally thin beam of light that defines a ray.

In Section 1.2 we will introduce Fermat's principle as an extremum principle from which one can trace the rays in a general medium. This principle is the optical analog of Hamilton's variational principle in classical mechanics. We will give examples in which the optical path of the ray between two points is a maximum, a minimum, or stationary.

From Fermat's principle, we can develop two parallel approaches: the Lagrangian approach (Section 1.3) and the Hamiltonian approach

* To give an example, the reflection and refraction of plane electromagnetic waves by a plane dielectric surface is not a difficult problem to solve; on the other hand, the reflection of plane waves by a curved surface is fairly difficult. (The generalized laws of reflection by a curved cylindrical surface have been discussed by Snyder and Mitchell, 1974.)

(Section 1.4). The latter will be used in Chapter 2 for the discussion of aberrations. The Lagrangian approach will be shown to yield the ray equation. This equation will be solved to obtain the path of rays in inhomogeneous media, i.e., media characterized by a spatially dependent variation of refractive index.

In Section 1.6 we will solve the scalar wave equation in the small-wavelength approximation to obtain the eikonal equation; this approximation, called the eikonal approximation, is similar to the WKB approximation in quantum mechanics. We then transform the eikonal equation into the ray equation. A derivation of Fermat's principle from the eikonal equation will also be given. In Section 1.7, we will discuss the transition from geometrical optics to wave optics in a manner analogous to the transition from classical mechanics to quantum mechanics.

1.2. Fermat's Principle

Similar to Hamilton's principle of least action in classical mechanics [see, e.g., Goldstein (1950), Chapter 2] we have, in optics, Fermat's principle, from which all of geometrical optics can be derived. As in classical mechanics, one can derive from the variational principle two related approaches, one involving the Lagrangian and the other the Hamiltonian.

Before we introduce Fermat's principle, it is necessary to introduce the concept of optical pathlength. Given any two points P and Q and a curve C connecting them, one can define the geometrical pathlength between the two points as the length of the curve lying between the two points, $\int_P^Q ds$, where the integral is performed from P to Q along the curve C and ds represents an infinitesimal arclength. The optical pathlength is defined as

$$\text{optical pathlength} = \int_{P \underset{C}{\to} Q} n(x, y, z)\, ds \tag{1.2-1}$$

where $n(x, y, z)$ is the refractive index function and the integral is again performed along the curve C. In the simple case of a homogeneous medium, the optical pathlength is just the geometrical pathlength multiplied by the refractive index of the medium. In the general case, the optical pathlength divided by c (the velocity of light in free space) represents the time that would be required for light to travel from P to Q along the given curve.

We may now state Fermat's principle according to which, out of the many paths that can connect two given points P and Q, the light ray would follow that path for which the optical pathlength between the two points is an extremum, i.e.,

$$\delta \int_P^Q n(x, y, z)\, ds = 0 \tag{1.2-2}$$

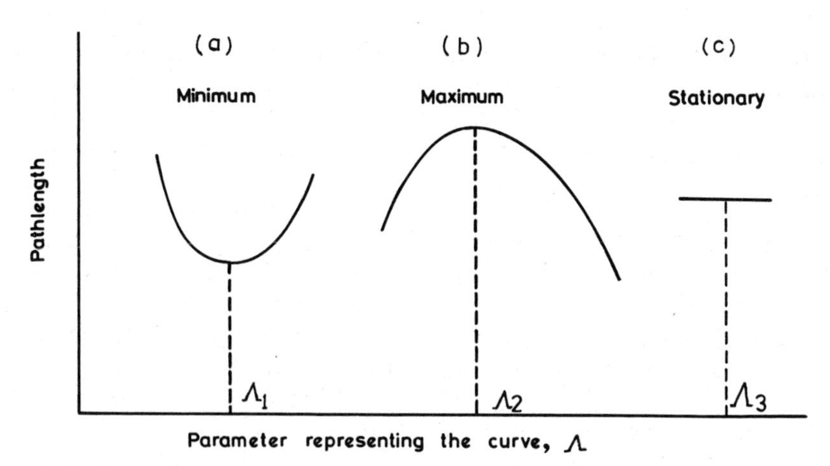

Fig. 1.1. The variation of optical pathlength between two points with Λ, which represents a parameter specifying the path. The three curves represent the cases when the actual ray is determined by a minimum, a maximum, and a stationary value of the optical pathlength between the two points. In (a) the actual ray path will correspond to $\Lambda = \Lambda_1$; in (b) the actual ray path will correspond to $\Lambda = \Lambda_2$; whereas in (c) all ray paths around the value $\Lambda = \Lambda_3$ are permissible.

where the δ variation of the integral means that it is a variation of the path of the integral such that the endpoints P and Q are fixed. It should be noted that Fermat's principle requires the optical pathlength to be an *extremum*, which may be a minimum (this is the case one most often encounters), a maximum, or stationary. It is at once clear that in a homogeneous medium, the rays are straight lines, since the shortest optical pathlength between two points is along a straight line.*

In Fig. 1.1 we have plotted the variation of the optical pathlength versus a parameter specifying a particular path. The first curve represents the case when the extremum is a minimum, the second when the extremum is a maximum, and the third when the extremum is a point of inflection. The comparison should be made only in the immediate neighborhood of the ray. The meaning of this is shown by the following example: Consider the simple case of finding the path of a ray from a point A to a point B when both of them lie on the same side of a mirror M (see Fig. 1.2a). It can be seen that the ray can go directly from A to B without suffering any reflection or it can go along the path APB after suffering a single reflection from the mirror. If Fermat's principle had asked for, say, an absolute minimum, then the path APB would be prohibited; but that is not the actual case. The path APB is also a minimum in the neighborhood involving paths like

* An excellent discussion on extremum principle has been given by Feynman *et al.* (1965).

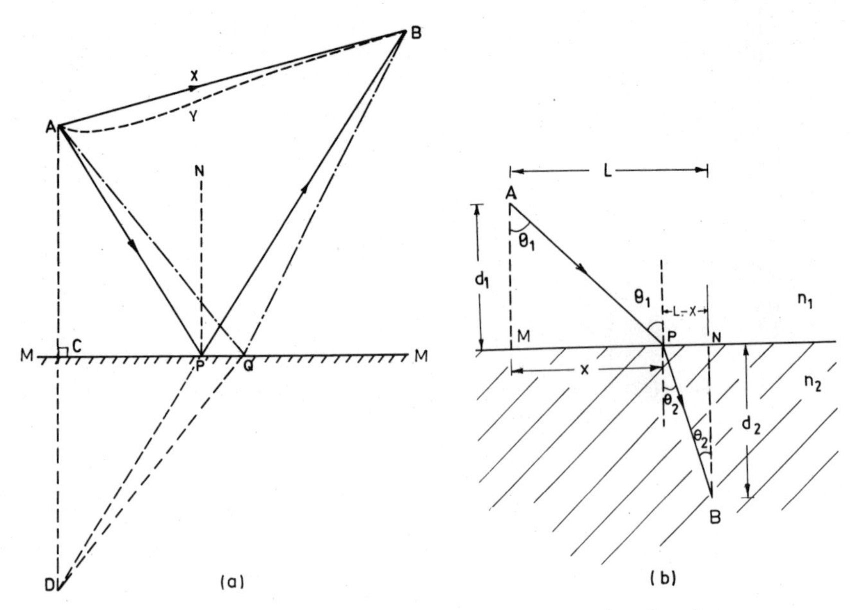

Fig. 1.2. (a) Reflection and (b) refraction of rays at a plane interface.

AQB. The phrase "immediate neighborhood of a path" would mean those paths that lie near the path under consideration and are similar to it. For example, the path *AQB* lies near *APB* and is similar to it; both paths suffer one reflection at the mirror. Thus Fermat's principle requires an extremum in the immediate neighborhood of the actual path, and in general, there may be more than one ray path connecting two points.

We give here representative examples of how the actual path may be a minimum, a maximum, or a point of inflection.

Problem 1.1. Using Fermat's principle derive the laws of reflection and refraction.

Solution. Let the ray path be *AQB*, where *Q* is an arbitrary point on the reflecting surface (see Fig. 1.2a). Drop a perpendicular from the point *A* on the surface of the mirror and extend this to the point *D* such that $AC = CD$; the point *C* is the foot of the perpendicular on the mirror surface. Clearly $AQ = QD$, and therefore the optical pathlength $AQ + QB$ is equal to $DQ + QB$. For $DQ + QB$ to be a minimum, the point *Q* must lie on the line joining the points *D* and *B*. Thus the ray path must be *APB*, where *P* is the point of intersection of the line *BD* with the plane mirror. Obviously, the point *P* will lie in the plane containing the points *A*, *C*, *D*, and *B*, and therefore the incident ray *AP*, the reflected ray *PB*, and the normal *PN* will lie in the same plane. Further, since $AP = PD$ and *DPB* is a straight line, the angle of incidence will be equal to the angle of reflection.

To obtain the laws of refraction, let MN be the surface separating two media of refractive indices n_1 and n_2 (see Fig. 1.2b). Let the ray start from A, intersect the surface at P, and proceed to B along PB. AP and PB must be straight lines since they are in homogeneous media. Let M and N represent the feet of the perpendiculars from A and B to the surface and let $MN = L$. Let x be the distance MP. We have to find the point P such that the optical pathlength APB is a minimum. The optical pathlength of APB is

$$\Lambda = n_1(d_1^2 + x^2)^{1/2} + n_2[d_2^2 + (L - x)^2]^{1/2} \tag{1.2-3}$$

For Λ to be an extremum with respect to x, we must have

$$\frac{d\Lambda}{dx} = n_1 \frac{x}{(d_1^2 + x^2)^{1/2}} - n_2 \frac{(L - x)}{[d_2^2 + (L - x)^2]^{1/2}} = 0 \tag{1.2-4}$$

If θ_1 and θ_2 are the angles defined as in Fig. 1.2b, then

$$\sin\theta_1 = \frac{x}{(d_1^2 + x^2)^{1/2}}, \qquad \sin\theta_2 = \frac{(L - x)}{[d_2^2 + (L - x)^2]^{1/2}}$$

Thus, for the optical path to be an extremum, we must have

$$n_1 \sin\theta_1 = n_2 \sin\theta_2 \tag{1.2-5}$$

which is Snell's law.

Problem 1.2. Generalize the above results when the surface of reflection or refraction is not plane but is given by an equation of the form

$$f(x, y, z) = 0 \tag{1.2-6}$$

Reference: Pegis (1961).

Solution. Let the surface given by Eq. (1.2-6) separate two media of refractive indices n_1 and n_2 as shown in Fig. 1.3. First consider the phenomenon of refraction. Let a ray starting from $A(x_1, y_1, z_1)$ reach the point $B(x_2, y_2, z_2)$ after getting refracted from the surface at a point $P(x, y, z)$. Let $\hat{\mathbf{u}}$ be the unit normal at P to the surface. Let $\hat{\mathbf{s}}$ and $\hat{\mathbf{s}}'$ be the unit vectors along AP and PB, respectively. We have to find a relation between $\hat{\mathbf{s}}$, $\hat{\mathbf{s}}'$, and $\hat{\mathbf{u}}$. The optical pathlength Λ between A and B along APB is

$$\Lambda(x, y, z) = n_1 d_1 + n_2 d_2 \tag{1.2-7}$$

where

$$d_1 = [(x - x_1)^2 + (y - y_1)^2 + (z - z_1)^2]^{1/2}$$

$$d_2 = [(x_2 - x)^2 + (y_2 - y)^2 + (z_2 - z)^2]^{1/2}$$

If the point $P(x, y, z)$ changes to $P'(x + \delta x, y + \delta y, z + \delta z)$ then the change in Λ

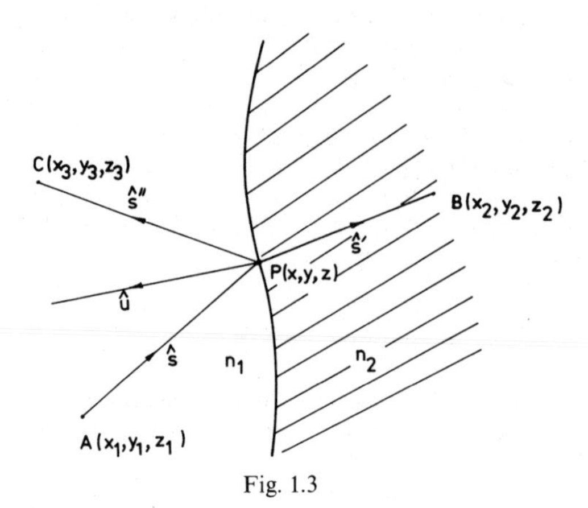

Fig. 1.3

is given by

$$\delta\Lambda = (n_1\alpha_1 - n_2\alpha_2)\,\delta x + (n_1\beta_1 - n_2\beta_2)\,\delta y + (n_1\gamma_1 - n_2\gamma_2)\,\delta z = 0 \qquad (1.2\text{-}8)$$

where $\alpha_1\,[=(x - x_1)/d_1]$, β_1, γ_1 and α_2, β_2, γ_2 represent the direction cosines of AP and PB, respectively, i.e., they are the x, y, and z components of $\hat{\mathbf{s}}$ and $\hat{\mathbf{s}}'$. Further, in Eq. (1.2-8), δx, δy, and δz cannot be varied arbitrarily because the changes δx, δy, and δz have to be such that the point P still lies on the surface given by Eq. (1.2-6); thus we must have

$$\frac{\partial f}{\partial x}\,\delta x + \frac{\partial f}{\partial y}\,\delta y + \frac{\partial f}{\partial z}\,\delta z = 0 \qquad (1.2\text{-}9)$$

If we substitute for δz from Eq. (1.2-9) into Eq. (1.2-8), we obtain an equation in which δx and δy can be varied arbitrarily. As such, the coefficients of δx and δy must be set equal to zero, which would lead to

$$\frac{n_1\alpha_1 - n_2\alpha_2}{\partial f/\partial x} = \frac{n_1\beta_1 - n_2\beta_2}{\partial f/\partial y} = \frac{n_1\gamma_1 - n_2\gamma_2}{\partial f/\partial z} = K \qquad (1.2\text{-}10)$$

where K is a constant. This yields

$$n_1\alpha_1 - n_2\alpha_2 = K\,\partial f/\partial x \qquad (1.2\text{-}11)$$

etc. Observing that the vector with components $\partial f/\partial x$, $\partial f/\partial y$, and $\partial f/\partial z$ represents*

* The normal to a surface $f(x,y,z) = \text{const}$ is given by ∇f, and hence the components are $(\partial f/\partial x, \partial f/\partial y, \partial f/\partial z)$.

the direction of the normal to the surface ($\hat{\mathbf{u}}$), we can write Eq. (1.2-11) as

$$n_1 \hat{\mathbf{s}} - n_2 \hat{\mathbf{s}}' = K_1 \hat{\mathbf{u}} \tag{1.2-12}$$

where K_1 is K times the magnitude of the vector with components ($\partial f/\partial x, \partial f/\partial y, \partial f/\partial z$). Equation (1.2-12) implies that $\hat{\mathbf{s}}$, $\hat{\mathbf{s}}'$, and $\hat{\mathbf{u}}$ are coplanar, i.e., the incident ray, the refracted ray, and the normal to the surface lie in the same plane. Cross-multiplying Eq. (1.2-12) with $\hat{\mathbf{u}}$ we get

$$n_1 \sin i_1 = n_2 \sin i_2 \tag{1.2-13}$$

where i_1 is the angle between $\hat{\mathbf{u}}$ and $\hat{\mathbf{s}}$ and i_2 is the angle between $\hat{\mathbf{u}}$ and $\hat{\mathbf{s}}'$. Equation (1.2-13) is Snell's law. In a similar manner, one can obtain the laws of reflection.

Problem 1.3. Consider a concave mirror. Let A and B be two points equidistant from the axis. The line AB passes through the center of curvature and is normal to the axis (see Fig. 1.4a). Show that the ray from A that reaches B after one reflection has a larger optical pathlength than any other path in the immediate neighborhood. This is an example where the ray path corresponds to the optical pathlength being a maximum.

Solution. Let P represent the vertex; clearly the ray will follow the path APB. If Q represents any other point on the mirror, then for a path like AQB the angle of incidence and the angle of reflection cannot be equal and hence AQB cannot represent a ray. We will now show that the length of AQB is indeed less than the length of APB. Construct an ellipse E passing through P with foci A and B. The ellipse has to lie *outside* the circle, since the radius of curvature at P is greater than PC. Produce AQ to R on the ellipse and join R to B, intersecting the circle M at S. It is clear that $QB < QR + RB$. Hence

$$AQ + QB < AQ + QR + RB = AR + RB \tag{1.2-14}$$

However, for any point P lying on an ellipse with foci A and B, $AP + PB = AR + RB$. Hence it follows that $AP + PB > AQ + QB$, i.e., the path of an actual ray is longer than a neighboring path.

Problem 1.4. Consider an elliptical reflector that reflects from the inner surface (see Fig. 1.4b). Let A and B represent the two foci. Show that a ray that travels from A to B through a single reflection has a stationary value of the optical pathlength with respect to variations in the point of reflection.

Solution. One can show that any ray starting from one of the foci A (say) after getting reflected from any point on the mirror must pass through B, with the condition that the angles of incidence and reflection be the same. Also the ellipse satisfies the property that the length of the path APB for any point P lying on the elliptical surface is the same. It follows then that the rays starting from A and passing through B after one reflection have the same pathlength.

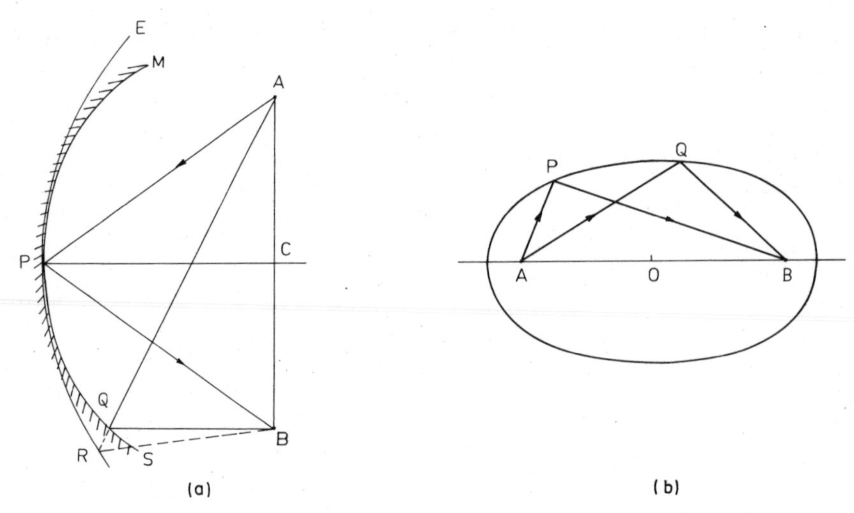

Fig. 1.4. (a) M represents a concave mirror and C its center of curvature. E is an ellipse whose foci are A and B. (b) An elliptical reflector with foci at A and B.

The above examples show that rays may proceed along paths whose optical pathlengths may be a maximum or stationary rather than a minimum.

1.3. Lagrangian Formulation

According to Hamilton's principle in classical mechanics, the trajectory of a particle between times t_1 and t_2 is such that

$$\delta \int_{t_1}^{t_2} \mathscr{L}\, dt = 0 \tag{1.3-1}$$

where $\mathscr{L}$ is called the Lagrangian and the integration is over time. In contrast, Fermat's principle [see Eq. (1.2-2)] has an integration over the space variable. The analogy can be made more explicit if one observes that the infinitesimal arclength ds can be written as

$$ds = [(dx)^2 + (dy)^2 + (dz)^2]^{1/2} = dz[1 + \dot{x}^2 + \dot{y}^2]^{1/2} \tag{1.3-2}$$

where dots represent differentiation with respect to z. Thus Eq. (1.2-2) can be written as

$$\delta \int_P^Q n(x, y, z)\,(1 + \dot{x}^2 + \dot{y}^2)^{1/2}\, dz = 0 \tag{1.3-3}$$

Comparing this with Eq. (1.3-1), one can define a corresponding optical Lagrangian as

$$L(x, y, \dot{x}, \dot{y}, z) = n(x, y, z)\,(1 + \dot{x}^2 + \dot{y}^2)^{1/2} \tag{1.3-4}$$

Here z may be assumed to play the same role as time in Lagrangian mechanics. The corresponding Lagrangian equations would be given by

$$\frac{d}{dz}\left(\frac{\partial L}{\partial \dot{x}}\right) = \frac{\partial L}{\partial x}, \qquad \frac{d}{dz}\left(\frac{\partial L}{\partial \dot{y}}\right) = \frac{\partial L}{\partial y} \tag{1.3-5}$$

(For the derivation of the Lagrangian equations from Hamilton's principle see Goldstein, 1950, Chapter 2.) The z direction is normally chosen in the direction along which the rays are approximately propagating. This direction in most cases coincides with the symmetry axis of the system. For example, for an optical system consisting of a system of coaxial lenses, the axis of the system is chosen as the z axis. Equations (1.3-4) and (1.3-5) form the fundamental equations of the Lagrangian formulation. If we substitute for L from Eq. (1.3-4) into (1.3-5), we find

$$\frac{d}{dz}\left[\frac{n\dot{x}}{(1 + \dot{x}^2 + \dot{y}^2)^{1/2}}\right] = (1 + \dot{x}^2 + \dot{y}^2)^{1/2}\,\frac{\partial n}{\partial x} \tag{1.3-6}$$

But from Eq. (1.3-2), we have

$$\frac{1}{(1 + \dot{x}^2 + \dot{y}^2)^{1/2}}\,\frac{d}{dz} = \frac{d}{ds}$$

Hence Eq. (1.3-6) reduces to

$$\frac{d}{ds}\left(n\,\frac{dx}{ds}\right) = \frac{\partial n}{\partial x} \tag{1.3-7}$$

Similarly the y and z components can be obtained:

$$\frac{d}{ds}\left(n\,\frac{dy}{ds}\right) = \frac{\partial n}{\partial y}, \qquad \frac{d}{ds}\left(n\,\frac{dz}{ds}\right) = \frac{\partial n}{\partial z} \tag{1.3-8}$$

Equations (1.3-7) and (1.3-8) can be combined into the following vector equation:

$$\frac{d}{ds}\left(n\,\frac{d\mathbf{r}}{ds}\right) = \nabla n \tag{1.3-9}$$

which is known as the *ray equation*, where $\mathbf{r}$ represents the position vector of any point on the ray.

Equation (1.3-9) is difficult to solve in most cases. However, if we restrict ourselves to rays that make small angles with the z axis, then we may write $ds \simeq dz$, and the ray equation would become

$$\frac{d}{dz}\left(n\,\frac{d\mathbf{r}}{dz}\right) = \nabla n \tag{1.3-10}$$

which is known as the paraxial ray equation.

Problem 1.5. Show that the z component of Eq. (1.3-9) can be obtained from the x and y components.

Solution. Consider the z component of the left-hand side of Eq. (1.3-9):

$$\frac{d}{ds}\left(n\,\frac{dz}{ds}\right) = \frac{dn}{ds}\frac{dz}{ds} + n\,\frac{d^2z}{ds^2} \tag{1.3-11}$$

From Eq. (1.3-2), we may write

$$\left(\frac{dz}{ds}\right)^2 = 1 - \left(\frac{dx}{ds}\right)^2 - \left(\frac{dy}{ds}\right)^2 \tag{1.3-12}$$

or

$$\begin{aligned} n\frac{dz}{ds}\frac{d^2z}{ds^2} &= -\frac{dx}{ds}\,n\,\frac{d^2x}{ds^2} - \frac{dy}{ds}\,n\,\frac{d^2y}{ds^2} \\ &= \frac{dn}{ds}\left[\left(\frac{dx}{ds}\right)^2 + \left(\frac{dy}{ds}\right)^2\right] - \frac{\partial n}{\partial x}\frac{dx}{ds} - \frac{\partial n}{\partial y}\frac{dy}{ds} \end{aligned} \tag{1.3-13}$$

where we have used the x and y components of Eq. (1.3-9). Substituting for $n\,d^2z/ds^2$ from Eq. (1.3-13) into Eq. (1.3-11) and using Eq. (1.3-12), one obtains the z component of the ray equation, where one must use

$$\frac{dn}{ds} = \frac{\partial n}{\partial x}\frac{dx}{ds} + \frac{\partial n}{\partial y}\frac{dy}{ds} + \frac{\partial n}{\partial z}\frac{dz}{ds} \tag{1.3-14}$$

Problem 1.6. Show, by solving the ray equation, that the rays in a homogeneous medium are straight lines.

Solution. In a homogeneous medium, the refractive index is constant and hence $\nabla n = 0$. Thus the ray equation becomes $d^2\mathbf{r}/ds^2 = 0$, the solution of which is simply $\mathbf{r} = \mathbf{a}s + \mathbf{b}$, which represents the equation of a straight line.

Problem 1.7. A Selfoc fiber* is characterized by the refractive-index variation

$$n^2(x, y, z) = n_0^2[1 - \alpha^2(x^2 + y^2)] \tag{1.3-15}$$

where n_0 and α are constants. Such a fiber is used in optical communications (see

---

* See Section 10.1.

Chapter 10). Find the general path of a paraxial ray in such a medium. Further, consider two cases, one in which the ray is confined to a plane and the other in which the ray propagates as a helix so that it is at a constant distance from the axis.

Solution. Since n is independent of z, the x component of Eq. (1.3-10) can be written as

$$\frac{d^2x}{dz^2} = \frac{1}{n}\frac{\partial n}{\partial x} = \frac{1}{2n^2}\frac{\partial n^2}{\partial x} \simeq -\alpha^2 x \tag{1.3-16}$$

where we have assumed that the factor $[1 - \alpha^2(x^2 + y^2)]$ can be replaced by unity, which is valid for paraxial rays. The solution of Eq. (1.3-16) is given by

$$x(z) = A\cos(\alpha z) + B\sin(\alpha z) \tag{1.3-17}$$

Similarly,

$$y(z) = C\cos(\alpha z) + D\sin(\alpha z) \tag{1.3-18}$$

where A, B, C, and D are constants to be determined by the initial conditions on the ray. Equations (1.3-17) and (1.3-18) determine the path of paraxial rays through the medium. These are sinusoidal with a period $2\pi/\alpha$, which is a constant and independent of the initial conditions on the ray (cf. next problem).

Since the medium is rotationally symmetric, we can always choose, without loss of generality, the point of incidence of the ray to be on the x axis. Thus $y(z = 0) = 0$ (where $z = 0$ has been chosen as the initial plane), which gives $C = 0$. For a ray incident at the point $(x_0, 0, 0)$ and subtending angles α_0, β_0, and γ_0 with the x, y, and z axes, respectively,

$$A = x_0, \qquad B = \cos\alpha_0/(\alpha\cos\gamma_0), \qquad D = \cos\beta_0/(\alpha\cos\gamma_0) \tag{1.3-19}$$

When $\beta_0 = \pi/2$, the ray is confined to the x-z plane (see Fig. 1.5a).

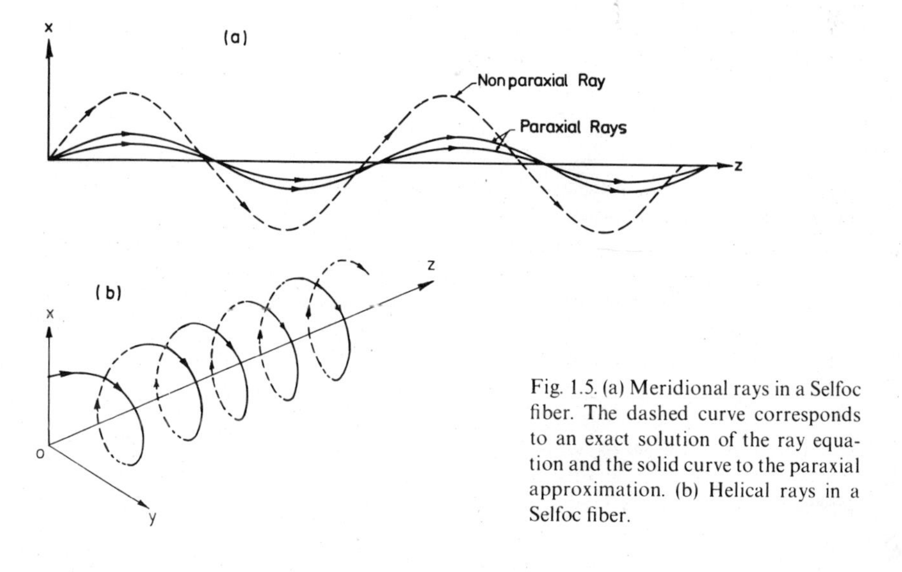

Fig. 1.5. (a) Meridional rays in a Selfoc fiber. The dashed curve corresponds to an exact solution of the ray equation and the solid curve to the paraxial approximation. (b) Helical rays in a Selfoc fiber.

Since a helical ray is always at a constant distance from the z axis (see Fig. 1.5b), $x_2(z) + y^2(z)$ must be a constant. For this to happen, we must have $B = 0$ (i.e., $\alpha_0 = \pi/2$) and $A = D$, i.e.,

$$x_0 = \cos\beta_0/(\alpha\cos\gamma_0) = \pm(1/\alpha)\tan\gamma_0 \tag{1.3-20}$$

where we have used the relation $\cos^2\alpha_0 + \cos^2\beta_0 + \cos^2\gamma_0 = 1$.

Problem 1.8. Show that if the refractive index is independent of z, then $x(z)$ rigorously satisfies the equation

$$\frac{d^2x}{dz^2} = \frac{1}{2C^2}\frac{\partial n^2}{\partial x} \tag{1.3-21}$$

where C is a constant. A similar equation can be written for $y(z)$.

Solution. Since the refractive index is independent of z, the z component of Eq. (1.3-9) becomes

$$\frac{d}{ds}\left(n\frac{dz}{ds}\right) = 0$$

Thus, $n\,dz/ds$ must be a constant, or we may write

$$n\,dz = C\,ds = C(1 + \dot{x}^2 + \dot{y}^2)^{1/2}\,dz \tag{1.3-22}$$

where the constant C is given by

$$C = n(x_0, y_0)\cos\gamma_0 \tag{1.3-23}$$

γ_0 being the angle that the ray makes with the z axis at the point (x_0, y_0). Equation (1.3-22) gives us $n/(1 + \dot{x}^2 + \dot{y}^2)^{1/2} = C$; if we use this in Eq. (1.3-6), we get Eq. (1.3-21).

Problem 1.9. Using Eq. (1.3-21), show that the path of a meridional ray in a Selfoc fiber [see Eq. (1.3-15)] is given by

$$x(z) = \frac{\sin\gamma_0}{\alpha}\sin\left(\frac{\alpha z}{\cos\gamma_0}\right) \tag{1.3-24}$$

where we have assumed the ray to be injected on the axis at $z = 0$. Notice that the period of oscillation of the ray depends on the angle of injection (see the dashed curve in Fig. 1.5a). For paraxial rays, $\cos\gamma_0 \simeq 1$.

Problem 1.10. Show that for a meridional ray [described by Eq. (1.3-24)] the time taken for a ray to travel a distance L is given by

$$T(\gamma_0) = \frac{n_0 L}{c}\left[1 + \frac{15}{192}\gamma_0^4 + \cdots\right] \tag{1.3-25}$$

Notice that for paraxial rays $\gamma_0 \simeq 0$, and the time taken is independent of γ_0. This is the reason for very small pulse dispersions in a Selfoc fiber (see Chapter 10). *Reference:* Bouillie *et al.* (1974).

Problem 1.11. Consider a medium whose refractive-index variation is given by

$$n(x) = n_0 \operatorname{sech}(\alpha x) \tag{1.3-26}$$

Solve Eq. (1.3-21) and show that the period of oscillation of the ray is independent of the initial conditions.

Solution. Multiplying Eq. (1.3-21) by $2(dx/dz)\,dz$ and integrating, we obtain

$$\left(\frac{dx}{dz}\right)^2 = \frac{n^2}{n^2(x_0)\cos^2\gamma_0} - 1 \tag{1.3-27}$$

where we have assumed that at $z = 0$, $x = x_0$, $y = 0$, and $dx/dz = \tan\gamma_0$. Using Eq. (1.3-26), Eq. (1.3-27) can be written in the form

$$\frac{1}{\alpha}\int\frac{d\Phi}{(1-\Phi^2)^{1/2}}\left(= \frac{1}{\alpha}\sin^{-1}\Phi\right) = \int dz = z + C \tag{1.3-28}$$

where

$$\Phi = \frac{1}{(A^2-1)^{1/2}}\sinh(\alpha x), \qquad A = \frac{n_0}{n(x_0)\cos\gamma_0} \tag{1.3-29}$$

and C is a constant of integration. Thus the ray path is

$$x(z) = \frac{1}{\alpha}\sinh^{-1}\{(A^2-1)^{1/2}\sin[\alpha(z+c)]\} \tag{1.3-30}$$

Typical ray paths are shown in Fig. 1.6. It is immediately evident that the rays are periodic in z with a period $2\pi/\alpha$ that is indeed independent of x_0 and γ_0.

Problem 1.12. Show that, in a medium possessing radial symmetry, i.e., where the refractive index is a function of radial coordinate r alone, the rays are confined to

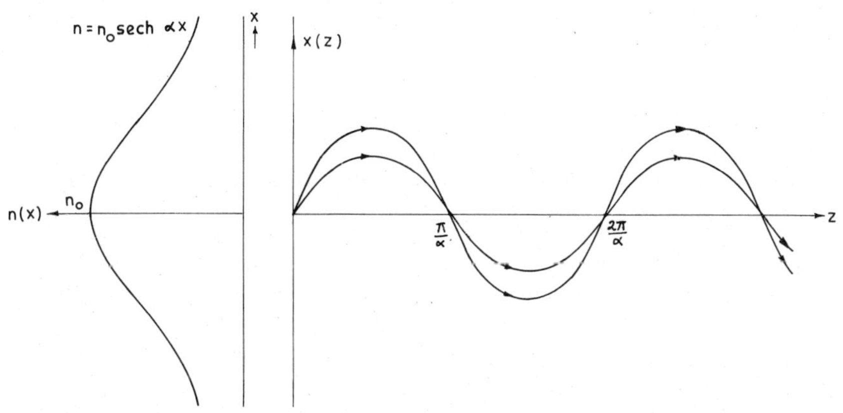

Fig. 1.6. Path of rays in a medium characterized by a refractive index of the form given by Eq. (1.3-26). Notice that all rays, irrespective of the angle of injection, have the same period of oscillation.

a single plane. Using this property, show that a medium with a refractive-index variation of the form

$$n(r) = n_0/(1 + r^2/a^2) \tag{1.3-31}$$

forms perfect images of point objects, i.e., all rays emanating from a point cross again at one point, the image point. This variation of refractive index is known as Maxwell's fish eye.

Solution. Equation (1.3-9) can be written in the form

$$\frac{d}{ds}(n\hat{\mathbf{s}}) = \nabla n \tag{1.3-32}$$

where $\hat{\mathbf{s}}\,(= d\mathbf{r}/ds)$ represents the unit vector along the tangent to the ray. For a medium possessing radial symmetry, ∇n will be along $\hat{\mathbf{r}}$. Thus

$$\hat{\mathbf{r}} \times \frac{d}{ds}(n\hat{\mathbf{s}}) = 0 \tag{1.3-33}$$

Now,

$$\frac{d}{ds}(n\hat{\mathbf{s}} \times \hat{\mathbf{r}}) = n\hat{\mathbf{s}} \times \frac{d\mathbf{r}}{ds} + \frac{d}{ds}(n\hat{\mathbf{s}}) \times \hat{\mathbf{r}} = 0 \tag{1.3-34}$$

Thus $n(\hat{\mathbf{s}} \times \hat{\mathbf{r}})$ is a constant, i.e., the rays are always confined to a plane. We choose this plane as the x-y plane for which $\theta = \pi/2$. Thus, along the path of the ray

$$ds = [(dr)^2 + r^2 \sin^2\theta (d\theta)^2 + r^2 (d\varphi)^2]^{1/2} = [1 + r^2\dot{\varphi}^2]^{1/2}\, dr \tag{1.3-35}$$

where dots here represent differentiation with respect to r. Hence the Lagrangian would be $n(r)\,(1 + r^2\dot{\varphi}^2)^{1/2}$ and the Lagrange equation in the variable φ will give us

$$n(r)\, r^2\dot{\varphi}/[1 + r^2\dot{\varphi}^2]^{1/2} = C \tag{1.3-36}$$

where C is some constant. Substituting for $n(r)$ from Eq. (1.3-31), after some rearrangement we obtain

$$\int d\varphi = \int \frac{\alpha(1 + \xi^2)\, d\xi}{\xi[\xi^2 - \alpha^2(1 + \xi^2)^2]^{1/2}} = \sin^{-1}\left[\frac{\alpha}{(1 - 4\alpha^2)^{1/2}} \frac{\xi^2 - 1}{\xi}\right] - \beta \tag{1.3-37}$$

where $\xi = r/a$, $\alpha = C/an_0$, and β is a constant of integration. Thus

$$\sin(\varphi + \beta) = \sin(\varphi_0 + \beta) \frac{r^2 - a^2}{r_0^2 - a^2} \frac{r_0}{r} \tag{1.3-38}$$

which represents the path of the ray in the medium. It can be seen from Eq. (1.3-38) that this is satisfied also for $\varphi = \varphi_0 + \pi$ and $r = a^2/r_0$. Thus all rays emanating from (r_0, φ_0) intersect again at $(a^2/r_0, \varphi_0 + \pi)$, and the imaging is perfect (see Fig. 1.7).

Problem 1.15. From Eq. (1.3-38), show that the rays are circles (see Fig. 1.7). [*Hint*: Transform to a Cartesian system.]

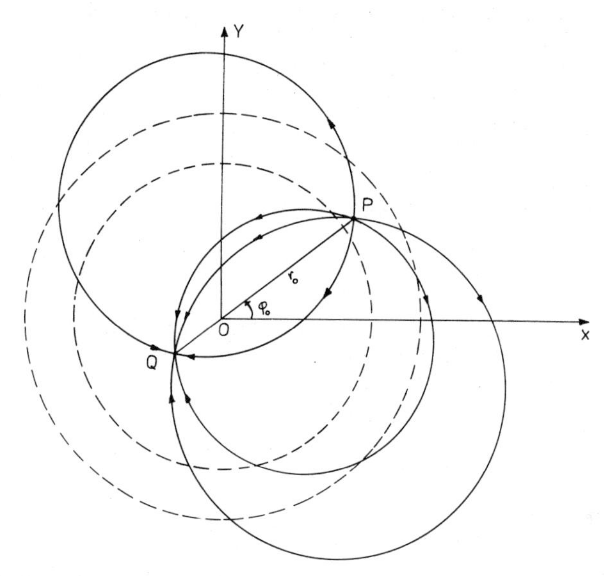

Fig. 1.7. Path of rays in Maxwell's fish eye [see Eq. (1.3-31)]. Loci of constant refractive index are shown as dashed circles.

1.4. Hamiltonian Formulation

Analogous to the case in classical mechanics, one can also develop the Hamiltonian formulation. This approach will be used in Chapter 2 to calculate explicit expressions for various aberration coefficients. In the Hamiltonian formulation, we have first to define the generalized momenta p and q by the relations*

$$p = \partial L/\partial \dot{x}, \qquad q = \partial L/\partial \dot{y} \tag{1.4-1}$$

where, as before, dots represent differentiation with respect to z and L is the Lagrangian. On substituting the value of L from Eq. (1.3-4) we find

$$p = \frac{n\dot{x}}{(1 + \dot{x}^2 + \dot{y}^2)^{1/2}} = n\,\frac{dx}{ds}, \qquad q = \frac{n\dot{y}}{(1 + \dot{x}^2 + \dot{y}^2)^{1/2}} = n\,\frac{dy}{ds} \tag{1.4-2}$$

where we have used Eq. (1.3-2). Since dx/ds and dy/ds represent the direction cosines of the ray at the point (x, y, z) along the x and y directions, $p\,(= n\,dx/ds)$ and $q\,(= n\,dy/ds)$ are termed the optical direction cosines of

* In classical mechanics q's represent the generalized coordinates and p's the corresponding generalized momenta. Notice that here both p and q are canonical momenta corresponding to x and y, respectively.

the ray. We now define the optical Hamiltonian H, in terms of L, through the relation

$$H = p\dot{x} + q\dot{y} - L(x, y, \dot{x}, \dot{y}, z) \tag{1.4-3}$$

Thus

$$\begin{aligned} dH &= \left(p - \frac{\partial L}{\partial \dot{x}}\right) d\dot{x} + \left(q - \frac{\partial L}{\partial \dot{y}}\right) d\dot{y} + \dot{x}\, dp + \dot{y}\, dq \\ &\quad - \frac{\partial L}{\partial x} dx - \frac{\partial L}{\partial y} dy - \frac{\partial L}{\partial z} dz \\ &= \dot{x}\, dp + \dot{y}\, dq - \dot{p}\, dx - \dot{q}\, dy - \frac{\partial L}{\partial z} dz \end{aligned} \tag{1.4-4}$$

where we have used Eqs. (1.4-1) and (1.3-5). From Eq. (1.4-4) it is clear that H is a function of x, y, p, q, and z and the following equations, Hamilton's equations, also follow readily:

$$\dot{x} = \partial H/\partial p, \qquad \dot{y} = \partial H/\partial q \tag{1.4-5}$$

$$\dot{p} = -\,\partial H/\partial x, \qquad \dot{q} = -\partial H/\partial y \tag{1.4-6}$$

$$\partial H/\partial z = -\,\partial L/\partial z \tag{1.4-7}$$

These equations form the basic equations of the Hamiltonian formulation. Given a Hamiltonian H, i.e., given a refractive index function $n(x, y, z)$, the above equations allow us to calculate the ray path. If we substitute the expressions for p, q, and L from Eqs. (1.4-2) and (1.3-4) into Eq. (1.4-3), we obtain

$$\begin{aligned} H &= \frac{n\dot{x}}{(1 + \dot{x}^2 + \dot{y}^2)^{1/2}}\dot{x} + \frac{n\dot{y}}{(1 + \dot{x}^2 + \dot{y}^2)^{1/2}}\dot{y} - n(1 + \dot{x}^2 + \dot{y}^2)^{1/2} \\ &= -\frac{n}{(1 + \dot{x}^2 + \dot{y}^2)^{1/2}} \end{aligned}$$

We also note that

$$\begin{aligned} (n^2 - p^2 - q^2)^{1/2} &= \left[n^2 - \frac{n^2\dot{x}^2}{1 + \dot{x}^2 + \dot{y}^2} - \frac{n^2\dot{y}^2}{1 + \dot{x}^2 + \dot{y}^2}\right]^{1/2} \\ &= \frac{n}{(1 + \dot{x}^2 + \dot{y}^2)^{1/2}} \end{aligned}$$

Thus

$$H = -[n^2(x, y, z) - p^2 - q^2]^{1/2} \tag{1.4-8}$$

1.5. Application of the Hamiltonian Formulation to the Study of Paraxial Lens Optics

Using Hamilton's equations, we will trace the rays in a rotationally symmetric optical system. By rotational symmetry, we imply that the properties of the system are the same on the circumference of any circle whose center lies on an axis; this axis is known as the symmetry axis of the system. A simple example is a coaxial system of lenses. Even for a rotationally symmetric system, since the Hamiltonian H is an irrational function of x, y, p, q, and z, it is difficult to solve Hamilton's equations and one has to look for approximate solutions. The lowest-order approximation would lead us to paraxial optics or Gaussian optics, which is concerned with rays that travel close to the axis of the system and make small angles with it. Under this approximation it will be shown that perfect images can be formed. Deviations from these determine the aberrations of the system. In this section we will show how the Hamiltonian formulation can be used to yield simple results for refracting surfaces and lenses, which are normally obtained by application of Snell's law. We will consider a few representative examples to show the applicability of this method. In Chapter 2 we will use the same formulation to calculate explicit expressions for the aberrations introduced by rotationally symmetric systems.

Since for a rotationally symmetric system the refractive index depends on the value of $(x^2 + y^2)$ (rather than on x and y independently) we may write the Hamiltonian in the form

$$H = -\left[n^2(u, z) - v\right]^{1/2} \tag{1.5-1}$$

where

$$u = x^2 + y^2, \qquad v = p^2 + q^2 \tag{1.5-2}$$

In the paraxial approximation, u and v are small quantities and hence we may make a Taylor series expansion of the Hamiltonian in ascending powers of u and v, retaining only first-order terms:

$$H(u, v, z) = H_0(z) + \left[H_1(z)\,u + H_2(z)\,v\right] + \cdots \tag{1.5-3}$$

where

$$H_1(z) = \left.\frac{\partial H}{\partial u}\right|_{u=0,\,v=0}, \qquad H_2 = \left.\frac{\partial H}{\partial v}\right|_{u=0,\,v=0} \tag{1.5-4}$$

In order to calculate the aberrations, higher-order terms in Eq. (1.5-3) have to be retained (see Chapter 2).

Since our system possesses rotational symmetry, we need consider only the set of equations in x and p; the equations for y and q would follow from analogy. From Eq. (1.5-2), we obtain

$$\frac{\partial}{\partial x} = 2x\,\frac{\partial}{\partial u}, \qquad \frac{\partial}{\partial p} = 2p\,\frac{\partial}{\partial v}, \qquad \text{etc.} \tag{1.5-5}$$

Thus, Hamilton's equations [Eqs. (1.4-5) and (1.4-6)] become

$$\dot{x} = \frac{dx}{dz} = 2p\,\frac{\partial H}{\partial v} = 2H_2 p, \qquad \dot{p} = \frac{dp}{dz} = -2x\,\frac{\partial H}{\partial u} = -2H_1 x \tag{1.5-6}$$

where we have used Eq. (1.5-3). Here x and p correspond to the paraxial approximations. Using Eqs. (1.4-8) and (1.5-4), we obtain

$$H_1 = -\left.\frac{\partial n}{\partial u}\right|_{u=0,v=0}, \qquad H_2 = \frac{1}{2n(0, z)} \tag{1.5-7}$$

where $n(0, z)$ is the refractive index along the axis. We will now make use of the above formulation to study the imaging properties of some simple optical systems, like a single refracting surface, a thin lens, and a thick lens. Other complicated optical systems can be analyzed by using these formulas in conjunction.

1.5.1. A Single Refracting Surface

Let us first consider a spherical refracting surface (of radius of curvature R) separating two homogeneous media of refractive indices n_1 and n_2 (see Fig. 1.8). The point C represents the center of curvature of the surface and the z axis passes through the point C. In order to write an equation describing the refractive-index variation, we must first find the equation of the spherical surface. Let (x, y, z) be the coordinates of an arbitrary point on the surface; the origin is assumed to be at the point O. Since the refracting surface is a portion of a sphere, we must have

$$x^2 + y^2 = z(2R - z) \tag{1.5-8}$$

Thus

$$z = R\left[1 \pm \left(1 - \frac{u^2}{R^2}\right)^{1/2}\right] \simeq \frac{u}{2R} + \frac{u^2}{8R^3} + \cdots = f(u) \tag{1.5-9}$$

where we have chosen the solution that makes z go to zero as $u \to 0$. (Positive values of R correspond to a convex surface and negative values of R correspond to a concave surface.) Clearly, a point whose z coordinate satisfies the inequality $z > f(u)$ lies to the right of the surface. Similarly

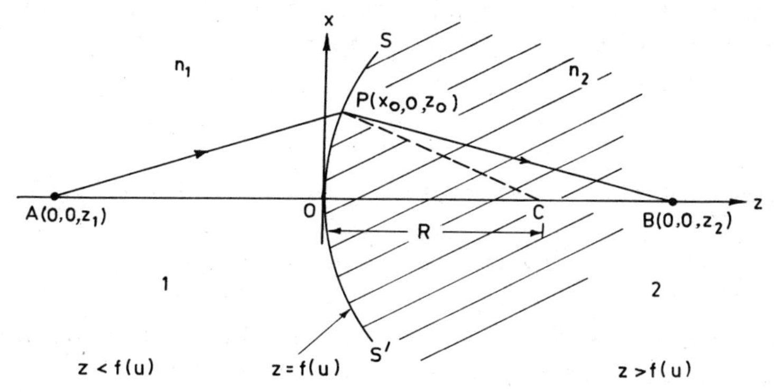

Fig. 1.8. SS' is a spherical surface of radius R separating two media of refractive indices n_1 and n_2. C represents the center of curvature of the surface.

if $z < f(u)$ then the point will lie to the left of the surface. Hence the refractive-index variation can be written in the form

$$n(x, y, z) = n_1\Theta\big(f(u) - z\big) + n_2\Theta\big(z - f(u)\big) \tag{1.5-10}$$

where $\Theta(x)$ is the unit step function, defined by the equation

$$\Theta(x) = \begin{cases} 1 & x > 0 \\ 0 & x < 0 \end{cases} \tag{1.5-11}$$

Thus

$$H_1 = -\left.\frac{\partial n}{\partial u}\right|_{u=0} = \frac{n_2 - n_1}{2R}\,\delta(z) \tag{1.5-12}$$

where we have used

$$d\Theta(x)/dx = \delta(x) \tag{1.5-13}$$

and $\delta(x)$ is the Dirac delta function (see Appendix A). Thus

$$\frac{dp}{dz} = -2H_1x = -(n_2 - n_1)\,\frac{\delta(z)}{R}\,x \tag{1.5-14}$$

It can immediately be seen that for $z < 0$ or for $z > 0$, $dp/dz = 0$, which shows that the rays connecting any two points lying in the same medium are straight lines. However, when the ray hits the refracting surface, it undergoes an abrupt change in its slope as can be seen by integrating Eq.

(1.5-14) from $z = -\varepsilon$ to $z = +\varepsilon$:

$$\int_{-\varepsilon}^{\varepsilon} \frac{dp}{dz} dz = -\int_{-\varepsilon}^{\varepsilon} \frac{n_2 - n_1}{R} \delta(z)\, x\, dz$$

or

$$p_2 - p_1 = -\frac{n_2 - n_1}{R} x_0 \tag{1.5-15}$$

where x_0 is the value of the x coordinate of the ray calculated at the refracting surface and p_1 and p_2 are the optical direction cosines in media 1 and 2, respectively. Further,

$$p = \frac{1}{2H_2} \dot{x} = n(0, z) \frac{dx}{dz} = \begin{cases} n_1\, dx/dz & \text{in medium 1} \\ n_2\, dx/dz & \text{in medium 2} \end{cases} \tag{1.5-16}$$

Let us consider a ray that starts from the point $A(0, 0, z_1)$ and gets refracted at the point $P(x_0, 0, z_0)$ as shown in Fig. 1.8; obviously z_1 is a negative quantity. (We are using the analytical geometry convention.) Thus, using Eq. (1.5-16) we get

$$p_1 = n_1 \frac{x_0}{z_0 - z_1} \simeq -n_1 \frac{x_0}{z_1}, \qquad p_2 \simeq -n_2 \frac{x_0}{z_2} \tag{1.5-17}$$

Substituting the above values in Eq. (1.5-15), we obtain

$$\frac{n_2}{z_2} - \frac{n_1}{z_1} = \frac{n_2 - n_1}{R} \tag{1.5-18}$$

which is the required formula for a single refracting surface. It can be seen from the above equation that in the realm of paraxial optics, for a given object point, the position of the image point is dependent only on n_1, n_2, and R. Thus the image formed under paraxial optics is an ideal image. The primary and the secondary focal lengths are $-n_1R/(n_2 - n_1)$ and $n_2R/(n_2 - n_1)$.

1.5.2. Thin Lens

A lens is called thin if a ray striking the first surface emerges at approximately the same height from the other surface. Consider a thin lens of refractive index n_1 placed between two media of refractive indices n_0 and n_2 as shown in Fig. 1.9a. Let the radii of curvatures of the two surfaces forming the lens be R_1 and R_2, respectively. If p_0 is the direction cosine of

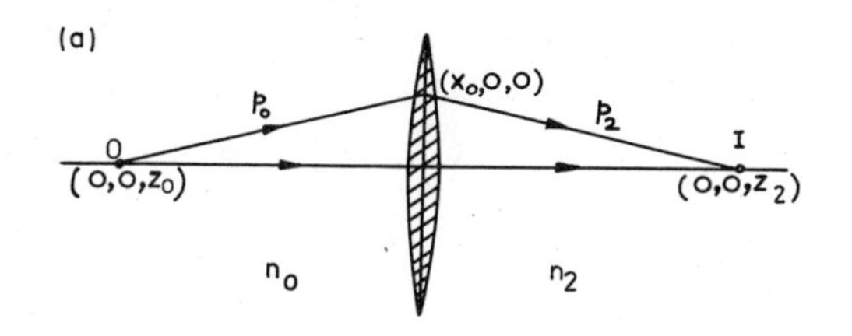

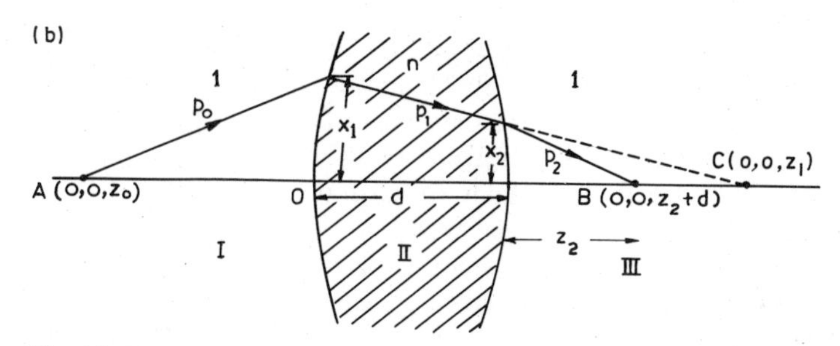

Fig. 1.9. (a) A thin lens made of a material of refractive index n_1 placed between media of refractive indices n_0 and n_2. (b) A thick lens made of a material of refractive index n placed in air.

the incident ray, then

$$p_1 - p_0 = -\frac{n_1 - n_0}{R_1} x_0, \qquad p_2 - p_1 = \frac{n_1 - n_2}{R_2} x_0 \tag{1.5-19}$$

where p_1 and p_2 are the optical direction cosines of the ray in media 1 and 2 and x_0 is the height at which the ray strikes the lens. Thus

$$p_2 - p_0 = \left(\frac{n_1 - n_2}{R_2} - \frac{n_1 - n_0}{R_1} \right) x_0 \tag{1.5-20}$$

Since

$$p_0 = -n_0 \frac{x_0}{z_0}, \qquad p_2 = -n_2 \frac{x_0}{z_2} \tag{1.5-21}$$

[see Eq. (1.5-17)], we obtain

$$\frac{n_0}{z_0} - \frac{n_2}{z_2} = -\left(\frac{n_1 - n_0}{R_1} - \frac{n_1 - n_2}{R_2} \right) \tag{1.5-22}$$

which is the thin-lens formula. In most cases, the two media on both sides of the lens are the same, i.e., $n_2 = n_0$, and if $n = n_1/n_0$ we get from Eq. (1.5-22)

$$\frac{1}{z_2} - \frac{1}{z_0} = (n-1)\left(\frac{1}{R_1} - \frac{1}{R_2}\right) = \frac{1}{f} \tag{1.5-23}$$

where f represents the focal length of the lens.

1.5.3. Thick Lens

If the thickness of the lens is not negligible compared to other parameters then one has a thick lens. Such a lens is shown in Fig. 1.9b. Let R_1 and R_2 be the radii of curvature of the two surfaces and let d be the thickness of the lens on the axis. For simplicity let us consider the media surrounding the lens to be of refractive-index unity and let n represent the refractive index of the material of the lens. Let the vertex of the first surface represent the origin of the coordinate system. Consider a ray that starts from point A with coordinates $(0, 0, z_0)$ and intersects the first surface at a height x_1 from the axis. After refraction let the ray intersect the second surface at a distance x_2 from the axis. The emerging ray passes through the point $B\,(0, 0, z_2 + d)$, where z_2 and d have been defined in Fig. 1.9b. If p_0, p_1, and p_2 represent the values of p in the three regions I, II, and III, then corresponding to Eq. (1.5-19) we have at the first and second surface

$$p_1 - p_0 = -\frac{n-1}{R_1}x_1, \qquad p_2 - p_1 = \frac{n-1}{R_2}x_2 \tag{1.5-24}$$

(For a thin lens we would have $x_1 = x_2$.) From the definition of optical direction cosine, we can write

$$x_2 - x_1 = \frac{p_1}{n}d \tag{1.5-25}$$

Using Eqs. (1.5-24) and (1.5-25) we obtain

$$p_0\left(1 + \frac{n-1}{n}\frac{d}{R_2}\right) - p_2 = \frac{n-1}{R_1}x_1\left(1 + \frac{n-1}{n}\frac{d}{R_2}\right) - \frac{n-1}{R_2}x_1 \tag{1.5-26}$$

Let us consider the case when $p_2 = 0$, i.e., the ray emerging from the lens is parallel to the axis. This will correspond to a particular object position shown as F_1 in Fig. 1.10a. A plane normal to the axis and passing through the point of intersection of the incident and the emergent rays is known as the primary principal plane and is shown as P_1 in the figure. The point F_1 is known as the primary focal point and its distance from

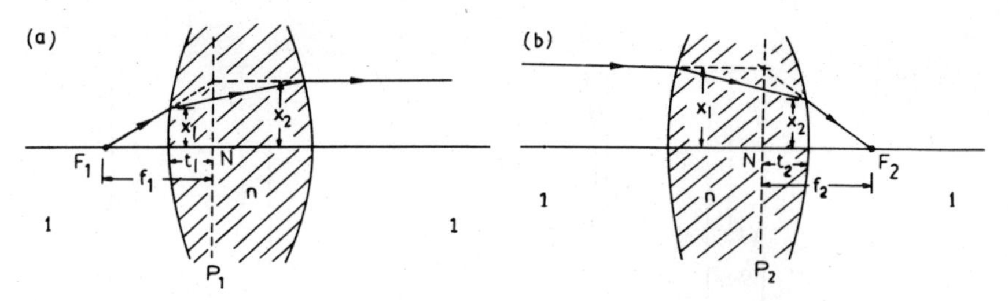

Fig. 1.10. (a) P_1 and (b) P_2 represent the two principal planes of the lens. The focal lengths are measured from the respective principal planes. The index of refraction of the lens is n and that of the surrounding medium is 1.

the principal plane P_1 (which is denoted by f_1) is known as the primary focal length. It can easily be seen from the figure that

$$p_0 = \frac{x_2}{F_1 N} = \frac{x_2}{(-f_1)}$$

and using the fact that $p_2 = 0$, we obtain $f_1 = -nR_1R_2/D$, where

$$D = (n-1)[n(R_2 - R_1) + (n-1)d] \tag{1.5-27}$$

Similarly, the second focal length f_2 (see Fig. 1.10b) is nR_1R_2/D. The magnitudes of the two focal lengths happen to be equal because of the medium being the same on both sides of the lens. The two planes P_1 and P_2 shown in the figure are called the principal planes of the lens. They are also called unit magnification planes.

The distances of the principal planes from the respective vertices can be easily determined. From Fig. 1.10a, we find by simple application of geometry

$$t_1 = f_1\left(1 - \frac{x_1}{x_2}\right) = -(n-1)\frac{dR_1}{D}$$

Similarly we can obtain (see Fig. 1.10b)

$$t_2 = (n-1)\frac{dR_2}{D}$$

The positions of the principal planes for some common types of lenses are shown in Fig. 1.11.

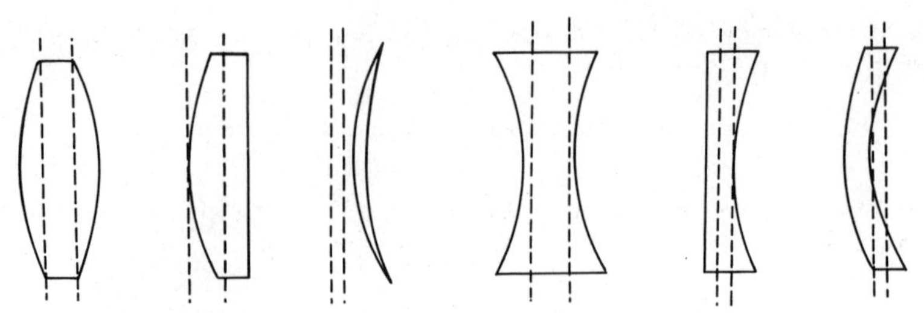

Fig. 1.11. Positions of principal planes for some common types of lenses.

1.6. Eikonal Approximation

1.6.1. Derivation of the Eikonal Equation

We start with the scalar wave equation*

$$\nabla^2 \psi + k_0^2 n^2 \psi = 0 \tag{1.6-1}$$

where ψ is taken to represent a component of the electric field, n is the refractive-index variation, and $k_0 = \omega/c = 2\pi/\lambda_0$ represents the free-space wave number, ω being the frequency of the wave, λ_0 the free-space wavelength, and c the speed of light in free space. When the medium is homogeneous, then Eq. (1.6-1) has solutions of the form $\exp(ikz)$, where z is the direction of propagation. Thus, when the medium is inhomogeneous one is led to assume a solution of Eq. (1.6-1) of the form

$$\psi = \psi_0 \exp[ik_0 S(x, y, z)] \tag{1.6-2}$$

where ψ_0 and S are real functions of x, y, and z and are assumed to be independent of k_0. The rapid variations of the optical field are represented by the exponential term. It is also assumed that ψ_0 and S vary slowly in space. Under these conditions we substitute for ψ from Eq. (1.6-2) into Eq. (1.6-1) to get

$$k_0^2[n^2 - (\nabla S)^2]\psi_0 + ik_0(2\nabla S \cdot \nabla\psi_0 + \psi_0 \nabla^2 S) + \nabla^2\psi_0 = 0 \tag{1.6-3}$$

Since ψ_0 and S are assumed to be real, after equating the real part separately to zero, we obtain

$$(\nabla S)^2 = n^2 + \frac{1}{k_0^2 \psi_0} \nabla^2 \psi_0 \tag{1.6-4}$$

* This equation can be derived from Maxwell's equations for an isotropic, charge-free, nonabsorbing homogeneous medium (see Section 10.3). For an inhomogeneous medium the scalar wave equation is valid as long as $\lambda \nabla n/n \ll 1$ (see Chapter 10).

Also since ψ_0 and S are assumed to be independent of k_0, we have in the limit $\lambda_0 \to 0$, i.e., $k_0 \to \infty$,

$$(\nabla S)^2 = n^2 \tag{1.6-5}$$

This equation is called the eikonal equation.* The quantity S is called the eikonal. We can see from Eq. (1.6-2) that the surfaces

$$S(x, y, z) = \text{const} \tag{1.6-6}$$

represent phase fronts, i.e., surfaces along which the phase remains constant. In isotropic media, the rays that may be defined as the direction along which energy propagates coincide with normals to the phase fronts.

The fact that the rays in a homogeneous medium are straight lines (see Problem 1.6) can be seen from the following consideration: In a homogeneous medium it follows from the eikonal equation that the magnitude of ∇S is independent of position. Thus, the position of the phase front at a later time would be "parallel" to the original position. This yields at once the fact that the rays in a homogeneous medium are straight lines. This is shown in Fig. 1.12a. On the contrary, in an inhomogeneous medium the rate of change of phase depends on position and the phase front after a certain interval of time is not parallel to the original one, resulting in the rays being curved. This shown in Fig. 1.12b.

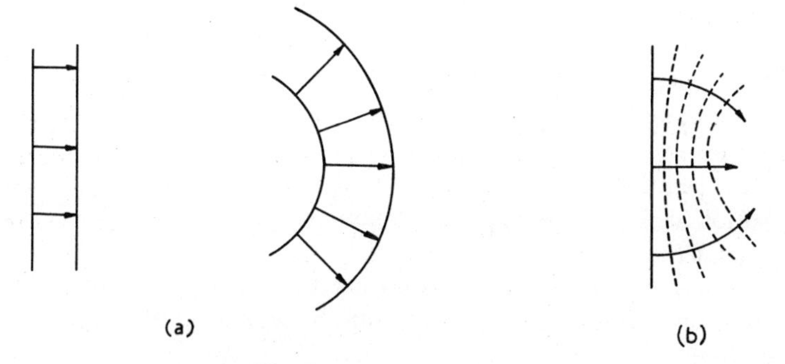

Fig. 1.12. (a) The phase front at any time in a homogeneous medium is "parallel" to the phase front at an earlier time. Thus the rays defined as normal to the phase front are straight lines in homogeneous media. (b) In an inhomogeneous medium, a plane phase front gets curved due to the different velocities at different points along its phase front. This leads to the fact that rays in an inhomogeneous medium are curved.

---

* For a more thorough discussion see Felsen and Marcuvitz (1973).

Problem 1.16. Derive the ray equation from the eikonal equation.

Solution. From the eikonal equation, we obtain

$$\nabla S = n\hat{\mathbf{s}} = n\, d\mathbf{r}/ds \tag{1.6-7}$$

where $\hat{\mathbf{s}}$ is the unit vector normal to the phase fronts, i.e., along the ray. Hence

$$\frac{d}{ds}\left(n\frac{d\mathbf{r}}{ds}\right) = \frac{d}{ds}(\nabla S) = (\hat{\mathbf{s}}\cdot\nabla)\,\nabla S \tag{1.6-8}$$

Taking the gradient of the eikonal equation [Eq. (1.6-5)] we obtain

$$\nabla\cdot[(\nabla S)^2] = \nabla n^2 = 2n\nabla n \tag{1.6-9}$$

since

$$\begin{aligned}\nabla[(\nabla S)^2] &= \nabla[\nabla S\cdot\nabla S] = 2\nabla\times(\nabla\times\nabla S) + 2(\nabla S\cdot\nabla)\,\nabla S\\ &= 2(\nabla S\cdot\nabla)\,\nabla S = 2n(\hat{\mathbf{s}}\cdot\nabla)\,\nabla S = 2n\frac{d}{ds}\left(n\frac{d\mathbf{r}}{ds}\right)\end{aligned} \tag{1.6-10}$$

where we have used Eqs. (1.6-7) and (1.6-8). Using the above equation in Eq. (1.6-9), one obtains the ray equation [Eq. (1.3-9)].

1.6.2. The Eikonal Equation and Fermat's Principle

If we take the curl of Eq. (1.6-7), and integrate over a surface A, we obtain

$$0 = \int_A \nabla\times(n\hat{\mathbf{s}})\,da = \oint_C n\hat{\mathbf{s}}\cdot d\mathbf{r} \tag{1.6-11}$$

where use has been made of Stokes' theorem and $\oint_C$ represents a line integral over a closed path C bounding the surface A. Let $PRSQ$ be any path obtained as a solution of the eikonal equation (see Fig. 1.13). The ray is normal to the phase front all along its path and in the figure we have shown the position of the phase front at different times. Let $PABQ$ be another curve connecting the two points P and Q. Let W_1 and W_2 be two infinitesimally close positions of the phase fronts. Let AC represent the normal to the phase front at the point A. We now apply Eq. (1.6-11) to the closed curve ACB and get

$$(n\hat{\mathbf{s}}\cdot d\mathbf{r})_{AC} + (n\hat{\mathbf{s}}\cdot d\mathbf{r})_{CB} + (n\hat{\mathbf{s}}\cdot d\mathbf{r})_{BA} = 0 \tag{1.6-12}$$

Since $\hat{\mathbf{s}}$ is normal to the phase front and BC is a part of the phase front, we have $(n\hat{\mathbf{s}}\cdot d\mathbf{r})_{BC} = 0$.

Since $(n\hat{\mathbf{s}}\cdot d\mathbf{r})_{BA} = -(n\hat{\mathbf{s}}\cdot d\mathbf{r})_{AB}$, we obtain

$$(n\hat{\mathbf{s}}\cdot d\mathbf{r})_{AC} = (n\hat{\mathbf{s}}\cdot d\mathbf{r})_{AB} \tag{1.6-13}$$

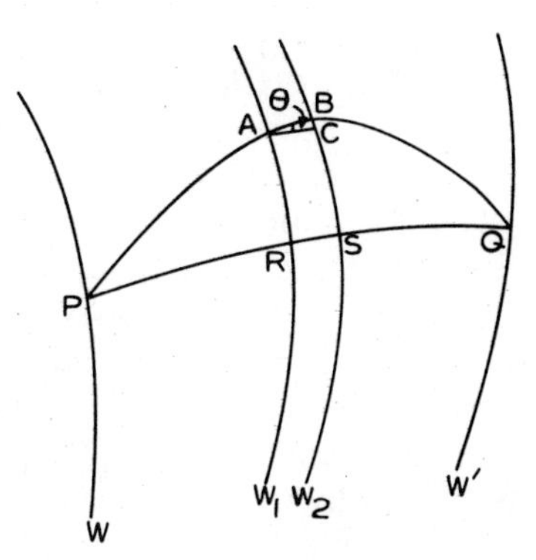

Fig. 1.13. W and W' represent the positions of the same phase front at different times. W_1 and W_2 represent the position of the phase front at two infinitesimally close intervals of time.

Since W_1 and W_2 represent two adjacent positions of the same phase front $(n\hat{\mathbf{s}}\cdot d\mathbf{r})_{AC} = (n\,ds)_{RS}$. If θ is the angle shown in the figure, then

$$(n\hat{\mathbf{s}}\cdot d\mathbf{r})_{AB} = (n\,ds\cos\theta)_{AB} \leq (n\,ds)_{AB} \tag{1.6-14}$$

Thus,

$$(n\,ds)_{RS} \leq (n\,ds)_{AB}$$

and hence

$$\int_{PRSQ} n\,ds \leq \int_{PABQ} n\,ds \tag{1.6-15}$$

If it is assumed that only one ray passes through any point then the equality sign in Eq. (1.6-15) can hold only when the path $PABQ$ is also a ray, which does not satisfy the above assumption. Hence we get

$$\int_{PRSQ} n\,ds < \int_{PABQ} n\,ds \tag{1.6-16}$$

which is Fermat's principle. From the above result, it seems as if the rays follow that path for which the optical pathlength is a minimum rather than an extremum. This is due to the fact that the above result has been derived under the assumption that between the points P and Q there always exist well-defined single-valued phase fronts with definite normals. Situations arise when there exists a point where all the rays focus and then emerge before reaching the point Q. In such situations, the ray is not determined by a minimum value of the optical pathlength but rather by an extremum value.

1.7. Wave Optics as Quantized Geometrical Optics*

Geometrical optics can be obtained as a limiting case of wave optics when the optical wavelength goes to zero. On the other hand, classical mechanics can be obtained as a limiting case of wave mechanics when the de Broglie wavelength goes to zero (see, e.g., Powell and Craseman, 1961). If we want to derive physical optics from geometrical optics we can use the analogy of the transition from classical mechanics to quantum mechanics. The quantum mechanical equations can be obtained from the classical equations by replacing the variables of classical theory by linear operators. For example, the classical momentum is replaced in quantum mechanics by a corresponding momentum operator, e.g., for the x component, the momentum operator is

$$p_x = \frac{\hbar}{i} \frac{\partial}{\partial x} \tag{1.7-1}$$

where $\hbar = h/2\pi$, h being Planck's constant, and $i = \sqrt{-1}$. In analogy, in going from geometrical optics to wave optics, we write the conjugate momenta (which represent the optical direction cosines) in terms of the operators

$$p = \frac{\kappa}{i} \frac{\partial}{\partial x}, \qquad q = \frac{\kappa}{i} \frac{\partial}{\partial y} \tag{1.7-2}$$

where κ is an unknown constant, corresponding to $\hbar$ in quantum mechanics. Since p and q are dimensionless quantities, κ should have the dimension of length. Since the classical results follow in the limit of $\hbar \to 0$, we expect that geometrical optics should follow in the limit $\kappa \to 0$. Further, in quantum mechanics, the energy corresponds to the operator $i\hbar\, \partial/\partial t$, which in this case will be $i\kappa\, \partial/\partial z$.

Now, using the optical Hamiltonian [see Eq. (1.4-8)] we may write the corresponding Schrödinger equation as

$$H\psi = i\kappa \frac{\partial \psi}{\partial z}$$

or

$$HH\psi = i\kappa \frac{\partial}{\partial z} H\psi = -\kappa^2 \frac{\partial^2 \psi}{\partial z^2}$$

i.e.,

$$(n^2 - p^2 - q^2)\psi = -\kappa^2 \frac{\partial^2 \psi}{\partial z^2} \tag{1.7-3}$$

* See Gloge and Marcuse (1969), Eichmann (1971), and Marcuse (1972).

which using Eq. (1.7-2) becomes

$$\left(\frac{\partial^2}{\partial x^2}+\frac{\partial^2}{\partial y^2}+\frac{\partial^2}{\partial z^2}\right)\psi+\frac{n^2}{\kappa^2}\psi=0 \tag{1.7-4}$$

where ψ is the wave function. Comparing this with the scalar wave equation [Eq. (1.6-1)] one can see that $\kappa = \lambda_0/2\pi$, where λ_0 is the free-space wavelength. The fact that geometrical optics should follow as a limiting case of the quantized theory when $\kappa \to 0$ is easily understood, since as $\kappa \to 0$, $\lambda_0 \to 0$; this, as we already know, is the limit under which geometrical optics is strictly valid.

Equation (1.7-4) corresponds to the relativistic Schrödinger equation or Klein–Gordon equation:

$$-\hbar^2\,\partial^2\psi/\partial t^2 = -\hbar^2 c^2\nabla^2\psi + m_0^2 c^4\psi \tag{1.7-5}$$

where m_0 is the rest mass of the particle. The Schrödinger equation can be obtained by making the approximation that $\mathbf{p}\,(= -i\hbar\boldsymbol{\nabla})$ is small compared to m_0c^2, thus obtaining the energy in nonrelativistic conditions. Analogously, in optics one would expect a Schrödinger type of equation under the approximation $p/n, q/n \ll 1$. Since p/n and q/n represent the direction cosines of the rays (see Section 1.3), the Schrödinger type of equation

$$i\frac{\lambda_0}{2\pi}\frac{\partial\psi}{\partial z} = -n\psi - \frac{\lambda_0^2}{8\pi^2 n_0}\left(\frac{\partial^2\psi}{\partial x^2}+\frac{\partial^2\psi}{\partial y^2}\right) \tag{1.7-6}$$

should correspond to the paraxial approximation.

We may conclude by noting that an uncertainty relation of the form

$$\Delta x\,\Delta p \geq \kappa/2 = \lambda_0/4\pi \tag{1.7-7}$$

can be physically interpreted in the following manner: If the spatial extent of the wavefront is Δx then the beam will undergo divergence because of diffraction and the spreading will be qualitatively determined by Eq. (1.7-7). Thus if d represents the slit width, then $\Delta x \sim d$ and $\Delta p \gtrsim \lambda_0/4\pi d$. If α is the angle made by the diffracted ray with the z axis, then $p = n\sin\alpha$ and Eq. (1.7-7) gives

$$\Delta\alpha \gtrsim \lambda_0/4\pi d \tag{1.7-8}$$

where we have assumed α to be small and $n = 1$. The above equation is consistent with the results obtained using diffraction theory.

2

Geometrical Theory of Third-Order Aberrations

2.1. Introduction

In Section 1.4 we used the Hamiltonian formulation to trace rays in some rotationally symmetric optical systems. They were derived under the paraxial approximation, i.e., the rays forming the image were assumed to lie infinitesimally close to the axis and to make infinitesimally small angles with it. It was found that the images of point objects were perfect, i.e., all rays starting from a given object point were found to intersect at *one* point, which is the image point. Such an image is called an ideal image. In general, rays that make large angles with the z axis or travel at large distances from the z axis do not intersect at one point. This phenomenon is known as aberration. In this chapter we will use the Hamiltonian formulation as developed by Luneberg (1964) to derive explicit expressions for aberrations in rotationally symmetric systems.

In Section 2.2, we derive explicit expressions for third-order aberrations in terms of two specific paraxial rays. In Section 2.3, we discuss how the different aberration coefficients obtained in Section 2.2 describe the five Seidel aberrations, namely, spherical aberration, coma, astigmatism, curvature of field, and distortion. After obtaining the explicit dependence of H_i and H_{ij} in terms of the refractive-index function in Section 2.4, we obtain explicit expressions for the aberrations of a rotationally symmetric graded-index medium, whose paraxial property we have already studied in Section 1.3. In Section 2.6, we obtain explicit expressions for the aberration coefficients in optical systems possessing finite discontinuities in refractive index, the refractive index between two discontinuities being constant. Examples of such optical systems are systems made up of lenses. Finally, in Section 2.7 we discuss the chromatic aberrations of optical systems.

2.2. Expressions for Third-Order Aberrations

For a rotationally symmetric system, the Hamiltonian is given by [see Eq. (1.4-8)]

$$H = -[n^2(U, z) - V]^{1/2} \tag{2.2-1}$$

where $U = X^2 + Y^2$, $V = P^2 + Q^2$, (X, Y, z) represents the position coordinate of any point on the ray, and (P, Q) are the optical direction cosines of the ray.* Hamilton's equations are [see Eqs. (1.4-5) and (1.4-6)]

$$\frac{dX}{dz} = 2P\frac{\partial H}{\partial V}, \qquad \frac{dP}{dz} = -2X\frac{\partial H}{\partial U} \tag{2.2-2}$$

To specify a ray one has to specify either the values of X, Y, P, and Q in any plane or the values of X and Y in two planes. We will choose the second type of boundary conditions (similar results can also be obtained by using the first set of boundary conditions). Let $z = z_0$ represent the object plane and $z = \zeta$ any other judiciously chosen reference plane; we will choose the plane $z = \zeta$ to contain the exit pupil. In general, a ray is completely specified if we know $X(z_0)$, $Y(z_0)$, $X(\zeta)$, and $Y(\zeta)$. For a particular ray, let these be given by

$$X(z_0) = x_0, \qquad X(\zeta) = \xi \tag{2.2-3}$$

$$Y(z_0) = y_0, \qquad Y(\zeta) = \eta \tag{2.2-4}$$

If we solve Eq. (2.2-2) under these boundary conditions, then X, Y, P, and Q will be general functions of x_0, y_0, ξ, and η. Paraxial rays would correspond to infinitesimal values of x_0, y_0, ξ, and η. For any general ray, we expand X, Y, P, and Q in ascending powers of x_0, y_0, ξ, and η to get

$$X = X_1 + X_2 + X_3 + \cdots, \qquad P = P_1 + P_2 + P_3 + \cdots \tag{2.2-5}$$

and similar equations for Y and Q. The subscripts represent the degree of the polynomial; for example, X_1 is linear in x_0, y_0, ξ, η; X_2 is quadratic in x_0, y_0, ξ, η. Note that there is no term like X_0 or P_0; this is because of the fact that a ray traveling along the symmetry axis (i.e., $X = Y = 0$) of a rotationally symmetric system travels undeviated. Hence if x_0, y_0, ξ, and η are zero, then (X, Y, P, Q) are zero everywhere. For infinitesimal values of x_0, y_0, ξ, and η, i.e., for a paraxial ray, X_2, X_3, etc., can be neglected in Eq. (2.2-5). Hence X_1 represents the paraxial value of X. Similarly P_1, Y_1, and Q_1 represent paraxial values of P, Y, and Q, respectively.

* The expression for the Hamiltonian is derived in Section 1.4. Here we are using capital letters for the transverse coordinates of a general ray to differentiate it from paraxial rays.

We will now show how the rotational symmetry of the system prohibits the even-order terms in the expansions given by Eq. (2.2-5). Since the system is rotationally symmetric, if a ray specified by the coordinates (x_0, y_0, ξ, η), has values $X(z_1)$, $Y(z_1)$, $P(z_1)$, $Q(z_1)$ in some plane $z = z_1$, then another ray specified by $(-x_0, -y_0, -\xi, -\eta)$ should have values $-X(z_1)$, $-Y(z_1)$, $-P(z_1)$, $-Q(z_1)$ in the same plane $z = z_1$ (see Fig. 2.1). If we try to impose this property in Eq. (2.2-5), we find that the quantities $X_1, X_3, \ldots$ change sign, while the quantities $X_2, X_4, \ldots$ do not. Hence it follows that the even-order terms should be absent from the expansions; thus Eq. (2.2-5) becomes

$$X = X_1 + X_3 + \cdots, \qquad P = P_1 + P_3 + \cdots \tag{2.2-6}$$

X_3 represents the third-order correction to X_1 and is called the third-order aberration. It is an aberration term because if it were absent (and $X_5, X_7, \ldots$, etc., were also absent) then as will be shown shortly, the images of every point object would be an exact point, and hence would represent an ideal image. Similarly, X_5 is called the fifth-order aberration, X_7 the seventh-order aberration, etc.

In order to calculate the aberrations, we make a Taylor expansion of H given by Eq. (2.2-1) in ascending powers of U and V about $U = 0$, $V = 0$, to get

$$H = H_0 + H_1 U + H_2 V + \tfrac{1}{2}(H_{11} U^2 + 2H_{12} UV + H_{22} V^2) + \cdots \tag{2.2-7}$$

where H_i and H_{ij} are the Taylor coefficients and are all functions of z. Equation (2.2-7) may be compared with Eq. (1.5-3), where terms up to first order in U and V were retained to obtain paraxial optics.

Substituting the expansions of X, P, and H from Eqs. (2.2-6) and (2.2-7) into Eq. (2.2-2), we get

$$(\dot{X}_1 + \dot{X}_3 + \cdots) = 2(P_1 + P_3 + \cdots)(H_2 + H_{12} U + H_{22} V + \cdots) \tag{2.2-8}$$

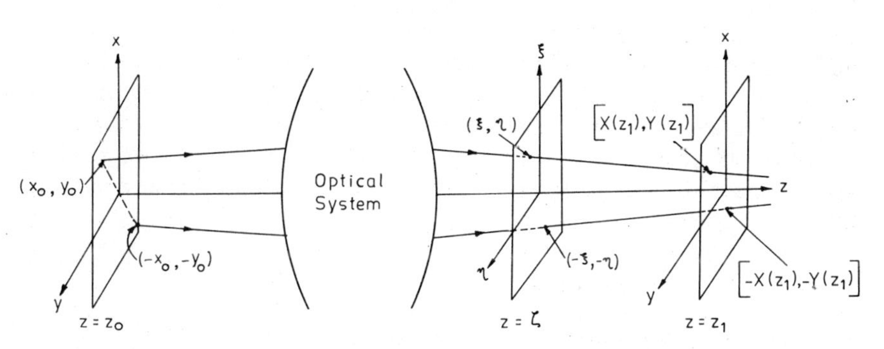

Fig. 2.1. A rotationally symmetric optical system.

where as before dots represent derivatives with respect to z. Now,

$$U = X^2 + Y^2 = (X_1 + X_3 + \cdots)^2 + (Y_1 + Y_3 + \cdots)^2$$
$$= U_1 + 2(X_1X_3 + Y_1Y_3) + \text{higher-order terms} \quad (2.2\text{-}9)$$

where $U_1 = X_1^2 + Y_1^2$. Similarly,

$$V = V_1 + 2(P_1P_3 + Q_1Q_3) + \text{higher-order terms} \quad (2.2\text{-}10)$$

where $V_1 = P_1^2 + Q_1^2$. Thus

$$(\dot{X}_1 + \dot{X}_3 + \cdots) = 2(P_1 + P_3 + \cdots)$$
$$\times \{H_2 + H_{12}[U_1 + 2(X_1X_3 + Y_1Y_3) + \cdots]$$
$$+ H_{22}[V_1 + 2(P_1P_3 + Q_1Q_3) + \cdots]\} \quad (2.2\text{-}11)$$

Equating terms of equal order we obtain

$$\dot{X}_1 = 2H_2P_1 \quad (2.2\text{-}12)$$

$$\dot{X}_3 = 2H_2P_3 + 2(H_{12}U_1 + H_{22}V_1)P_1 \quad (2.2\text{-}13)$$

Similarly,

$$\dot{P}_1 = -2H_1X_1 \quad (2.2\text{-}14)$$

$$\dot{P}_3 = -2H_1X_3 - 2(H_{11}U_1 + H_{12}V_1)X_1 \quad (2.2\text{-}15)$$

Equations (2.2-12) and (2.2-14) represent paraxial equations. Equations (2.2-13) and (2.2-15) are equations that determine the third-order aberrations in the system. The corresponding equations for Y_1, Q_1, Y_3, and Q_3 would be

$$\begin{aligned} \dot{Y}_1 &= 2H_2Q_1, & \dot{Y}_3 &= 2H_2Q_3 + 2(H_{12}U_1 + H_{22}V_1)Q_1 \\ \dot{Q}_1 &= -2H_1Y_1, & \dot{Q}_3 &= -2H_1Y_3 - 2(H_{11}U_1 + H_{12}V_1)Y_1 \end{aligned} \quad (2.2\text{-}16)$$

Given an optical system, we know the refractive index n as a function of U and z. Knowing n, the coefficients H_i and H_{ij} can be calculated. Then the paraxial equations (2.2-12) and (2.2-14) can be solved. This solution can then be substituted in Eqs. (2.2-13) and (2.2-15) to get the expression for X_3. An analogous procedure can be adopted for Y_3.

We will now use two specific paraxial rays to get explicit expressions for X_3 and Y_3 directly in terms of the system parameters. The first paraxial ray, called the axial ray, satisfies the boundary conditions

$$X_1(z_0) = 0, \quad Y_1(z_0) = 0, \quad X_1(\zeta) = 1, \quad Y_1(\zeta) = 0 \quad (2.2\text{-}17)$$

The solution of the paraxial equations under this boundary condition would be represented by $X_1(z) \equiv g(z)$ and $P_1(z) \equiv \vartheta(z)$ such that $g(z_0) = 0$, $g(\zeta) = 1$ (see Fig. 2.2). The second paraxial ray, called the field ray, satisfies

draw

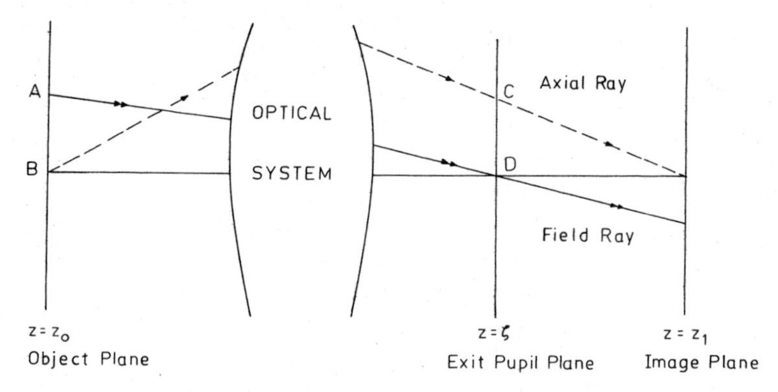

Fig. 2.2. The dashed and solid curves correspond to the axial and field rays, respectively. *AB* and *CD* are of unit length.

the boundary conditions

$$X_1(z_0) = 1, \qquad Y_1(z_0) = 0, \qquad X_1(\zeta) = 0, \qquad Y_1(\zeta) = 0 \tag{2.2-18}$$

The solution of the paraxial equations under this set of boundary conditions would be represented by $X_1(z) \equiv G(z)$, $P_1(z) \equiv \theta(z)$ such that $G(z_0) = 1$, $G(\zeta) = 0$ (see Fig. 2.2).

Thus $g(z)$, $G(z)$, $\vartheta(z)$, and $\theta(z)$ form two linearly independent solutions of the set of paraxial equations. Hence solutions of Eqs. (2.2-12) and (2.2-14) satisfying any general boundary condition can be expressed as linear combinations of these. This is so because we have a set of coupled first-order differential equations, which is equivalent to a second-order differential equation. Thus for a ray that satisfies the boundary conditions specified by Eqs. (2.2-3) and (2.2-4), the paraxial solution is given by

$$\begin{aligned} X_1(z) &= x_0 G(z) + \xi g(z) \\ Y_1(z) &= y_0 G(z) + \eta g(z) \end{aligned} \tag{2.2-19}$$

This can be verified by noting that $G(z_0) = 1$, $g(z_0) = 0$, and $G(\zeta) = 0$, $g(\zeta) = 1$. We also have

$$dg/dz = 2H_2\vartheta, \qquad d\vartheta/dz = -2H_1 g \tag{2.2-20}$$

$$dG/dz = 2H_2\theta, \qquad d\theta/dz = -2H_1 G \tag{2.2-21}$$

The above equations arise because (g, ϑ) and (G, θ) are special solutions of the paraxial equation [Eqs. (2.2-12) and (2.2-14)]. Now,

$$P_1(z) = \frac{1}{2H_2}\frac{dX_1}{dz} = \frac{1}{2H_2}(x_0\dot{G} + \xi\dot{g}) = x_0\theta + \xi\vartheta \tag{2.2-22}$$

Similarly,

$$Q_1(z) = y_0\theta + \eta\vartheta \tag{2.2-23}$$

We are interested in calculating the aberration (i.e., the value of X_3) in the paraxial image plane defined in the following manner: Let us consider an axial point in the object plane $z = z_0$. Then a ray leaving this point and traveling along the axis of symmetry of the system would not be deviated. Any other paraxial ray starting from this point would be represented by a multiple of $g(z)$. The image would be formed at the intersection of these two rays, i.e., at a value of z, say $z = z_1$, such that $g(z_1) = 0$, $z_1 \neq z_0$. This would then represent the paraxial image plane (see Fig. 2.2).

Since $g(z_1) = 0$, therefore in the image plane the paraxial solution [see Eq. (2.2-19)] would be simply

$$X_1(z_1) = x_0 G(z_1) \tag{2.2-24}$$

Since $G(z_1)$ is independent of x_0, y_0, ξ, and η, Eq. (2.2-24) says that the paraxial image of a point (which is at a distance x_0 from the z axis) in the object plane would be formed at the point that is at a distance $X_1(z_1)$ from the z axis in the image plane $z = z_1$, i.e., all rays emanating from the point $(x_0, 0)$ in the plane $z = z_0$ would meet at $[X_1(z_1), 0]$, irrespective of the values of ξ and η. Thus the image would be an ideal image.

Let us now calculate the aberrations X_3 and Y_3 of a ray satisfying the boundary conditions given by Eqs. (2.2-3) and (2.2-4) in terms of the functions $g(z)$, $\vartheta(z)$, $G(z)$, and $\theta(z)$. If we solve the paraxial equations with the boundary conditions

$$X_1(z_0) = x_0, \qquad X_1(\zeta) = \xi \tag{2.2-25}$$

$$Y_1(z_0) = y_0, \qquad Y_1(\zeta) = \eta \tag{2.2-26}$$

then clearly we have

$$X_3(z_0) = X_5(z_0) = \cdots = 0, \qquad Y_3(z_0) = Y_5(z_0) = \cdots = 0 \tag{2.2-27}$$

The solution of the paraxial equations under the boundary conditions given by Eqs. (2.2-25) and (2.2-26) is indeed given by Eqs. (2.2-19), (2.2-22), and (2.2-23). Substituting for $2H_2$ from Eq. (2.2-20) into Eq. (2.2-13) we get

$$\dot{X}_3\vartheta - \dot{g}P_3 = 2(H_{12}U_1 + H_{22}V_1)P_1\vartheta \tag{2.2-28}$$

Similarly, substituting for $2H_1$ from Eq. (2.2-20) in Eq. (2.2-15) we get

$$X_3\dot{\vartheta} - g\dot{P}_3 = 2(H_{11}U_1 + H_{12}V_1)X_1 g \tag{2.2-29}$$

Adding Eqs. (2.2-28) and (2.2-29) we get

$$\frac{d}{dz}(X_3\vartheta - gP_3) = 2[(H_{12}U_1 + H_{22}V_1)P_1\vartheta + (H_{11}U_1 + H_{12}V_1)X_1g] \tag{2.2-30}$$

Integrating the above equation from the object plane $z = z_0$ to the image plane $z = z_1$, we get

$$(X_3\vartheta - gP_3)\Big|_{z_0}^{z_1} = 2\int_{z_0}^{z_1} [(H_{11}U_1 + H_{12}V_1)X_1g + (H_{12}U_1 + H_{22}V_1)P_1\vartheta]\,dz \tag{2.2-31}$$

Using the fact that $g(z_0) = 0$, $X_3(z_0) = 0$, $g(z_1) = 0$, we get

$$X_3(z_1) = \frac{2}{\vartheta(z_1)}\int_{z_0}^{z_1} [(H_{11}U_1 + H_{12}V_1)X_1g + (H_{12}U_1 + H_{22}V_1)P_1\vartheta]\,dz \tag{2.2-32}$$

which represents the aberration of the ray along the x direction in the paraxial image plane. Similarly,

$$Y_3(z_1) = \frac{2}{\vartheta(z_1)}\int_{z_0}^{z_1} [(H_{11}U_1 + H_{12}V_1)Y_1g + (H_{12}U_1 + H_{22}V_1)Q_1\vartheta]\,dz \tag{2.2-33}$$

which represents the aberration of the ray along the y direction in the paraxial image plane. We write X_3 and Y_3 explicitly in terms of g, G, ϑ, and θ by observing that

$$\begin{aligned} U_1 = X_1^2 + Y_1^2 &= [x_0G(z) + \xi g(z)]^2 + [y_0G(z) + \eta g(z)]^2 \\ &= rG^2 + 2tGg + sg^2 \end{aligned} \tag{2.2-34}$$

where

$$r = x_0^2 + y_0^2, \qquad t = x_0\xi + y_0\eta, \qquad s = \xi^2 + \eta^2$$

Similarly,

$$\begin{aligned} V_1 = P_1^2 + Q_1^2 &= [x_0\theta(z) + \xi\vartheta(z)]^2 + [y_0\theta(z) + \eta\vartheta(z)]^2 \\ &= r\theta^2 + 2t\theta\vartheta + s\vartheta^2 \end{aligned} \tag{2.2-35}$$

We next substitute the expressions for U_1, V_1, X_1, P_1 from Eqs. (2.2-34), (2.2-35), (2.2-19), and (2.2-22) into Eq. (2.2-32). For example, the term

$\int (H_{11}U_1 + H_{12}V_1) X_1 g\, dz$ becomes

$$\int [H_{11}(rG^2 + 2tGg + sg^2) + H_{12}(r\theta^2 + 2t\theta\vartheta + s\vartheta^2)](x_0 G + \xi g) g\, dz$$

$$= \left[\int (H_{11}G^3 g + H_{12}\theta^2 G g)\, dz\right] r x_0 + \left[\int (H_{11}G^2 g^2 + H_{12}\theta^2 g^2)\, dz\right] r\xi$$

$$+ \left[\int (2H_{11}G^2 g^2 + 2H_{12}\theta\vartheta G g)\, dz\right] t x_0$$

$$+ \left[\int (2H_{11}G g^3 + 2H_{12}\theta\vartheta g^2)\, dz\right] t\xi$$

$$+ \left[\int (H_{11}g^3 G + H_{12}\vartheta^2 G g)\, dz\right] s x_0$$

$$+ \left[\int (H_{11}g^4 + H_{12}\vartheta^2 g^2)\, dz\right] s\xi \tag{2.2-36}$$

In a similar manner the other terms can also be calculated. The final result can be put in the form

$$\begin{aligned} X_3(z_1) &= [As + 2Bt + (C + D)r]\,\xi + [Bs + 2Ct + Er]\,x_0 \\ Y_3(z_1) &= [As + 2Bt + (C + D)r]\,\eta + [Bs + 2Ct + Er]\,y_0 \end{aligned} \tag{2.2-37}$$

where A, B, C, D, and E are known as the aberration coefficients and are defined by the relations

$$\begin{aligned} A &= \frac{2}{\vartheta(z_1)} \int_{z_0}^{z_1} [H_{11}g^4 + 2H_{12}g^2\vartheta^2 + H_{22}\vartheta^4]\, dz \\ B &= \frac{2}{\vartheta(z_1)} \int_{z_0}^{z_1} [H_{11}Gg^3 + H_{12}g\vartheta(g\theta + G\vartheta) + H_{22}\theta\vartheta^3]\, dz \\ C &= \frac{2}{\vartheta(z_1)} \int_{z_0}^{z_1} [H_{11}G^2g^2 + 2H_{12}Gg\theta\vartheta + H_{22}\theta^2\vartheta^2]\, dz \\ D &= \frac{2}{\vartheta(z_1)} \Gamma^2 \int_{z_0}^{z_1} H_{12}\, dz \\ E &= \frac{2}{\vartheta(z_1)} \int_{z_0}^{z_1} [H_{11}G^3 g + H_{12}G\theta(g\theta + G\vartheta) + H_{22}\theta^3\vartheta]\, dz \end{aligned} \tag{2.2-38}$$

where Γ, called the paraxial optical invariant of the system (see Problem

2.1), is defined by

$$\Gamma = G\vartheta - g\theta \tag{2.2-39}$$

The coefficients A, B, C, D, and E are called the aberration coefficients, and their physical significance will be discussed in the next section. They depend only on the system parameters and the object and image planes.

Problem 2.1. Show that $d\Gamma/dz = 0$. [*Hint:* Use the differential equations satisfied by g, ϑ, G, and θ.]

Problem 2.2. If instead of choosing the boundary conditions followed in the text we had specified X, Y, P, and Q at the plane $z = z_0$ (say), then state the boundary conditions for the two fundamental paraxial rays (corresponding to g, ϑ, G, and θ) and obtain any general solution in terms of these.

2.3. *Physical Significance of the Coefficients A, B, C, D, and E*

In this section we study what kind of image pattern is produced separately by each of the terms containing A, B, C, D, and E. They correspond, respectively, to the five Seidel aberrations: spherical aberration, coma, astigmatism, curvature of field, and distortion.

Since the system is rotationally symmetric, we can without loss of generality choose our object point to lie on the x axis, i.e., $y_0 = 0$, and the object would have coordinates $(x_0, 0, z_0)$ with some chosen origin of reference. We will study the image pattern produced by the system when all rays passing through the system are considered. Let $z = \zeta$ be the exit pupil plane. It is easier to consider the image formed due to sets of rays striking the exit pupil plane at a constant distance ρ from the z axis. Then this distance can be varied from zero to the radius of the exit pupil to get the complete set of rays forming the image.

The values of ξ and η for rays hitting the exit pupil plane $z = \zeta$ at a constant distance ρ from the z axis are given by (see Fig. 2.3)

$$\xi = \rho \cos \varphi, \qquad \eta = \rho \sin \varphi \tag{2.3-1}$$

Different values of φ would correspond to the different rays striking the plane $z = \zeta$ along different points on the circle of radius ρ centered at $\xi = \eta = 0$. Since $y_0 = 0$ we have

$$r = x_0^2, \qquad t = x_0\rho \cos \varphi, \qquad s = \rho^2 \tag{2.3-2}$$

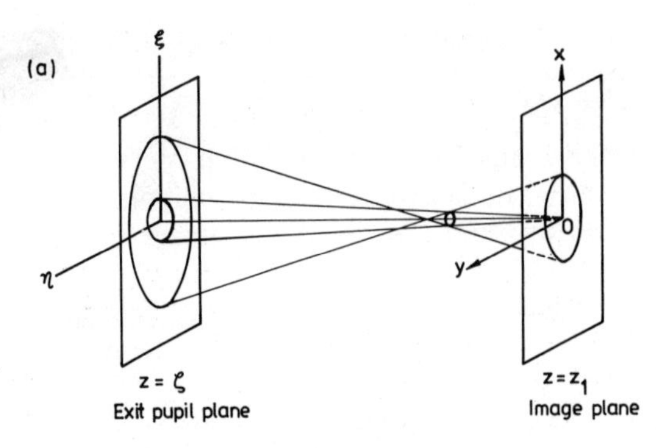

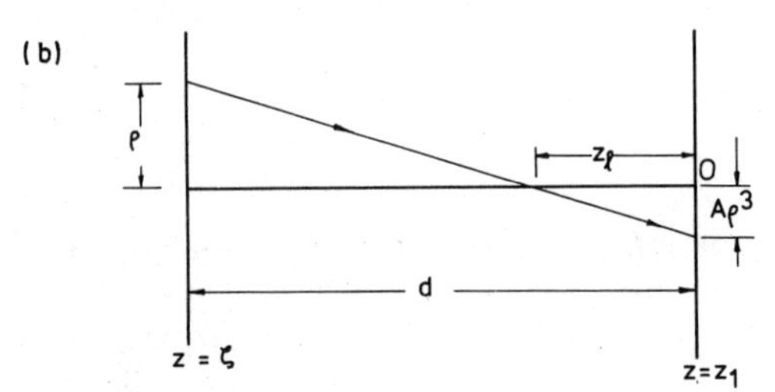

Fig. 2.3. (a) In the presence of spherical aberration, rays striking the exit pupil plane $z = \zeta$, at a constant distance ρ from the axis, intersect the image plane $z = z_1$ in a circle of radius $A\rho^3$ centered at the paraxial image point 0. The figure shows the combined effect of defocusing and spherical aberration, leading to the formation of a circle of least confusion. (b) Relation between lateral spherical aberration and longitudinal spherical aberration.

and the aberrations X_3 and Y_3 reduce to

$$X_3 = A\rho^3 \cos\varphi + Bx_0\rho^2(2 + \cos 2\varphi) + (3C + D)\,x_0^2\rho\cos\varphi + Ex_0^3$$

$$Y_3 = A\rho^3 \sin\varphi + Bx_0\rho^2 \sin 2\varphi + (C + D)\,x_0^2\rho\sin\varphi \tag{2.3-3}$$

2.3.1. Spherical Aberration

If A is the only nonzero coefficient, then the aberrations along the x and y directions are

$$X_3 = A\rho^3\cos\varphi, \qquad Y_3 = A\rho^3\sin\varphi \tag{2.3-4}$$

It should be noted that even in the presence of other aberrations, for an axial object (i.e., $x_0 = 0$, $y_0 = 0$) X_3 and Y_3 are rigorously given by Eq. (2.3-4), and the other aberrations do not make any contribution. From Eq. (2.3-4) we get

$$(X_3^2 + Y_3^2) = (A\rho^3)^2 \tag{2.3-5}$$

X_3 and Y_3 represent the deviations of the point of intersection of the ray from the paraxial image point, along the x and y directions, respectively. Consequently, rays that strike the exit pupil plane at a constant distance ρ

from the axis intersect the paraxial image plane $z = z_1$ in a circle of radius $A\rho^3$ whose center lies at the paraxial image point O (see Fig. 2.3a). Thus in the presence of this aberration the image of a point object is a circular patch of light. If the radius of the exit pupil is ρ_0, then the image would be a circular patch of light of radius $A\rho_0^3$. The distance $A\rho_0^3$ represents the lateral spherical aberration. Thus, when a set of rays parallel to the axis hits a lens, the rays striking the edge of the lens focus to a point called the marginal focus, which is different from the paraxial focus. To calculate the longitudinal spherical aberration (i.e., the aberration along the axis of the system) we refer to Fig. 2.3b and find

$$A\rho^3/\rho = z_l/(d - z_l)$$

or

$$z_l \simeq A\rho^2 d \tag{2.3-6}$$

where z_l (known as the longitudinal spherical aberration) is the distance between the marginal focus and the paraxial focus and d is the distance between the exit pupil and the image plane. Thus the paraxial rays focus at the paraxial image plane and the rays coming from the edge of the exit aperture of radius ρ_0 focus at a distance z_l $(= A\rho_0^2 d)$ from the paraxial image plane. When a screen is placed at the paraxial image plane then the image is a patch of light of radius $A\rho_0^3$. When the screen is placed at the marginal focus, the image is again a patch of light, due to the defocusing of the paraxial rays. Hence somewhere in between, one expects the circular patch to have a minimum diameter. This image is called the circle of least confusion (see Fig. 2.3a). Figure 2.4 explicitly shows the spherical aberration of a lens.

Fig. 2.4. The spherical aberration of a lens (courtesy of K. K. Gupta and A. Sharma).

2.3.2. Coma

As already mentioned, the only aberration present in the image for an on-axis object point is spherical aberration. The first off-axis aberration from which the image suffers is coma. This aberration is represented by B. For very small values of x_0, as can be seen from Eq. (2.3-3) the aberration in the image is due to A and B only. Let us consider the effect of the term containing B on the image of a point object. The aberrations are, from Eq. (2.3-3),

$$\begin{aligned} X_3 &= Bx_0\rho^2(2 + \cos 2\varphi) \\ Y_3 &= Bx_0\rho^2 \sin 2\varphi \end{aligned} \tag{2.3-7}$$

Eliminating φ from Eq. (2.3-7), we get

$$(X_3 - 2Bx_0\rho^2)^2 + Y_3^2 = (Bx_0\rho^2)^2 \tag{2.3-8}$$

Equation (2.3-8) represents a circle of radius $Bx_0\rho^2$ whose center is at a distance $2Bx_0\rho^2$ from the paraxial image point. It can be seen from Eq. (2.3-7) that for a given ρ as φ varies from 0 to 2π, the image point moves twice over a circle of radius $Bx_0\rho^2$, and thus rays with values of φ differing by π intersect the image plane at the same point (see Fig. 2.5a). Thus as ρ increases, i.e., rays from larger and larger portions of the exit pupil are considered, the image radius increases and the center shifts away from

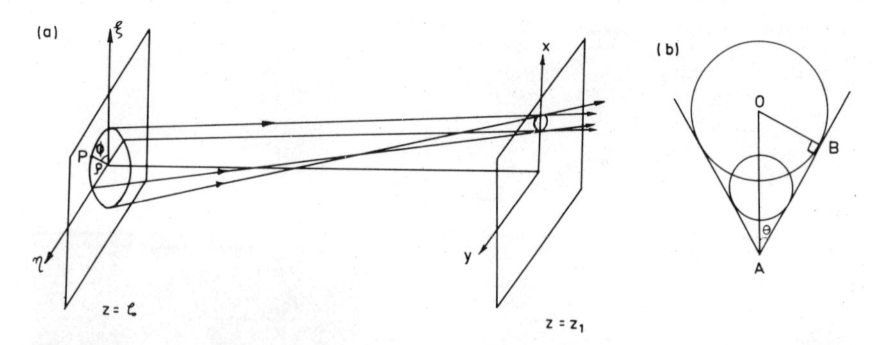

Fig. 2.5. (a) In the presence of coma, the rays passing through the circumference of a circle of radius ρ on the exit pupil plane ($z = \zeta$) intersect the image plane on a circle of radius $Bx_0\rho^2$ [$= OB$ (see Fig. b)] whose center is displaced from the paraxial image point A by $2Bx_0\rho^2$ [$= OA$ (see Fig. b)]. Notice that rays with the same value of ρ and values of φ differing by π intersect the image plane at the same point. (b) The image of a point object afflicted with coma is confined within two lines subtending angles $\theta = 30°$ with the x direction.

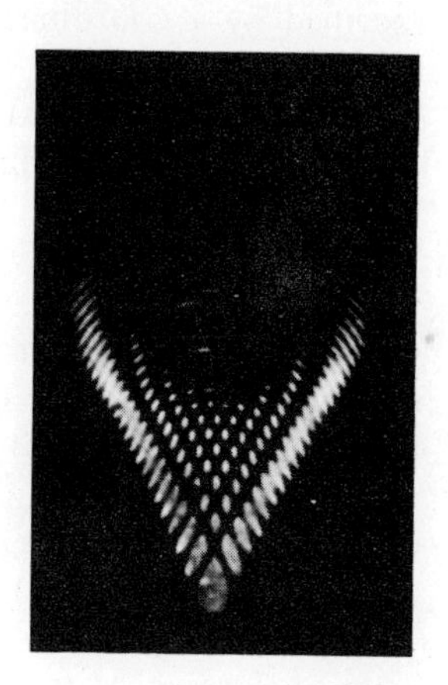

Fig. 2.6. The image of a point source by a system suffering from coma. The fringes are due to diffraction effects. (After Born and Wolf, 1975. Reprinted with permission.)

the paraxial image point. These circles have a common tangent and the angle subtended by the two tangents with the x axis is 30°, which can easily be seen from the following consideration. From Fig. 2.5b it follows that

$$\sin\theta = OB/OA = Bx_0\rho^2/2Bx_0\rho^2 = \tfrac{1}{2} \qquad (2.3\text{-}9)$$

which gives $\theta = 30°$. Thus the image is confined between two lines subtending angles of 30° with the x direction. As can be seen from Fig. 2.5b, the image pattern has a cometlike appearance, and hence this aberration is known as coma. Figure 2.6 shows the image of a point object in the presence of coma. One can see the cometlike appearance of the image; however, the fringes are due to diffraction (see, e.g., Born and Wolf, 1975, Chapter IX).

2.3.3. *Astigmatism and Curvature of Field*

It would be easier to consider these two phenomena simultaneously. If C and D are the only nonzero coefficients, then the aberrations are given by [see Eq. (2.3-3)]

$$X_3 = (3C + D)\, x_0^2\rho\cos\varphi, \qquad Y_3 = (C + D)\, x_0^2\rho\sin\varphi \qquad (2.3\text{-}10)$$

These equations can be combined to give

$$\frac{X_3^2}{[(3C + D)\, x_0^2\rho]^2} + \frac{Y_3^2}{[(C + D)\, x_0^2\rho]^2} = 1 \qquad (2.3\text{-}11)$$

which is the equation of an ellipse with semiaxes $(3C + D)\, x_0^2\rho$ and $(C + D)\, x_0^2\rho$, with its center at the paraxial image point. Thus rays emanating from a given value of ρ intersect the paraxial image plane in an ellipse (see Fig. 2.7a).

Let us consider a plane parallel to the paraxial image plane and specified by a coordinate z measured from the plane AB (see Fig. 2.7b). Let $(\bar{x}, \bar{y})$ represent the variables in this plane (see Fig. 2.7b). Let d be the distance between the exit pupil plane and the paraxial image plane. Then by consideration of similar triangles we can obtain*

$$\bar{x} = -\frac{z}{d}\xi + \left(1 + \frac{z}{d}\right)x_1 \tag{2.3-12}$$

where (x_1, y_1) represent the coordinates of the point of intersection of the ray with the paraxial image plane, and

$$x_1 = X_1 + (3C + D)\, x_0^2 \xi \tag{2.3-13}$$

where the point (X_1, Y_1) represents the paraxial image point. Substituting for x_1 from Eq. (2.3-13) into Eq. (2.3-12), we get

$$\bar{x} = \frac{d+z}{d} X_1 + \left[-\frac{z}{d} + \frac{d+z}{d}(3C + D)\, x_0^2 \right] \xi \tag{2.3-14}$$

Similarly, we can obtain

$$\bar{y} = \left[-\frac{z}{d} + \frac{d+z}{d}(C + D)\, x_0^2 \right] \eta \tag{2.3-15}$$

where we have used the fact that $Y_1 = 0$ since the object lies in the x-z plane. Since $\xi^2 + \eta^2 = \rho^2$, we have

$$\frac{[\bar{x} - (d + z) X_1/d]^2}{[-z/d + (d + z)(3C + D)\, x_0^2/d]^2} + \frac{\bar{y}^2}{[-z/d + (d + z)(C + D)\, x_0^2/d]^2} = \rho^2 \tag{2.3-16}$$

Equation (2.3-16) tells us that in a plane at a distance $-z$ from the paraxial image plane, a set of rays with a fixed ρ and varying φ form an ellipse centered at the point $\{[(d + z)/d]\, X_1, 0\}$. For certain values of z, the ellipse degenerates into a straight line. When $z = z_T$ such that

$$-\frac{z_T}{d} + \frac{d + z_T}{d}(3C + D)\, x_0^2 = 0$$

or

$$z_T \simeq (3C + D)\, x_0^2 d \tag{2.3-17}$$

then the ellipse degenerates into a straight line along the y direction (shown as a in Fig. 2.7a). This is called the tangential focus. Similarly, when $z = z_S$

* The space between the exit pupil and the image plane is assumed to be homogeneous.

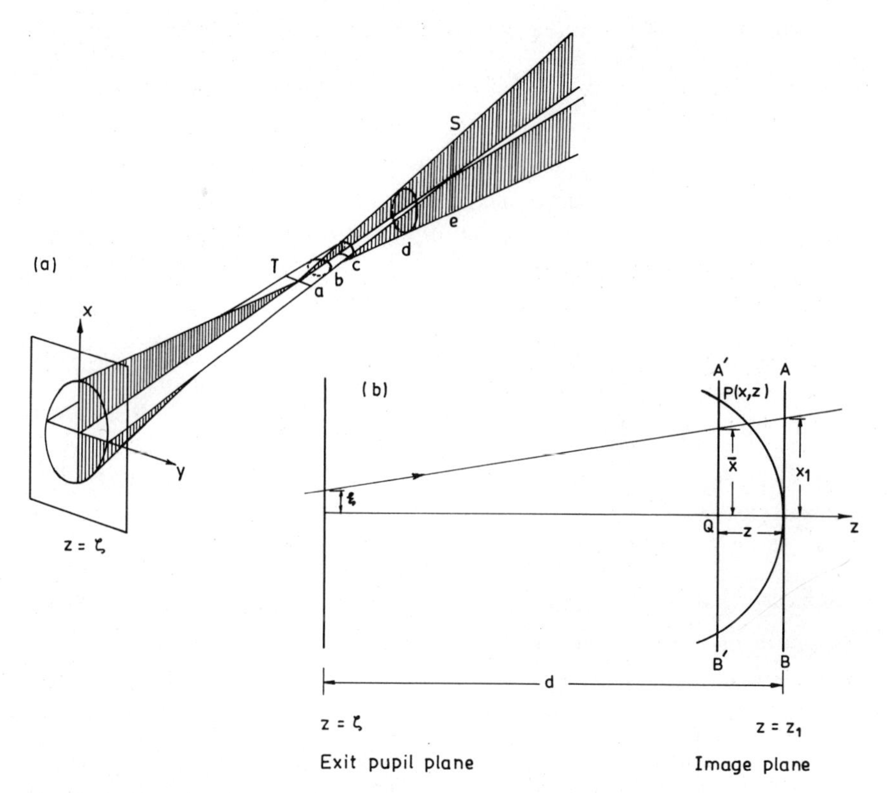

Fig. 2.7. (a) Focusing of rays in the presence of astigmatism. *a* represents the tangential focus and *e* the sagittal focus. (b) *AB* represents the paraxial image plane and *A′B′* any other nearby plane parallel to *AB*, at a distance *z* from it.

such that

$$-\frac{z_S}{d} + \frac{d + z_S}{d}(C + D)\,x_0^2 = 0$$

or

$$z_S \simeq (C + D)\,x_0^2 d \tag{2.3-18}$$

the ellipse degenerates into a straight line along the x direction (shown as *e* in Fig. 2.7a). This is called the sagittal focus. The distance between these two foci, namely,

$$z_T - z_S = 2Cx_0^2 d \tag{2.3-19}$$

is a measure of astigmatism. Thus the two foci would coincide when $C = 0$.

The ellipse degenerates into a circle when

$$\left[-\frac{z}{d}+\frac{d+z}{d}(3C+D)\,x_0^2\right]=\pm\left[-\frac{z}{d}+\frac{d+z}{d}(C+D)\,x_0^2\right] \tag{2.3-20}$$

If we use the positive sign, we obtain $C = 0$, which corresponds to no astigmatism. Thus in the absence of astigmatism, the image is circular at all planes, which would degenerate to a point at $z = Dx_0^2 d$. On the other hand, if we use the negative sign in Eq. (2.3-20), we get

$$z = (2C + D)\,x_0^2 d$$

i.e., the image is circular only at a fixed plane. This image is called the circle of least confusion and is shown as c in Fig. 2.7a. Notice that this plane coincides with the paraxial image plane when $C = -D/2$, which will be shown to correspond to zero curvature of field. If M is the magnification of the system, then to the third order of approximation, we can substitute

$$x_0 = x_1/M \tag{2.3-21}$$

in Eqs. (2.3-17) and (2.3-18) and get

$$z_T = \frac{(3C+D)}{M^2}\,dx_1^2, \qquad z_S = \frac{(C+D)}{M^2}\,dx_1^2 \tag{2.3-22}$$

The above equations represent equations of a circle for small x_1; this can be seen as follows. Consider a circle of radius R and choose the origin to lie on the circumference of the circle as shown in Fig. 2.7b. Let $P(x, z)$ be any point on the circle. If PQ represents the perpendicular from P to the z axis, then from geometry one obtains rigorously $-z(-2R + z) = x^2$, the solution of which (that goes to 0 as $x \to 0$) is

$$z = R - R\left(1 - \frac{x^2}{R^2}\right)^{1/2} \simeq \frac{x^2}{2R} \tag{2.3-23}$$

Thus the curves given by Eq. (2.3-22) represent circles of radii R_T and R_S given by

$$\frac{1}{R_T} = \frac{2d}{M^2}(3C + D), \qquad \frac{1}{R_S} = \frac{2d}{M^2}(C + D) \tag{2.3-24}$$

These are called the tangential and sagittal field curvatures, respectively. The mean of these two,

$$\frac{1}{R} = \frac{1}{2}\left(\frac{1}{R_T} + \frac{1}{R_S}\right) = \frac{2d}{M^2}(2C + D) \tag{2.3-25}$$

is called the field curvature.

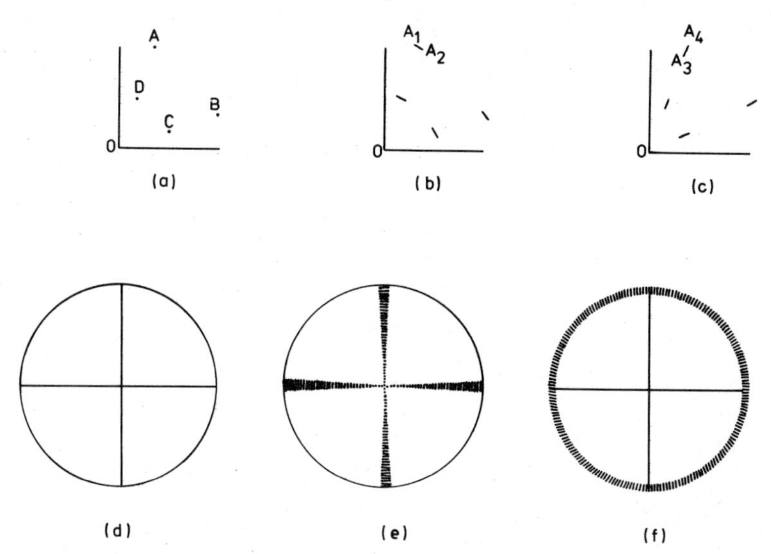

Fig. 2.8. If the object is of the form shown in (a) [or (d)], then the image formed in the tangential plane will be of the form (b) [or (e)] and the image in the sagittal plane will be of the form (c) [or (f)].

Let us consider four object points A, B, C, and D in the object plane (see Fig. 2.8a). In the tangential plane, each point will be imaged as a straight line, which will be perpendicular to the line joining the object point and the axial point O. For example, the point A will be imaged along a line A_1A_2, which is in a direction perpendicular to OA. Similarly, in the sagittal plane the point A will be imaged along a line A_3A_4, which is in a direction parallel to OA. Thus, if we have a spoked wheel in the object plane (see Fig. 2.8d) then the image in the tangential and sagittal planes will be similar to that shown in Figs. 2.8e and 2.8f.

2.3.4. Distortion

The coefficient E represents the distortion in the image. In order to study its physical significance, we consider an object point specified by the coordinates (x_0, y_0) in the plane $z = z_0$. Neglecting the effect of other aberrations (i.e., assuming $A = B = C = D = 0$), Eq. (2.2-37) becomes

$$X_3 = E(x_0^2 + y_0^2)\,x_0, \qquad Y_3 = E(x_0^2 + y_0^2)\,y_0 \tag{2.3-26}$$

Let us consider a square grid as shown in Fig. 2.9a. Let a represent the spacing between the lines. We need consider only one of the quadrants in detail; the others follow from symmetry. The coordinates of various intersection

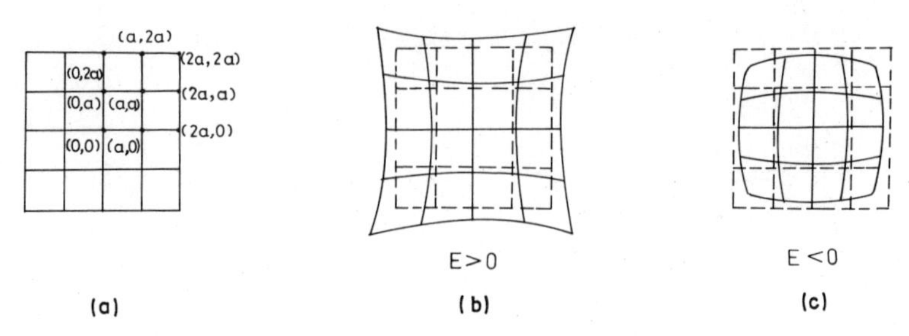

Fig. 2.9. Imaging of a grid in the presence of distortion.

points would be $(0, 0)$, $(a, 0)$, $(2a, 0)$, $(0, a)$, $(0, 2a)$, (a, a), $(a, 2a)$, $(2a, a)$, $(2a, 2a)$ (see Fig. 2.9a). The coordinates of the image of an object point (x_0, y_0) in the presence of distortion would be

$$\begin{aligned} x_1 &= Mx_0 + E(x_0^2 + y_0^2)\,x_0 \\ y_1 &= My_0 + E(x_0^2 + y_0^2)\,y_0 \end{aligned} \tag{2.3-27}$$

where M is the magnification of the system. Considering, for simplicity, the system to have unit positive magnification, the image of the above points would have coordinates

$$\begin{gathered} (0, 0), \quad (a + Ea^3, 0), \quad (2a + 4Ea^3, 0), \quad (0, a + Ea^3), \quad (0, 2a + 4Ea^3) \\ (a + 2Ea^3, a + 2Ea^3), \qquad (a + 5Ea^3, 2a + 10Ea^3) \\ (2a + 10Ea^3, a + 5Ea^3), \quad (2a + 16Ea^3, 2a + 16Ea^3) \end{gathered}$$

These have been plotted in Figs. 2.9b and 2.9c for positive and negative E, respectively. The dashed lines show the ideal image. It follows at once how positive E modifies the image into a form called pincushion distortion and negative E into a form called barrel distortion.

Physically, if we have a pinhole on the axis at the exit aperture, then the only aberration present in the image plane is distortion. This can be seen easily if we assume $\xi = \eta = 0$; then Eqs. (2.2-37) reduce to Eqs. (2.3-26). What the pinhole does is to allow only those rays that satisfy $\xi = \eta = 0$ (see Fig. 2.10). Further, when there is a pinhole, then corresponding to a point on the object plane only one ray will reach the image plane. Consequently there cannot be any aberration present except distortion of the object due to nonuniform magnification.

We now consider the case when all the third-order aberrations are present simultaneously. The resultant image shape will be a combination of all the aberration image shapes we have so far considered.

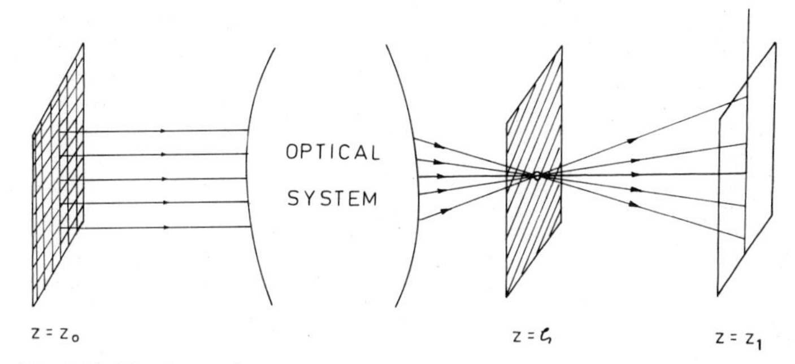

Fig. 2.10. The image formed when the exit pupil plane is a pinhole on the axis suffers only from distortion.

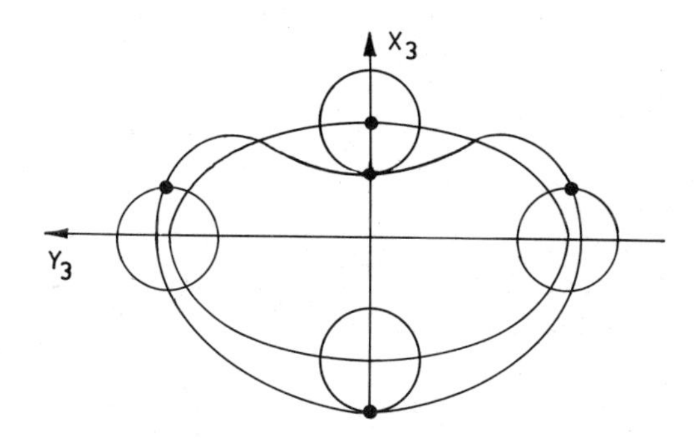

Fig. 2.11. Form of the image of a point object in the presence of all the aberrations (after Luneburg, 1964; reprinted with permission).

Equations (2.3-3) can be combined to yield

$$\{X_3 - 2Bx_0\rho^2 - Ex_0^3 - [A\rho^3 + (3C + D)\,x_0^2\rho]\cos\varphi\}^2 + \{Y_3 - [A\rho^3 + (C + D)\,x_0^2\rho]\sin\varphi\}^2 = (Bx_0\rho^2)^2 \quad (2.3\text{-}28)$$

Equation (2.3-28) is the equation of a circle of radius $Bx_0\rho^2$ whose center lies on an ellipse with axes $A\rho^3 + (3C + D)\,x_0^2\rho$ and $A\rho^3 + (C + D)\,x_0^2\rho$. The center of the ellipse in turn lies at $(2Bx_0\rho^2 + Ex_0^3, 0)$. The composite shape is shown in Fig. 2.11. Every time the center of the circle moves around the ellipse once, any point on the circumference of the circle moves around twice.

2.4. The Coefficients H_{ij} in Terms of Refractive-Index Variation

Since the series on the right-hand side of Eq. (2.2-7) is essentially a Taylor series expansion of the Hamiltonian H around $U = 0$, $V = 0$, we may write the coefficients H_1, H_2, etc., in the following form:

$$H_1 = \left.\frac{\partial H}{\partial U}\right|_{U=0,V=0}, \qquad H_2 = \left.\frac{\partial H}{\partial V}\right|_{U=0,V=0}$$
$$H_{11} = \left.\frac{\partial^2 H}{\partial U^2}\right|_{U=0,V=0}, \qquad H_{12} = \left.\frac{\partial^2 H}{\partial U\,\partial V}\right|_{U=0,V=0}, \qquad H_{22} = \left.\frac{\partial^2 H}{\partial V^2}\right|_{U=0,V=0} \tag{2.4-1}$$

Using the expression $H = -[n^2(U, z) - V]^{1/2}$, we obtain

$$H_1 = -\left.\frac{\partial[n^2(U, z) - V]^{1/2}}{\partial U}\right|_{U=0,V=0}$$
$$= -\left.\frac{n(U, z)}{[n^2(U, z) - V]^{1/2}}\frac{\partial n(U, z)}{\partial U}\right|_{U=0,V=0} = -\left.\frac{\partial n(U, z)}{\partial U}\right|_{U=0} \tag{2.4-2}$$

Similarly

$$H_2 = \frac{1}{2n(0, z)}, \qquad H_{11} = -\left.\frac{\partial^2 n(U, z)}{\partial U^2}\right|_{U=0}$$
$$H_{12} = -\left.\frac{1}{2n^2(0, z)}\frac{\partial n(U, z)}{\partial U}\right|_{U=0}, \qquad H_{22} = \frac{1}{4n^3(0, z)} \tag{2.4-3}$$

Thus given an index distribution $n(U, z)$, the above coefficients can be calculated. Using the values of H_1 and H_2, the paraxial equations have to be solved under specific boundary conditions to obtain the axial and the field rays. Once they have been obtained, the explicit values for the aberration coefficients can be calculated easily.

In Section 2.5, we make an explicit calculation of the third-order aberration in a continuously varying inhomogeneous medium. In Section 2.6 the aberrations of systems consisting of lenses formed of homogeneous media will be considered. The general case of systems consisting of inhomogeneous media separated by surfaces has been discussed by Thyagarajan and Ghatak (1976) and will not be dealt with here.

2.5. Aberrations of Graded-Index Media

In this section we will calculate the aberrations in a medium characterized by the following refractive-index variation:

$$n(U, z) = n_0(1 - \tfrac{1}{2}\alpha^2 U + \tfrac{1}{2}\beta\alpha^4 U^2) \tag{2.5-1}$$

A general rotationally symmetric inhomogeneous medium contains higher-order terms also, like U^3, U^4. However, we need not consider these higher-order terms since they would be important only for fifth- and higher-order aberrations.* This is because of the fact that the coefficients of the cubic and higher powers of U and V in the expansion of the Hamiltonian H [see Eq. (2.2-7)] do not contribute to third-order aberration. It should be mentioned that fibers possessing such a refractive-index variation have recently been fabricated (see, e.g., Kitano *et al.*, 1970). Such media exhibit focusing properties (see Section 1.3) and can be used in the form of image relays and as microlenses of ultrashort focal lengths. As such the study of aberrations introduced by such media is important. (See Marchand, 1973.) These fibers can also be used in optical communications (see Chapter 10).

Although n_0, α, and β may in general be continuous functions† of z, we consider them to be independent of z, which is almost true in actual practical fibers. For such a refractive-index distribution,

$$H_1 = \frac{n_0\alpha^2}{2}, \qquad H_2 = \frac{1}{2n_0}$$
$$H_{11} = -n_0\beta\alpha^4, \qquad H_{12} = \frac{\alpha^2}{4n_0}, \qquad H_{22} = \frac{1}{4n_0^3} \tag{2.5-2}$$

and hence the paraxial equations [Eqs. (2.2-12) and (2.2-14)], from which the axial and field rays may be derived, reduce to

$$\frac{dX_1}{dz} = \frac{1}{n_0}P_1, \qquad \frac{dP_1}{dz} = -n_0\alpha^2 X_1 \tag{2.5-3}$$

The general solution of Eq. (2.5-3) when n_0 and α are independent of z is

$$X_1 = A\sin\alpha z + B\cos\alpha z$$
$$P_1 = n_0\alpha(A\cos\alpha z - B\sin\alpha z) \tag{2.5-4}$$

We choose the plane $z = \zeta$ such that

$$\sin\alpha\zeta = 1 \tag{2.5-5}$$

* An analysis of fifth-order aberrations of inhomogeneous media has recently been carried out by Gupta *et al.* (1976).

† Conical Selfoc fibers are characterized by a refractive-index variation of the form

$$n(U, z) = n_0[1 - \tfrac{1}{2}\alpha^2\kappa(z)U + \tfrac{1}{2}\beta\alpha^4U^2]$$

where $\kappa(z) = (1 - \sigma z)^{-2}$, σ being a constant. Third-order aberrations of such media have been obtained by Thyagarajan *et al.* (1976).

Under such a condition, the axial (g, ϑ) and the field (G, θ) rays are given by

$$
\begin{aligned}
g(z) &= \sin \alpha z, \qquad \vartheta(z) = n_0 \alpha \cos \alpha z \\
G(z) &= \cos \alpha z, \qquad \theta(z) = -n_0 \alpha \sin \alpha z
\end{aligned} \tag{2.5-6}
$$

where the object plane has been chosen to be $z = 0$. Hence we have a series of paraxial image planes at values of z satisfying

$$\sin \alpha z_1 = 0 \tag{2.5-7}$$

i.e., $\alpha z_1 = m\pi$, $m = 1, 2, 3, \ldots$. Choosing $m = 2$ for simplicity, we have

$$z_1 = 2\pi/\alpha \tag{2.5-8}$$

Having obtained g, ϑ, G, θ, H_{11}, H_{12}, H_{22}, z_0, and z_1, we have to substitute these in Eqs. (2.2-38) to get explicit values of A, B, C, D, and E. For example, the coefficient of spherical aberration is given by

$$
\begin{aligned}
A &= \frac{2}{n_0 \alpha} \int_0^{z_1} \left[-n_0 \beta \alpha^4 \sin^4 \alpha z + \frac{\alpha^2}{2n_0} n_0^2 \alpha^2 \cos^2 \alpha z \sin^2 \alpha z \right. \\
&\qquad \left. + \frac{1}{4n_0^3} n_0^4 \alpha^4 \cos^4 \alpha z \right] dz \\
&= z_1 \alpha^3 \left(\tfrac{5}{16} - \tfrac{3}{4}\beta\right)
\end{aligned} \tag{2.5-9}
$$

where use has been made of the fact that if

$$I_n \equiv \int_0^{z_1} (\sin \alpha z)^n (\cos \alpha z)^{4-n}\, dz \tag{2.5-10}$$

then

$$I_0 = \frac{3z_1}{8}, \qquad I_1 = 0, \qquad I_2 = \frac{z_1}{8}, \qquad I_3 = 0, \qquad I_4 = \frac{3z_1}{8} \tag{2.5-11}$$

Similarly

$$B = 0, \qquad C = -\frac{z_1 \alpha^3}{4}\left(\beta + \tfrac{1}{4}\right), \qquad D = \tfrac{1}{2} z_1 \alpha^3, \qquad E = 0 \tag{2.5-12}$$

Thus the system is free from coma ($B = 0$) and distortion ($E = 0$). The total aberrations along the x and y directions can now be written as [see Eq. (2.2-37)]

$$
\begin{aligned}
X_3 &= \left(\tfrac{5}{16} - \tfrac{3}{4}\beta\right) z_1 \alpha^3 \xi (x_0^2 + \xi^2 + \eta^2) \\
Y_3 &= z_1 \alpha^3 \left[\left(\tfrac{7}{16} - \tfrac{1}{4}\beta\right) x_0^2 + \left(\tfrac{5}{16} - \tfrac{3}{4}\beta\right)(\xi^2 + \eta^2) \right] \eta
\end{aligned} \tag{2.5-13}
$$

where, as before, the object point is specified by $(x_0, 0)$.

Let us first consider the case when the image is formed by meridional rays only. These rays satisfy the condition that they always remain in a plane containing the optical axis, which in the present case is the z axis. Since the object point lies in the x-z plane, the condition for a ray to be meridional would be $\eta = 0$. In such a case, from Eq. (2.5-13) we get

$$X_3 = (\tfrac{5}{16} - \tfrac{3}{4}\beta)\, z_1 \alpha^3 \xi (x_0^2 + \xi^2), \qquad Y_3 = 0 \tag{2.5-14}$$

It can be seen from Eq. (2.5-14) that when $\beta = 5/12$, the third-order aberrations in meridional rays can be made identically zero. In such a case the index distribution given by Eq. (2.5-1) reduces to

$$n(U) = n_0(1 - \tfrac{1}{2}\alpha^2 U + \tfrac{5}{24}\alpha^4 U^2) \tag{2.5-15}$$

This can be seen to be the first three terms of an index function

$$n(U) = n_0 \operatorname{sech}(\alpha U^{1/2}) \tag{2.5-16}$$

We have here been able to prove that such an index distribution is free from third-order aberrations as far as meridional rays are concerned, but it can be proved by other detailed considerations that such an index distribution is indeed free from all orders of aberrations as far as meridional rays are concerned (see Problem 1.11). Meridional rays are not the only type of rays that take part in image formation. Of course, if the object point happened to be axial, all rays emanating from the object would be meridional and such an index distribution would produce ideal images. This can be seen from the fact (as already observed in Section 2.3) that the only type of aberration present for an axial object is the spherical aberration, and for an index distribution given by Eq. (2.5-16) the spherical aberration is identically zero.

Rays that do not remain in one plane containing the optical axis are called skew rays. A special kind of these are helical rays, which maintain a constant distance from the optical axis as they travel through the medium; they trace a helix and hence the name (see Fig. 1.5b). For such helical rays (see Problem 1.7)

$$X_1(z) = x_0 \cos \alpha z, \qquad Y_1(z) = x_0 \sin \alpha z \tag{2.5-17}$$

At the reference plane, $z = \zeta = \pi/2\alpha$, and $X_1(\zeta) = \xi = 0$ and $Y_1(\zeta) = \eta = x_0$. Thus using Eq. (2.5-13), we obtain

$$X_3 = 0, \qquad Y_3 = z_1 \alpha^3 \eta^3 (\tfrac{3}{4} - \beta) \tag{2.5-18}$$

Hence for helical rays to be free from third-order aberrations we must have $\beta = \frac{3}{4}$. Under such a condition the refractive-index distribution given by Eq. (2.5-1) becomes

$$n(U) = n_0(1 - \tfrac{1}{2}\alpha^2 U + \tfrac{3}{8}\alpha^4 U^2) \tag{2.5-19}$$

which can be seen to represent the first three terms of

$$n(U) = n_0/(1 + \alpha^2 U)^{1/2} \tag{2.5-20}$$

This refractive-index distribution can also be derived from other considerations (see Problem 2.3), and it can be shown that for such an index variation the helical rays travel without aberrations of any order.

Thus it is clear from the above analysis how certain specific index distributions are free from aberrations for certain specific subclasses of rays, meridional and helical. Since all kinds of rays, including skew rays, are involved in image formation, we can see that any particular index distribution cannot be completely free from aberrations for all kinds of rays. In the above analysis we have not considered any surfaces to be present in the system. These would also add to the aberrations. Here lies the importance of inhomogeneous media, because the aberrations can be controlled by changing the parameters of the inhomogeneous medium in a system.

Problem 2.3. Starting from Fermat's principle, obtain the refractive-index distribution, which is free from aberration for helical rays.

Solution. The arclength in the cylindrical system of coordinates is given by [cf. Eq. (1.3-2)]

$$ds = \left[1 + \left(\frac{dr}{dz}\right)^2 + r^2\left(\frac{d\theta}{dz}\right)^2\right]^{1/2} dz \tag{2.5-21}$$

Since we are concerned at present with helical rays, i.e., rays that maintain a constant distance from the z axis, we have for such rays

$$dr/dz = 0 \tag{2.5-22}$$

and therefore the Lagrangian is given by [cf. Eq. (1.3-4)]

$$L = n(r)\left[1 + r^2\left(\frac{d\theta}{dz}\right)^2\right]^{1/2} \tag{2.5-23}$$

The Lagrangian equation for the variable r is

$$\frac{d}{dz}\left(\frac{\partial L}{\partial \dot{r}}\right) = \frac{\partial L}{\partial r} \tag{2.5-24}$$

(where dots represent derivatives with respect to z). Using Eq. (2.5-23), Eq. (2.5-24) reduces to

$$\frac{dn}{dr} = -\frac{nr\dot{\theta}^2}{(1 + r^2\dot{\theta}^2)} \tag{2.5-25}$$

If the index distribution has to be such that all helical rays have the same period, $d\theta/dz$ should be independent of r. If L is the axial distance traveled by the ray in completing an angle 2π, we have

$$d\theta/dz = 2\pi/L \tag{2.5-26}$$

where L is required to be independent of r. Thus we have to determine the index distribution that satisfies Eqs. (2.5-25) and (2.5-26) simultaneously. From Eq. (2.5-25) we get

$$\int \frac{dn}{n} = -\int \frac{\dot{\theta}^2 r\, dr}{1 + r^2\dot{\theta}^2} = -\frac{1}{2}\ln(1 + r^2\dot{\theta}^2) \tag{2.5-27}$$

which gives

$$n = n_0 \Big/ \left[1 + \left(\frac{2\pi}{L}\right)^2 r^2\right]^{1/2} \tag{2.5-28}$$

where n_0 is the refractive index on the axis. Notice that we have not made use of the paraxial approximation.

2.6. *Aberrations in Systems Possessing Finite Discontinuities in Refractive Index*

In Section 2.5, we considered the aberrations of media possessing continuous variation of refractive index with position. In this section we modify the formulas for A, B, C, D, and E into a form suitable for direct application to systems composed of lenses.

We will first derive formulas for a single surface formed between two homogeneous media. This will later be extended to systems consisting of any number of surfaces.

The refractive-index variation of a system composed of two homogeneous media of refractive indices n_1 and n_2 separated by a surface with a radius of curvature R is [see Eq. (1.5-10)]

$$n(U, z) = n_1\Theta(\Lambda) + n_2\Theta(-\Lambda) \tag{2.6-1}$$

where $\Lambda = f(U) - z$, $U = X^2 + Y^2$, $\Theta(x)$ is the unit step function [see Eq. (1.5-11)], and

$$z = f(U) \simeq \frac{U}{2R} + \frac{U^2}{8R^3} \tag{2.6-2}$$

represents the equation of the spherical surface. It immediately follows from Eq. (2.6-1) that for $z < f(U)$, $n = n_1$, and for $z > f(U)$, $n = n_2$.

For evaluating the quantities H_i and H_{ij} we have to determine the first and second derivatives of H with respect to U. Thus from Eq. (2.6-1) we have,

$$\frac{dn}{dU} = n_1 \frac{d\Theta(\Lambda)}{d\Lambda} \frac{df}{dU} - n_2 \frac{d\Theta(-\Lambda)}{d(-\Lambda)} \frac{df}{dU}$$

$$= -\Delta n_1 \delta(\Lambda) \frac{df}{dU} \tag{2.6-3}$$

where use has been made of Eq. (1.5-13) and $\Delta n_1 = n_2 - n_1$. Differentiating again, we obtain

$$\frac{d^2n}{dU^2} = -\Delta n_1 \delta(\Lambda) \frac{d^2f}{dU^2} - \Delta n_1 \frac{d\delta(\Lambda)}{d\Lambda} \left(\frac{df}{dU} \right)^2 \tag{2.6-4}$$

Thus, using Eqs. (2.6-3) and (2.6-4) in Eqs. (2.4-2) and (2.4-3) we get

$$H_1 = \frac{\Delta n_1}{2R} \delta(z), \qquad H_2 = \frac{1}{2n_0(z)}, \qquad H_{12} = \frac{\Delta n_1}{4n_0^2(z)\, R} \delta(z)$$
$$H_{11} = \frac{\Delta n_1}{4R^3} \delta(z) - \frac{\Delta n_1}{4R^2} \delta'(z), \qquad H_{22} = \frac{1}{4n_0^3(z)} \tag{2.6-5}$$

where the prime denotes differentiation with respect to the argument and we have used the fact that $\delta(-z) = \delta(z)$. Further, $n_0(z) = n_1\Theta(-z) + n_2\Theta(z)$, which represents the refractive-index variation along the z axis. If we substitute for H_{11}, H_{12}, and H_{22} in Eq. (2.2-38), we will obtain the expressions for the various aberration coefficients. For example,

$$A = \frac{1}{2\vartheta(z_1)} \int_{z_0}^{z_1} \left\{ \left[\delta(z) g^4 - R\delta'(z) g^4 + 2\frac{R^2}{n_0^2} \delta(z) g^2 \vartheta^2 \right] \frac{\Delta n_1}{R^3} + \frac{\vartheta^4}{n_0^3} \right\} dz$$

$$= \frac{1}{2\vartheta(z_1)} \int_{z_0}^{z_1} \left\{ \left[g^4 + 4Rg^3 \frac{dg}{dz} + \frac{2R^2}{n_0^2} g^2 \vartheta^2 \right] \frac{\Delta n_1}{R^3} \delta(z) + \frac{1}{n_0^3} \vartheta^4 \right\} dz \tag{2.6-6}$$

where we have used the following property of the delta function:

$$\int \delta'(z)\, f(z)\, dz = -\int f'(z)\, \delta(z)\, dz \tag{2.6-7}$$

Since

$$\frac{dg}{dz} = 2H_2 \vartheta = \frac{1}{n_0} \vartheta \tag{2.6-8}$$

we obtain

$$\int_{z_0}^{z_1} \frac{1}{2n_0^3} \vartheta^4 \, dz = \int_{z_0}^{z_1} \frac{1}{2n_0^2} \vartheta^3 \frac{dg}{dz} dz$$

$$= \frac{\vartheta^3}{2n_0^2} g \Bigg|_{z_0}^{z_1} - \int_{z_0}^{z_1} \left\{ \frac{1}{2n_0^2} 3\vartheta^2 \left(-\frac{\Delta n_1}{R} \delta(z) g \right) - \frac{\vartheta^3}{n_0^3} \Delta n_1 \delta(z) \right\} g \, dz$$

$$= \int_{z_0}^{z_1} \left(\frac{3g^2\vartheta^2}{2n_0^2 R} + g \frac{\vartheta^3}{n_0^3} \right) \dot{n}_0 \, dz \tag{2.6-9}$$

where we have used

$$\dot{n}_0 \equiv \frac{dn_0}{dz} = \frac{d}{dz} [n_1 \Theta(-z) + n_2 \Theta(z)] = \Delta n_1 \delta(z) \tag{2.6-10}$$

$$\frac{d\vartheta}{dz} = -2H_1 g = -\frac{\Delta n_1}{R} \delta(z) g = -\dot{n}_0 \frac{g}{R} \tag{2.6-11}$$

and the fact that

$$g(z_0) = g(z_1) = 0 \tag{2.6-12}$$

(see Section 2.2). Using Eqs. (2.6-9) and (2.6-11) in Eq. (2.6-6), we obtain

$$\vartheta(z_1) A = \frac{1}{2} \int_{z_0}^{z_1} \left(\frac{g^3}{R^3} + 4 \frac{g^2}{R^2} \frac{\vartheta}{n_0} + 5 \frac{g}{R} \frac{\vartheta^2}{n_0^2} + 2 \frac{\vartheta^3}{n_0^3} \right) \dot{n}_0 g \, dz \tag{2.6-13}$$

One can use Eqs. (2.6-8), (2.6-10), and (2.6-11) to prove that

$$\left(\frac{g^3}{R^3} + 4 \frac{g^2}{R^2} \frac{\vartheta}{n_0} + 5 \frac{g}{R} \frac{\vartheta^2}{n_0^2} + 2 \frac{\vartheta^3}{n_0^3} \right) \dot{n}_0$$

$$= - \left[\frac{d}{dz} \left(\frac{\vartheta}{n_0} \right) \right]^2 \left[\frac{d}{dz} \left(\frac{\vartheta}{n_0^2} \right) \right] \Big/ \left[\frac{d}{dz} \left(\frac{1}{n_0} \right) \right]^2 \tag{2.6-14}$$

Thus Eq. (2.6-13) reduces to

$$\vartheta(z_1) A = -\frac{1}{2} \int_{z_0}^{z_1} g \left\{ \left[\frac{d}{dz} \left(\frac{\vartheta}{n_0} \right) \right]^2 \left[\frac{d}{dz} \left(\frac{\vartheta}{n_0^2} \right) \right] \Big/ \left[\frac{d}{dz} \left(\frac{1}{n_0} \right) \right]^2 \right\} dz \tag{2.6-15}$$

The above formula can be extended to systems consisting of any number of surfaces. Between two points in a homogeneous medium, n_0 and ϑ remain constant and the integrand in Eq. (2.6-15) vanishes. Thus, the contribution to A from a homogeneous medium is zero. Consequently, the above integral may be evaluated around the discontinuities only. For a discontinuity at $z = 0$, the integral should be evaluated from $z = -\epsilon$ to $z = +\epsilon$. Now, for an arbitrary function $G_1(z)$ that is discontinuous at $z = 0$, we may write (around $z = 0$)

$$dG_1/dz = \delta(z)\Delta G_1 \tag{2.6-16}$$

where ΔG_1 represents the change in G_1 at the discontinuity. Since ϑ and n_0 have a discontinuity at $z = 0$, we immediately obtain

$$\vartheta(z_1)A = -\frac{1}{2}g_s\left[\Delta\left(\frac{\vartheta_1}{n_1}\right)\right]^2\left[\Delta\left(\frac{\vartheta_1}{n_1^2}\right)\right]\Bigg/\left[\Delta\left(\frac{1}{n_1}\right)\right]^2 \tag{2.6-17}$$

where g_s is the x coordinate of the axial ray on the surface and as before ΔX represents the change in X on refraction, i.e.,

$$\left[\Delta\left(\frac{1}{n_1}\right)\right] = \left(\frac{1}{n_2} - \frac{1}{n_1}\right) = \frac{1}{n_1}(k-1) \tag{2.6-18}$$

$$\Delta\left(\frac{\vartheta_1}{n_1}\right) = \left(\frac{\vartheta_2}{n_2} - \frac{\vartheta_1}{n_1}\right) \tag{2.6-19}$$

where ϑ_1 and ϑ_2 are the values of ϑ before and after refraction and $k = n_1/n_2$. Integrating Eq. (2.6-11) from a point to the left of the surface to a point to the right of the surface, we have

$$\vartheta_2 - \vartheta_1 = -\frac{\Delta n_1}{2R}g_s \tag{2.6-20}$$

Using Eq. (2.6-20), Eq. (2.6-19) reduces to

$$\Delta\left(\frac{\vartheta_1}{n_1}\right) = (k-1)\left(\frac{g_s}{R} + \frac{\vartheta_1}{n_1}\right) \tag{2.6-21}$$

We also have

$$\Delta\left(\frac{\vartheta_1}{n_1^2}\right) = \left(\frac{\vartheta_2}{n_2^2} - \frac{\vartheta_1}{n_1^2}\right) = \frac{k-1}{n_2}\left(\frac{g_s}{R} + \frac{\vartheta_1}{n_1} + \frac{\vartheta_1}{n_1 k}\right) \tag{2.6-22}$$

Using Eqs. (2.6-18), (2.6-21), and (2.6-22), Eq. (2.6-17) reduces to

$$A_s = -\frac{1}{2\vartheta(z_1)}n_1(k-1)g_s\left(\frac{g_s}{R} + \frac{\vartheta_1}{n_1}\right)^2\left(\frac{g_s}{R} + \frac{\vartheta_1}{n_1} + \frac{\vartheta_2}{n_2}\right) \tag{2.6-23}$$

This is the contribution to A from each surface of the system, where the quantities are referred to their values on the surface. Similarly we can obtain

$$B_s = qA_s, \qquad D_s = -\frac{1}{2\vartheta(z_1)}\Gamma^2\frac{1}{R}\Delta\left(\frac{1}{n_1}\right)$$
$$C_s = q^2A_s, \qquad E_s = q^3A_s + qD_s \tag{2.6-24}$$

where

$$q = \left(\frac{G_s}{R} + \frac{\theta_1}{n_1}\right)\Bigg/\left(\frac{g_s}{R} + \frac{\vartheta_1}{n_1}\right) \tag{2.6-25}$$

G and θ represent the coordinates of the field ray. Hence to calculate aberrations one has to first trace the axial and field rays and then calculate the values of A_s, B_s, C_s, D_s, and E_s due to each surface, and then add the contribution due to each surface to obtain the complete aberration present in the system. We will now consider some examples.

2.6.1. A Plane Glass Surface

Consider the image formed by a plane glass surface, of an object O as shown in Fig. 2.12. The image is virtual but the analysis that we have developed remains valid. The reference plane $z = \zeta$ is the plane of the glass surface and the relative index of refraction of the two media is μ; the point I represents the paraxial image point (see Fig. 2.12). To find the axial ray, consider a paraxial ray starting from O and hitting the plane $z = \zeta = 0$ at a unit height. Hence

$$g_s = g(z = \zeta) = 1, \qquad \vartheta = 1 \cdot \sin\alpha_1 \simeq \alpha_1 \simeq -1/z_0 \tag{2.6-26}$$

$$\vartheta' = \mu\sin\alpha_2 \simeq \mu\alpha_2 \simeq -\mu/z_1 \tag{2.6-27}$$

Observe that both z_0 and z_1 are negative quantities. Since the rays satisfy Snell's law at the boundary ($z = \zeta$), we have $\sin\alpha_1 = \mu\sin\alpha_2$, which for paraxial rays gives $z_1 = \mu z_0$, which gives the position of the paraxial image point. Thus, in the paraxial approximation, from Eq. (2.6-27) we obtain

$$\vartheta' = -1/z_0 \tag{2.6-28}$$

Substituting these values in the expression for the coefficient of spherical aberration [scc Eq. (2.6-23)], we obtain

$$A = -\frac{1}{2(-1/z_0)}1 \times \left(\frac{1}{\mu} - 1\right) \times 1 \times \left(0 - \frac{1}{z_0}\right)^2 \times \left(0 - \frac{1}{z_0} - \frac{1}{\mu z_0}\right)$$
$$= \frac{\mu^2 - 1}{2z_0^2\mu^2} \tag{2.6-29}$$

where we have used the fact that for a plane glass surface $R = \infty$. Thus for a ray striking the surface at a height l from the z axis, the lateral spherical aberration is

$$X_3 = Al^3 = \frac{\mu^2 - 1}{2z_0^2\mu^2} l^3 \tag{2.6-30}$$

The longitudinal spherical aberration Δz is given by [see Eq. (2.3-6)]

$$\Delta z = Al^2 z_1 = \frac{\mu^2 - 1}{2z_0\mu} l^2 \tag{2.6-31}$$

Problem 2.4. Show that an image formed by a plane mirror is free from aberration. [*Hint:* Formulas for a reflecting surface follow from those of a refracting surface by substituting $\mu = -1$; see Snell's law.]

Problem 2.5. Obtain the longitudinal spherical aberration as given by Eq. (2.6-31) from application of Snell's laws.

2.6.2. *Aberration of a Thin Lens*

We will now obtain explicit expressions for the various aberrations present in a thin lens. To determine the aberrations of a thin lens we have first to trace the axial and field rays. Let z_0 be the distance of the object plane from the lens (since the lens is assumed to be thin, z_0 is the distance from either the first or the second surface) and let the distance of the paraxial image plane from the lens (see Fig. 2.13) be z_1. Let R_1 and R_2 be the radii of curvatures of the two surfaces and let μ be the refractive index of the medium of the lens. Let g_0, g_1, g_2, and g_i represent the height of the axial

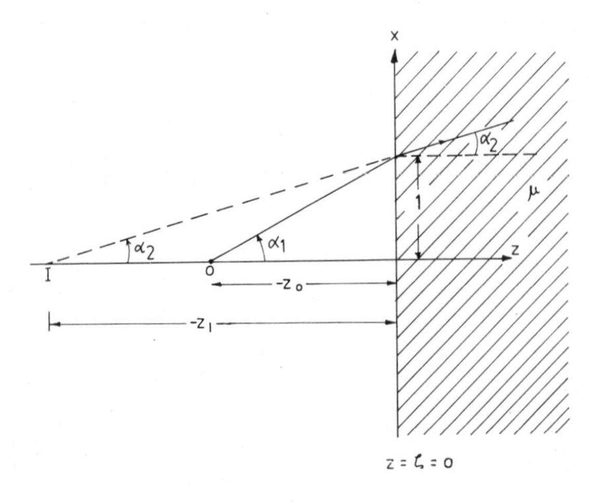

Fig. 2.12. Aberration introduced by a plane refracting surface separating two media of relative refractive index μ. The surface is taken to be the exit pupil plane.

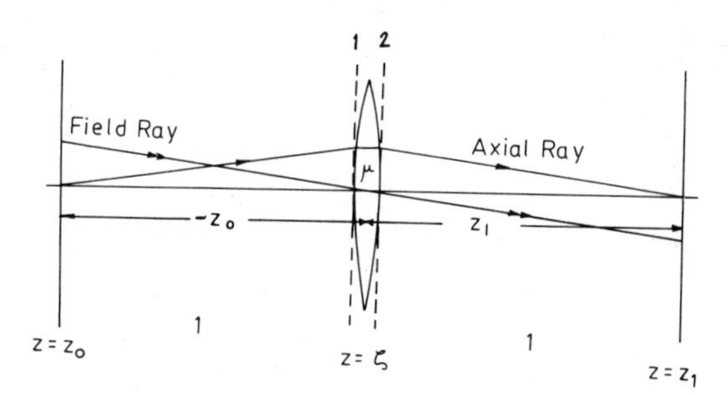

Fig. 2.13. The configuration for calculating the aberrations of a thin lens. The axial ray and the field ray are also shown.

ray in the object plane, the first and second refracting surfaces, and the image plane, respectively; clearly

$$g_0 = 0, \qquad g_1 = 1, \qquad g_2 = 1, \qquad g_i = 0 \tag{2.6-32}$$

Further using the analysis of Section 1.5, we obtain the following expressions for the optical direction cosines in the three media:

$$\vartheta_1 = -\frac{1}{z_0} \tag{2.6-33}$$

$$\vartheta_2 = \vartheta_1 - \frac{1}{R_1}(\mu - 1) = -\frac{1}{z_0} - \frac{\mu - 1}{R_1} \tag{2.6-34}$$

$$\vartheta_3 = \vartheta_2 + \frac{\mu - 1}{R_2} = -\frac{1}{z_0} - (\mu - 1)\left(\frac{1}{R_1} - \frac{1}{R_2}\right) = -\frac{1}{z_1} \tag{2.6-35}$$

Similarly, for the field ray we obtain

$$G_0 = 1, \qquad G_1 = 0, \qquad G_2 = 0, \qquad G_i = -z_1/z_0 \tag{2.6-36}$$

and

$$\theta_1 = 1/z_0, \qquad \theta_2 = 1/z_0, \qquad \theta_3 = 1/z_0 \tag{2.6-37}$$

Further,

$$q_1 = \left[z_0\left(\frac{1}{R_1} - \frac{1}{z_0}\right)\right]^{-1}, \qquad q_2 = \left[\mu z_0\left(\frac{1}{R_2} - \frac{1}{\mu z_0} - \frac{\mu - 1}{\mu R_1}\right)\right]^{-1} \tag{2.6-38}$$

$$\mathscr{P} = (\mu - 1)\left(\frac{1}{R_1} - \frac{1}{R_2}\right) \tag{2.6-39}$$

Using the above expressions, we can evaluate the aberration coefficients A, B, C, D, and E. For example, the contribution from the first surface to the spherical aberration is [see Eq. (2.6-23)]

$$A_{s1} = -\frac{1}{2(-1/z_1)} 1 \times \left(\frac{1}{\mu} - 1\right) \times 1 \times \left(\frac{1}{R_1} + \frac{1}{z_0}\right)^2$$

$$\times \left[\frac{1}{R_1} + \frac{1}{z_0} + \frac{1}{\mu z_0} - \frac{(\mu - 1)}{\mu}\frac{1}{R_1}\right]$$

$$= \frac{(\mu - 1) z_1}{2\mu}\left(\frac{1}{R_1} - \frac{1}{z_0}\right)^2\left(\frac{\mu + 1}{\mu z_0} - \frac{1}{\mu R_1}\right) \tag{2.6-40}$$

The contribution from the second surface is

$$A_{s2} = \frac{z_1 \mu}{2}(\mu - 1)\left[\frac{1}{R_2} + \frac{1}{\mu}\left(\frac{1}{z_0} - \frac{\mu - 1}{R_1}\right)\right]^2$$

$$\times \left[\frac{1}{R_2} + \frac{1}{\mu}\left(\frac{1}{z_0} - \frac{\mu - 1}{R_1}\right) - \frac{1}{z_1}\right]$$

$$= \frac{\mu(\mu - 1) z_1}{2}\left\{\left[\frac{1}{\mu}\left(-\frac{1}{z_0} + \frac{1}{R_1}\right) - \frac{\mathscr{P}}{\mu - 1}\right]^3\right.$$

$$\left. - \frac{1}{z_1}\left[\frac{1}{\mu}\left(-\frac{1}{z_0} + \frac{1}{R_1}\right) - \frac{\mathscr{P}}{\mu - 1}\right]^2\right\} \tag{2.6-41}$$

Thus the coefficient of spherical aberration A for the complete lens is the sum of Eqs. (2.6-40) and (2.6-41), which after some algebraic simplifications can be put in the form

$$A = A_{s1} + A_{s2}$$

$$= \frac{z_1(\mu - 1)}{2\mu^2}\left\{\left[\left(\frac{1}{R_2} - \mathscr{P}\right)^2\left(\frac{1}{R_2} - \mathscr{P}(\mu + 1)\right) - \frac{1}{R_1^3}\right]\right.$$

$$- \frac{1}{z_0}\left[(\mu + 3)\left(\frac{1}{R_2^2} - \frac{1}{R_1^2} + \mathscr{P}^2 - \frac{2\mathscr{P}}{R_2}\right) - 2\mu\mathscr{P}\left(\frac{1}{R_2} - \mathscr{P}\right)\right]$$

$$\left. + \frac{1}{z_0^2}\left[(2\mu + 3)\left(\frac{1}{R_2} - \frac{1}{R_1} - \mathscr{P}\right) - \mu\mathscr{P}\right]\right\} \tag{2.6-42}$$

Similarly, the other aberration coefficients can be obtained:

$$B = \frac{z_1(\mu - 1)}{2(-z_0)}\left[\frac{(\mu - 1)(2\mu + 1)}{\mu R_1 R_2} - \frac{\mu^2 - \mu - 1}{\mu^2 R_1^2} - \frac{\mu}{R_2^2}\right] + \frac{(2\mu + 1)}{z_0^2 \mu(\mu - 1)}\mathscr{P} \tag{2.6-43}$$

$$C = \frac{z_1 \mathscr{P}}{2z_0^2}, \qquad D = -\frac{z_1 \mathscr{P}}{2\mu z_0^2}, \qquad E = 0 \tag{2.6-44}$$

It is interesting to note that the thin lens is free from distortion ($E = 0$). A simple case of this analysis is when the object is at infinity. Then for rays incident parallel to the axis one can use these formulas by taking z_0 to $-\infty$. When $z_0 \to -\infty$, as is evident, B, C, and D also go to zero and A reduces to

$$A = \frac{f(\mu - 1)}{2\mu^2}\left[\left(\frac{1}{R_2} - \mathscr{P}\right)^2\left(\frac{1}{R_2} - \mathscr{P}(\mu + 1)\right) - \frac{1}{R_1^3}\right] \tag{2.6-45}$$

In this case the image is at the focus, and the aberration determines the difference in paraxial and marginal foci. It should be borne in mind that the aberration obtained by the above calculation is the lateral spherical aberration, i.e., the aberration in a plane normal to the axis. Thus the lateral spherical aberration is

$$S_{\text{lat}} = -\frac{(\mu - 1) f}{2\mu^2}\left[\frac{1}{R_1^3} - \left(\frac{1}{R_2} - \frac{\mu + 1}{f}\right)\left(\frac{1}{R_2} - \frac{1}{f}\right)^2\right]\rho^3 \tag{2.6-46}$$

where f is the focal length of the thin lens ($=1/\mathscr{P}$) and ρ is the distance from the axis at which the ray strikes the lens. From here one can easily obtain the longitudinal spherical aberration, i.e., the difference between the paraxial and marginal foci [see Eq. (2.3-6); in the present case $d = f$]:

$$S_{\text{long}} = -\frac{(\mu - 1) f^2 \rho^2}{2\mu^2}\left[\frac{1}{R_1^3} - \left(\frac{1}{R_2} - \frac{\mu + 1}{f}\right)\left(\frac{1}{R_2} - \frac{1}{f}\right)^2\right] \tag{2.6-47}$$

This represents the difference in foci between rays striking the lens at a distance ρ from the axis and the paraxial rays.

The image formed by a set of rays parallel to the axis of the lens suffers only from spherical aberration. If we consider a set of parallel rays inclined to the axis, then the image suffers from other aberrations also. To calculate the coma in an image formed by a set of parallel rays inclined to the axis at an angle φ, we consider an object at a height x_0 from the axis at a distance $-z_0$ from the lens. Then the coma present in the image is $3Bx_0\rho^2$. Using Eq.

(2.6-43), this becomes

$$\text{coma} = \frac{3(\mu - 1)}{2} z_1 \left(\frac{x_0}{-z_0} \right) \rho^2 \left[\frac{(\mu - 1)(2\mu + 1)}{\mu R_1 R_2} - \frac{\mu^2 - \mu - 1}{\mu^2 R_1^2} - \frac{\mu}{R_2^2} \right]$$
$$+ \frac{3(2\mu + 1)\mathscr{P} z_1}{\mu(\mu - 1)} \rho^2 \left(\frac{x_0}{z_0^2} \right) \tag{2.6-48}$$

In the limit of $x_0 \to -\infty$ and $z_0 \to -\infty$ such that $x_0/z_0 = \tan\varphi$ remains constant, we get a parallel set of rays inclined at an angle φ to the axis. Thus the amount of coma in the image formed by a parallel bundle of rays inclined at an angle φ to the z axis is given by

$$\text{coma} = \frac{3(\mu - 1)}{2} f\rho^2 \tan\varphi \left[\frac{(\mu - 1)(2\mu + 1)}{\mu R_1 R_2} - \frac{\mu^2 - \mu - 1}{\mu^2 R_1^2} - \frac{\mu}{R_2^2} \right] \tag{2.6-49}$$

Problem 2.6. Consider a thin lens made up of a material of refractive index 1.5. Let the radii of curvatures of the two surfaces be -40 and -8 cm. Calculate the longitudinal spherical aberration of a ray incident parallel to the axis at a height 1 cm from the axis. Also calculate the coma when a ray is incident at a height 1 cm from the axis at an angle $\varphi = \tan^{-1}(\frac{1}{5})$.

Solution. The focal length of the lens can be calculated to be $f = 20$ cm. Then substituting the given values of the parameters in Eqs. (2.6-47) and (2.6-49), we obtain

$$S_{\text{long}} = -0.34 \text{ cm}, \qquad \text{coma} = -0.03 \text{ cm}$$

Problem 2.7. For every lens, one can define a parameter q called the shape factor, which is defined by

$$q = \frac{R_2 + R_1}{R_2 - R_1} \tag{2.6-50}$$

Consider a lens of focal length 20 cm and made of a medium of refractive index $n = 1.5$. Obtain the variation of spherical aberration and coma as a function of the shape factor (i.e., different combinations of R_1 and R_2 such that $f = 20$ cm) from $q = -2$ to $q = +2$. The values of other parameters are the same as in the previous problem.

Solution. Table 2.1 gives the values of R_1 and R_2 for different values of q. It can be seen that $q = -1$ represents a planoconvex lens with the plane side facing the incident rays, $q = +1$ represents again a planoconvex lens with the convex surface facing the incident rays, and $q = 0$ represents an equiconvex lens. Figure 2.14 shows the variation of spherical aberration and coma as a function of q. It may at once be observed that for a given focal length, a lens with $q \simeq 0.72$ has minimum (but not zero) spherical aberration and a lens with $q = +0.8$ has zero coma. The lens with $q = +1$ is seen to possess very little spherical aberration and coma.

Table 2.1

q	R_1 (cm)	R_2 (cm)
−2.0	−20	−6.666
−1.5	−40	−8.0
−1.0	∞	−10
−0.5	40	−13.333
0.0	20	−20
0.5	13.333	−40
1.0	10	∞
1.5	8	40
2.0	6.666	20

This is the reason why in different eyepieces one uses planoconvex lenses with their convex surfaces facing the incident light. The above process of decreasing the spherical aberration by changing q is called bending the lens.

Problem 2.8. Consider a thin biconvex lens with the following values of the parameters: $R_1 = 20$ cm, $R_2 = -20$ cm, $n = 1.5$. Consider an object point situated at a distance 30 cm from the lens and at a height 1 cm from the axis. Determine the different amounts of aberrations present in an image formed by rays striking the lens at a height of 1 cm.

Solution. The focal length of the lens can be calculated to be $f = 20$ cm. Hence the power $\mathscr{P} = 1/f = 0.05$ cm^{-1}. The image point lies at a distance of 60 cm

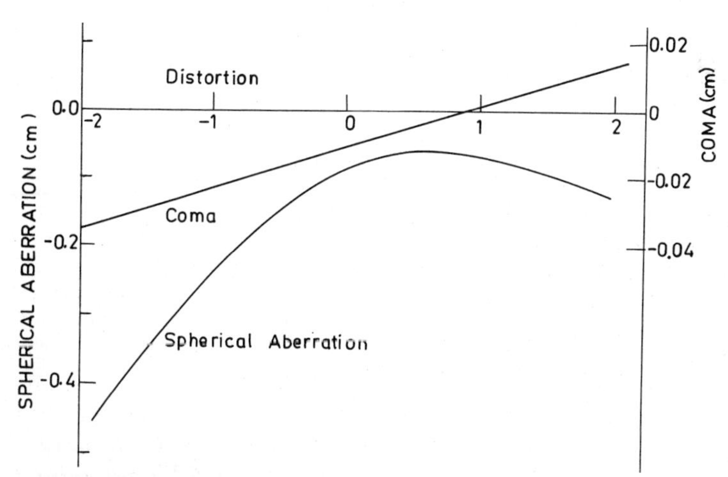

Fig. 2.14. The variation of spherical aberration and coma for a thin lens as a function of the shape factor q. A planoconvex lens with $q = +1$ has very little spherical aberration and coma.

from the lens. Then using Eqs. (2.6-42)–(2.6-44) one can obtain

$$A = -8.89 \times 10^{-3}\ \text{cm}^{-2}, \qquad B = -3.11 \times 10^{-3}\ \text{cm}^{-2}$$

$$C = 1.67 \times 10^{-3}\text{cm}^{-2}, \qquad D = -1.11 \times 10^{-3}\ \text{cm}^{-2}, \qquad E = 0$$

Thus, the longitudinal spherical aberration $A\rho^2 z_1 = 0.5334$ cm; the coma $3Bx_0\rho^2 = -0.009318$ cm, and the distance between sagittal and tangential foci $-2Cx_0^2 z_1 = -0.2$ cm. Further, the radius of curvature of the field $M^2/2d(2C + D) = 14.9$ cm, where $M = z_1/(-z_0)$ is the magnification of the system.

In general, for a combination of lenses, one must first trace out the axial and field rays and find the values of $g(z)$ and $G(z)$ at various refracting surfaces. The values of ϑ and θ in various media are also determined, from which one can obtain the aberration coefficients.

2.7. Chromatic Aberration

The refractive index of any material medium depends on the wavelength of the radiation under consideration. Since image formation is accompanied by refraction at refractive-index discontinuities or by propagation through continuously varying refractive-index media, the property of dispersion affects image formation. If the light is not monochromatic, then after refraction, different wavelength components proceed along different directions and hence form images at different points. This introduces aberration in the image and this aberration is called chromatic aberration. Since white light is normally used for image formations in most optical instruments, this aberration leads to colored images.

In this section we study the primary chromatic aberration of an optical system, where we assume the absence of any of the five Seidel aberrations (the five monochromatic aberrations discussed in Sections 2.2 and 2.3). The difference between the actual image point and the image point for any specified wavelength is the chromatic aberration. The component of this difference along the optical axis is called longitudinal chromatic aberration and the component normal to the optical axis is called lateral chromatic aberration.

For studying chromatic aberration, we must consider only the paraxial equations. The paraxial equations (as already discussed in Section 1.5) are

$$\dot{x} = 2H_2 p, \qquad \dot{p} = -2H_1 x \tag{2.7-1}$$

Let $z = z_0$ be the object plane with respect to some conveniently chosen origin. Since we are going to study the effect of wavelength on the image, we have to choose a particular wavelength and define aberrations with respect to the image formed by light of this wavelength. We define our axial ray (coordinates g, ϑ) and field ray (coordinates G, θ) with respect to this wavelength. (It is normally chosen to be $\lambda = 5893$ Å.) Let $z = z_1$ be the paraxial image plane with respect to this wavelength.

We will first obtain the chromatic aberration of a single refracting surface. It will be shown that the chromatic aberration for a system of surfaces can be readily obtained from the formula for a single refracting surface. Let R be the radius of curvature of the refracting surface separating two media of refractive indices n_1 and n_2. In such a case, as in Section 2.6, the refractive-index function for the system would be

$$n(u, z) = n_1 \Theta(\Lambda - z) + n_2 \Theta(z - \Lambda) \tag{2.7-2}$$

where $\Lambda = f(u) + z_2$ and z_2 is the pole of the surface. As before, $u = x^2 + y^2$ and $\Theta(x)$ is the unit step function. The equation of the surface is [see Eq. (1.5-9)]

$$z = z_2 + f(u) \simeq z_2 + \frac{u}{2R} + \frac{u^2}{8R^3} + \cdots \tag{2.7-3}$$

The coefficients H_1 and H_2 have already been determined in Section 2.6:

$$H_1 = \frac{1}{2R} \dot{n}_0, \qquad H_2 = \frac{1}{2n_0} \tag{2.7-4}$$

where dots represent differentiation with respect to z and n_0 is the refractive-index variation along the optical axis at the chosen reference wavelength. Thus Eq. (2.7-1) becomes

$$\dot{x} = \frac{1}{n_0} p, \qquad \dot{p} = -\frac{1}{R} \dot{n}_0 x \tag{2.7-5}$$

If δx represents the change in x for a change $\delta\lambda$ in λ, then from Eqs. (2.7-5) we obtain

$$\delta\dot{x} = \frac{1}{n_0} \delta p + p\delta\left(\frac{1}{n_0}\right), \qquad \delta\dot{p} = -\frac{1}{R} (\dot{n}_0 \delta x + x \delta \dot{n}_0) \tag{2.7-6}$$

Since (g, ϑ) is a solution of the paraxial equations, we have as before

$$\dot{g} = \frac{1}{n_0} \vartheta, \qquad \dot{\vartheta} = -\frac{1}{R} \dot{n}_0 g \tag{2.7-7}$$

We substitute the values of $1/n_0$ and $\dot{n}_0/R$ in Eq. (2.7-6) to obtain

$$\frac{d}{dz} (\vartheta\, \delta x - g\, \delta p) = p \vartheta\, \delta\left(\frac{1}{n_0}\right) + \frac{xg}{R} \delta \dot{n}_0 \tag{2.7-8}$$

Integrating Eq. (2.7-8) from z_0 to z_1, where $z = z_0$ and $z = z_1$ represent the object and the image planes, respectively, and using the fact that $g(z_0) = g(z_1) = 0$ and $(\delta x)_{z_0}$ (which represents the chromatic aberration in the object plane) is zero, we have

$$\vartheta(z_1)\,(\delta x)_{z_1} = \int_{z_0}^{z_1} \left[p\vartheta\,\delta\left(\frac{1}{n_0}\right) + \frac{xg}{R}\,\delta\dot{n}_0 \right] dz \tag{2.7-9}$$

Using

$$x = x_0 G(z) + \xi g(z), \qquad p = x_0\theta(z) + \xi\vartheta(z) \tag{2.7-10}$$

where x_0 is the x coordinate of the ray in the object plane $z = z_0$, and ξ is the x coordinate in the exit pupil plane $z = \zeta$, Eq. (2.7-9) reduces to

$$(\delta x)_{z_1} = K\xi + Lx_0 \tag{2.7-11}$$

where

$$K = \frac{1}{\vartheta(z_1)} \int_{z_0}^{z_1} \left[\vartheta^2\,\delta\left(\frac{1}{n_0}\right) + \frac{g^2}{R}\,\delta\dot{n}_0 \right] dz$$
$$L = \frac{1}{\vartheta(z_1)} \int_{z_0}^{z_1} \left[\theta\vartheta\,\delta\left(\frac{1}{n_0}\right) + \frac{gG}{R}\,\delta\dot{n}_0 \right] dz \tag{2.7-12}$$

We can simplify Eqs. (2.7-12) by using Eq. (2.7-7). For example,

$$K = \frac{1}{\vartheta(z_1)} \int_{z_0}^{z_1} \left[-\frac{\vartheta}{n_0}\,\delta n_0\,\frac{dg}{dz} - g\,\frac{\dot{\vartheta}}{\dot{n}_0}\,\delta\dot{n}_0 \right] dz \tag{2.7-13}$$

Integrating the first term in Eq. (2.7-13) by parts and using the fact that $g(z_0) = g(z_1) = 0$, we obtain

$$K = \frac{1}{\vartheta(z_1)} \int_{z_0}^{z_1} g \left[-\frac{\dot{\vartheta}}{\dot{n}_0}\,\delta\dot{n}_0 + \frac{\dot{\vartheta}}{n_0}\,\delta n_0 + \frac{\vartheta}{n_0}\,\delta\dot{n}_0 - \frac{\vartheta}{n_0^2}\,\dot{n}_0\delta n_0 \right] dz$$
$$= \frac{1}{\vartheta(z_1)} \int_{z_0}^{z_1} g \left[\frac{d}{dz}\left(\frac{\vartheta}{n_0}\right) \frac{d}{dz}\left(\frac{\delta n_0}{n_0}\right) \Big/ \frac{d}{dz}\left(\frac{1}{n_0}\right) \right] dz \tag{2.7-14}$$

Similarly,

$$L = \frac{1}{\vartheta(z_1)} \int_{z_0}^{z_1} g \left[\frac{d}{dz}\left(\frac{\theta}{n_0}\right) \frac{d}{dz}\left(\frac{\delta n_0}{n_0}\right) \Big/ \frac{d}{dz}\left(\frac{1}{n_0}\right) \right] dz \tag{2.7-15}$$

where $\delta n_0/n_0$ measures the dispersion of the medium. Dispersion is normally measured by the quantity

$$\frac{n_F - n_C}{n_D - 1}$$

where n_F, n_C, and n_D represent the refractive indices of the medium at wavelengths corresponding to the hydrogen F line ($\lambda = 4861$ Å), hydrogen C line ($\lambda = 6563$ Å), and sodium D line ($\lambda = 5893$ Å), respectively. For typical values of n_F, n_C, and n_D for various common optical materials, refer to Jenkins and White (1957).

For an optical system composed of homogeneous media separated by surfaces, the contribution to K and L is only at the surfaces, and the chromatic aberration for the complete system is given by Eq. (2.7-11), where

$$K = \frac{1}{\vartheta(z_1)} \sum_i g_i \left[\Delta_i \left(\frac{\vartheta}{n} \right) \right] \frac{\Delta_i \kappa}{\Delta_i (1/n)}$$
$$L = \frac{1}{\vartheta(z_1)} \sum_i g_i \left[\Delta_i \left(\frac{\theta}{n} \right) \right] \frac{\Delta_i \kappa}{\Delta_i (1/n)} \tag{2.7-16}$$

the summation being done over the surfaces of the system. Here $\kappa = \delta n / n$.

The normal procedure employed is to remove the chromatic aberration for the F and C lines with respect to the D line. Even after this correction, the system would still be suffering from chromatic aberration for other wavelengths.

Problem 2.9. Calculate the chromatic aberration of a thin lens made up of a medium of refractive index n and placed in air.

Solution. Let R_1 and R_2 represent the radii of curvature of the thin lens. We have already traced the axial and field rays through a thin lens (see Section 2.6). Thus the expressions for K and L as given by Eq. (2.7-16) can be obtained for a thin lens. For example, we have

$$K = -z_1 \left\{ 1 \times \left[\frac{1}{n} \left(-\frac{1}{z_0} - \frac{n-1}{R_1} \right) + \frac{1}{z_0} \right] \frac{\kappa}{(1/n) - 1} \right.$$
$$\left. + 1 \times \left[-\frac{1}{z_1} - \frac{1}{n} \left(-\frac{1}{z_0} - \frac{n-1}{R_1} \right) \right] \frac{(-\kappa)}{1 - (1/n)} \right\} \tag{2.7-17}$$

where the symbols have the same meaning as in Section 2.6. The total K is obtained as a sum of the contribution from each surface of the lens. On simplification, we obtain

$$K = -z_1 \mathscr{P} \frac{\delta n}{n-1} \tag{2.7-18}$$

In an exactly similar manner,

$$L = -z_1 \left[1 \times \left(\frac{1}{z_0 n} - \frac{1}{z_0} \right) \frac{\kappa}{(1/n) - 1} + 1 \times \left(\frac{1}{z_0} - \frac{1}{z_0 n} \right) \frac{(-\kappa)}{1 - (1/n)} \right] = 0 \tag{2.7-19}$$

Thus the lateral chromatic aberration is given by [see Eq. (2.7-11)]

$$\delta x = K\xi = -\mathscr{P} z_1 \frac{\delta n}{n-1} \xi \tag{2.7-20}$$

where ξ is the height at which the ray strikes the exit pupil, which is the lens itself in the present case.

The longitudinal chromatic aberration (δz) is given by

$$\delta z = \frac{z_1}{\xi}(\delta x) = -\mathscr{P} z_1^2 \frac{\delta n}{n-1} \tag{2.7-21}$$

For a parallel beam of light incident on the lens, $z_1 = f$, and the chromatic aberration is

$$\delta f = -f \frac{\delta n}{n-1} \tag{2.7-22}$$

It may be noted that for the simple case of a thin lens, the expression for chromatic aberration, namely, Eq. (2.7-22), could have been obtained in an alternative way as given below. The focal length of a thin lens is given by

$$\frac{1}{f} = (n-1)\left(\frac{1}{R_1} - \frac{1}{R_2}\right) \tag{2.7-23}$$

If a change of n by δn (the change of n is due to the change in the wavelength of the light) results in a change of f by δf, then by differentiating Eq. (2.7-23) we obtain

$$-\frac{\delta f}{f^2} = \delta n\left(\frac{1}{R_1} - \frac{1}{R_2}\right) = \frac{\delta n}{n-1}\frac{1}{f}$$

i.e.,

$$\delta f = -f \frac{\delta n}{n-1} \tag{2.7-24}$$

which is the same as Eq. (2.7-22).

3

Characteristic Functions

3.1. Introduction

In Chapter 1 we developed geometrical optics on the basis of Fermat's principle. In this chapter, we introduce certain characteristic functions, first studied by Hamilton, that characterize the properties of the system completely. The usefulness of this method lies in the fact that just by knowing certain symmetries possessed by the system, one can draw general conclusions about the performance of the system. For example, it will be shown that the third-order aberrations of any rotationally symmetric system, whether it is made up of homogeneous media separated by surfaces or has a continuous variation of refractive index, can be completely specified by five aberration coefficients, which are known as the Seidel aberrations. It will also be shown that using the characteristic functions it is possible to obtain the characteristics of optical systems with desired properties. For example, one can determine a surface that focuses all rays emanating from a single point onto another point (see Problem 3.1) or one can determine the equation of a surface that renders parallel all rays emanating from a single point, after a reflection (see Problem 3.5). At the same time it should be pointed out that it is, in general, difficult to calculate the characteristic functions for a given system exactly, although even approximate forms give good results. In fact, the lowest-order approximation yields the paraxial properties of the system, and higher and higher orders of approximations give the various orders of aberrations present in the system.

3.2. Point Characteristic Function

3.2.1. Definition and Properties

Fermat's principle tells us that out of all possible paths that can connect any two points, the ray follows that path for which the optical pathlength between the two points is an extremum. The point characteristic function V between the points $P_1(x_1, y_1, z_1)$ and $P_2(x_2, y_2, z_2)$ is defined through the relation

$$V(x_1, y_1, z_1; x_2, y_2, z_2) = \int_{P_1}^{P_2} n(x, y, z)\, ds \qquad (3.2\text{-}1)$$

where the integration is performed along the ray. As can be seen, V is simply the optical pathlength of the ray connecting the two points and is called the point characteristic function because it is a function of the coordinates of two points. If S represents the eikonal, then it satisfies the following equation [see Eq. (1.6-7)]:

$$\nabla S = n\hat{\mathbf{s}} \qquad (3.2\text{-}2)$$

where $\hat{\mathbf{s}}$ is a unit vector along the ray. The rate of variation of the eikonal S along the ray path is given by

$$\frac{dS}{ds} = \nabla S \cdot \frac{d\mathbf{r}}{ds} = n\hat{\mathbf{s}} \cdot \hat{\mathbf{s}} = n \qquad (3.2\text{-}3)$$

where we have used the fact that $d\mathbf{r}/ds$ represents the unit vector along the ray. Substituting for n from Eq. (3.2-3) into Eq. (3.2-1), we obtain

$$\begin{aligned} V(x_1, y_1, z_1; x_2, y_2, z_2) &= \int_{P_1(x_1,y_1,z_1)}^{P_2(x_2,y_2,z_2)} \frac{dS}{ds}\, ds \\ &= S(x_2, y_2, z_2) - S(x_1, y_1, z_1) \qquad (3.2\text{-}4) \end{aligned}$$

If

$$\nabla_i = \hat{\mathbf{i}}\frac{\partial}{\partial x_i} + \hat{\mathbf{j}}\frac{\partial}{\partial y_i} + \hat{\mathbf{k}}\frac{\partial}{\partial z_i}, \qquad i = 1, 2 \qquad (3.2\text{-}5)$$

then from Eq. (3.2-4) we obtain, using Eq. (3.2-2),

$$\nabla_1 V = -\nabla_1 S(x_1, y_1, z_1) = -n_1\hat{\mathbf{e}}_1 \qquad (3.2\text{-}6)$$

$$\nabla_2 V = \nabla_2 S(x_2, y_2, z_2) = n_2\hat{\mathbf{e}}_2 \qquad (3.2\text{-}7)$$

where n_1 and n_2 are the indices of refraction at the points P_1 and P_2, and $\hat{\mathbf{e}}_1$ and $\hat{\mathbf{e}}_2$ represent the unit vectors along the direction of the ray at P_1 and P_2, respectively. If (p_1, q_1, r_1) and (p_2, q_2, r_2) are the optical direction cosines of the ray at P_1 and P_2, respectively, then Eq. (3.2-6) can be written in the form

$$\hat{\mathbf{i}}\frac{\partial V}{\partial x_1} + \hat{\mathbf{j}}\frac{\partial V}{\partial y_1} + \hat{\mathbf{k}}\frac{\partial V}{\partial z_1} = -(\hat{\mathbf{i}}p_1 + \hat{\mathbf{j}}q_1 + \hat{\mathbf{k}}r_1) \qquad (3.2\text{-}8)$$

from which we obtain

$$p_1 = -\partial V/\partial x_1, \qquad q_1 = -\partial V/\partial y_1, \qquad r_1 = -\partial V/\partial z_1 \tag{3.2-9}$$

Similarly from Eq. (3.2-7) we obtain

$$p_2 = \partial V/\partial x_2, \qquad q_2 = \partial V/\partial y_2, \qquad r_2 = \partial V/\partial z_2 \tag{3.2-10}$$

Also, since the optical direction cosines satisfy the equations

$$p_1^2 + q_1^2 + r_1^2 = n_1^2, \qquad p_2^2 + q_2^2 + r_2^2 = n_2^2 \tag{3.2-11}$$

we obtain

$$\left(\frac{\partial V}{\partial x_1}\right)^2 + \left(\frac{\partial V}{\partial y_1}\right)^2 + \left(\frac{\partial V}{\partial z_1}\right)^2 = n_1^2 \tag{3.2-12}$$

$$\left(\frac{\partial V}{\partial x_2}\right)^2 + \left(\frac{\partial V}{\partial y_2}\right)^2 + \left(\frac{\partial V}{\partial z_2}\right)^2 = n_2^2 \tag{3.2-13}$$

Thus the point characteristic function satisfies two eikonal-type equations in variables (x_1, y_1, z_1) and (x_2, y_2, z_2). From Eqs. (3.2-9) and (3.2-10), it follows that for a given system, if the point characteristic function is obtained, then one can immediately determine the initial and final direction cosines of the rays. It also follows from Eqs. (3.2-9) and (3.2-10) that

$$dV = \frac{\partial V}{\partial x_1} dx_1 + \frac{\partial V}{\partial y_1} dy_1 + \frac{\partial V}{\partial z_1} dz_1 + \frac{\partial V}{\partial x_2} dx_2 + \frac{\partial V}{\partial y_2} dy_2 + \frac{\partial V}{\partial z_2} dz_2$$

$$= p_2\, dx_2 + q_2\, dy_2 + r_2\, dz_2 - p_1 dx_1 - q_1\, dy_1 - r_1\, dz_1 \tag{3.2-14}$$

We will now consider two examples, where the point characteristic function is used to determine the form of certain optical systems with specified properties.

Problem 3.1. Two points are said to be perfectly conjugate if all rays allowed by the system and emanating from one point intersect each other at the other point. The image formation in such a case is perfect for the given pair of object and image points. Determine the shape of the refracting surface separating two media of refractive indices n_1 and n_2, such that two specified points, each in one of the media, are perfectly conjugate.

Solution. For two points to be conjugate it is clear that the point characteristic function has to be a constant, i.e., the optical pathlength of all rays joining these points must be the same. In such a case, the optical pathlength has a stationary value with respect to variations in the path connecting the two points. Let the points $A(0, 0, 0)$ lying in medium I and $B(0, 0, z_2)$ lying in medium II be the two points shown in Fig. 3.1. Let $P(x, y, z)$ represent any point on the surface of which the

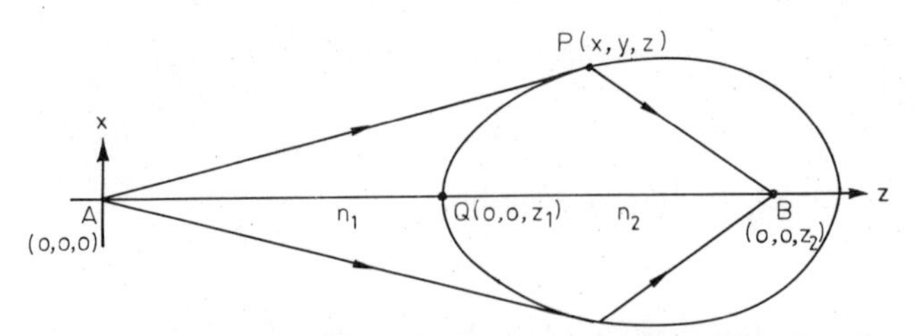

Fig. 3.1. A and B are two perfectly conjugate points (adapted from Luneberg, 1964; reprinted with permission).

equation is to be determined. Then the point characteristic function for the ray joining A to P and P to B is given by

$$V = n_1(x^2 + y^2 + z^2)^{1/2} + n_2[x^2 + y^2 + (z_2 - z)^2]^{1/2} \tag{3.2-15}$$

For A and B to be perfectly conjugate points, V should be a constant independent of the direction cosines of the rays. Thus

$$n_1(x^2 + y^2 + z^2)^{1/2} + n_2[x^2 + y^2 + (z_2 - z)^2]^{1/2} = C \tag{3.2-16}$$

which gives us the equation of the surface. To evaluate the constant C, we observe that if the coordinates of the vertex Q are $(0, 0, z_1)$, then for the straight path AQB, V should be equal to $n_1z_1 + n_2(z_2 - z_1)$. Thus

$$n_1(x^2 + y^2 + z^2)^{1/2} + n_2[x^2 + y^2 + (z_2 - z)^2]^{1/2} = n_1z_1 + n_2(z_2 - z_1) \tag{3.2-17}$$

Equation (3.2-17) gives us the equation of the surface such that the two points are conjugate. The surface, as expected, is symmetric in x and y and is obtained by revolving the following curve about the z axis:

$$n_1(x^2 + z^2)^{1/2} + n_2[x^2 + (z_2 - z)^2]^{1/2} = n_1z_1 + n_2(z_2 - z_1) \tag{3.2-18}$$

This curve is called a Cartesian oval and is shown in Fig. 3.1.

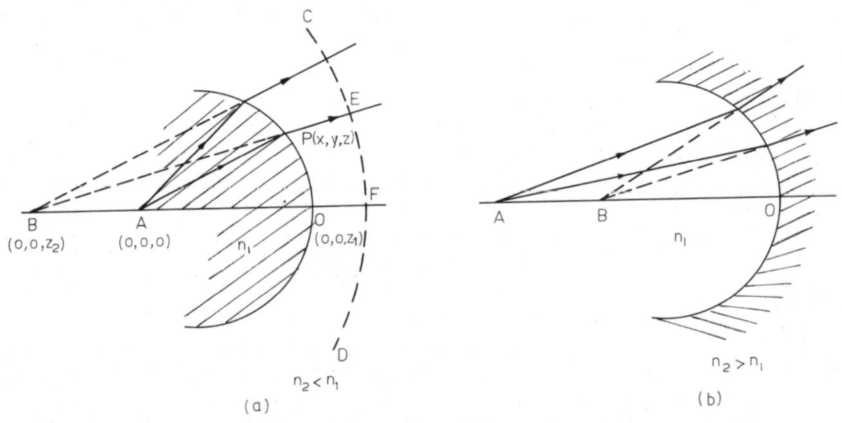

Fig. 3.2. The aplanatic points of a sphere; (a) and (b) correspond to $n_2 < n_1$ and $n_2 > n_1$, respectively.

Problem 3.2. Determine the equation of a surface separating two media of refractive indices n_1 and n_2 such that all rays emanating from one point, after refraction through the surface, appear to be emanating from another point.

Solution. The configuration is shown in Fig. 3.2. Rays from the point $A(0,0,0)$, after refraction from the surface at the point $P(x, y, z)$, appear to be diverging from the point $B(0, 0, z_2)$. The optical pathlength from A to B along the ray is

$$V = n_1(x^2 + y^2 + z^2)^{1/2} - n_2[x^2 + y^2 + (z - z_2)^2]^{1/2} \tag{3.2-19}$$

where the second term carries a negative sign because the point B is obtained by projecting the ray backward. Proceeding along the same argument as in the previous problem, we obtain*

$$n_1(x^2 + y^2 + z^2)^{1/2} - n_2[x^2 + y^2 + (z - z_2)^2]^{1/2} = n_1 z_1 - n_2(z_1 - z_2) \tag{3.2-20}$$

If z_1 is such that†

$$n_1 z_1 - n_2(z_1 - z_2) = 0 \tag{3.2-21}$$

then Eq. (3.2-20) reduces to

$$x^2 + y^2 + \left(z - \frac{n_2 z_1}{n_1 + n_2}\right)^2 = \left(\frac{n_1 z_1}{n_1 + n_2}\right)^2 \tag{3.2-22}$$

The above equation is the equation of a sphere of radius

$$R = \frac{n_1}{n_1 + n_2} z_1 \tag{3.2-23}$$

and centered at $C[0, 0, n_2 z_1/(n_1 + n_2)]$. Thus a sphere of radius R given by Eq. (3.2-23) produces a perfect virtual image of the point $(0, 0, 0)$ at the point $[0, 0, (n_2 - n_1) z_1/n_2]$, i.e., *all* rays emanating from $(0, 0, 0)$ after refraction from the surface of the sphere appear to diverge from a point $[0, 0, (n_2 - n_1) z_1/n_2]$. These points are called the aplanatic points of a sphere. Figures 3.2a and b show the two conjugate points when $n_2 < n_1$ and $n_2 > n_1$, respectively. The aplanatic points of a sphere are utilized in the construction of wide-aperture oil immersion microscope objectives.

* This can also be understood from the fact that for the rays to appear to be emanating from the point B, the refracted wavefront should be a perfect sphere whose center is at the point B. Let the dotted curve CD in Fig. 3.2a denote the position of the refracted wavefront (at any instant of time). Clearly

$$n_1 \cdot AP + n_2 \cdot PE = n_1 \cdot AO + n_2 \cdot OF$$

or

$$n_1 \cdot AP + n_2(R - BP) = n_1 \cdot AO + n_2(R - BO)$$

i.e.,

$$n_1 \cdot AP - n_2 \cdot BP = n_1 \cdot AO - n_2 \cdot BO$$

which is the same as Eq. (3.2-20).

† This is not a necessary condition, but we will see that under such a condition, the resulting surface has a simple form, namely, spherical.

3.2.2. *Abbe Sine Condition*

We saw in Problem 3.1 that for the formation of a perfect image of a point object, the point characteristic function has to be a constant. This constant will depend on the coordinates of the two points between which the point characteristic is calculated but it must be independent of the direction cosines of the ray. We will now consider a general optical system and, using the above property, determine the condition that rays have to satisfy in order that an off-axis point be imaged sharply. Let A be an axial point that is imaged perfectly to a point B as shown in Fig. 3.3a. This implies that the point characteristic function between the points A and B is a constant. Let $P(x_1, y_1, z_1)$ be an off-axis point near A, which is to be imaged sharply to the point $Q(x_2, y_2, z_2)$, P and Q being referred to the coordinate axes centered at A and B, respectively. This would imply that the point

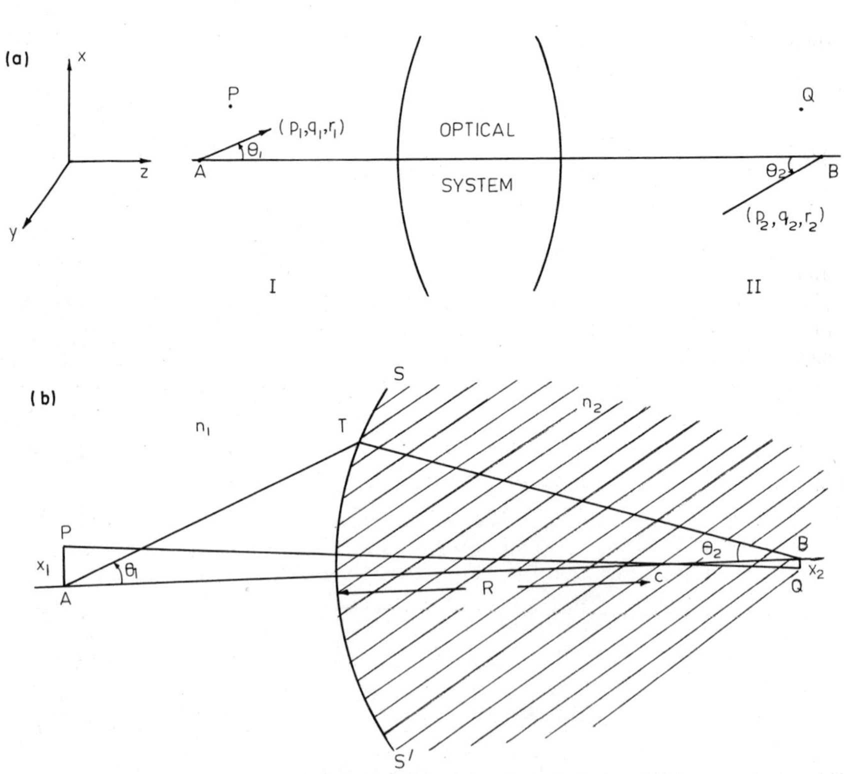

Fig. 3.3. (a) If the point characteristic function between the points A and B is a constant and if the Abbe sine condition is also satisfied, then the point characteristic function between the points P and Q is also a constant. (b) SS' is a spherical surface of radius of curvature R separating two media of refractive indices n_1 and n_2.

characteristic function between the points P and Q should again be a constant. Thus we can write

$$V(x_1, y_1, z_1; x_2, y_2, z_2) - V_{AB} = C(x_1, y_1, z_1; x_2, y_2, z_2) \quad (3.2\text{-}24)$$

where $V_{AB} = V(x_1 = 0, y_1 = 0, z_1 = 0; x_2 = 0, y_2 = 0, z_2 = 0)$ is the point characteristic function between the points A and B, and C is a function of only x_1, y_1, z_1, x_2, y_2, and z_2 and is independent of the direction cosines of the rays considered. If the points P and Q are very near the points A and B, then one can write

$$V(x_1, y_1, z_1; x_2, y_2, z_2) = V(x_1 = 0, y_1 = 0, z_1 = 0; x_2 = 0, y_2 = 0, z_2 = 0)$$
$$+ \left(\frac{\partial V}{\partial x_1}\right)_A x_1 + \left(\frac{\partial V}{\partial y_1}\right)_A y_1 + \left(\frac{\partial V}{\partial z_1}\right)_A z_1 + \left(\frac{\partial V}{\partial x_2}\right)_B x_2$$
$$+ \left(\frac{\partial V}{\partial y_2}\right)_B y_2 + \left(\frac{\partial V}{\partial z_2}\right)_B z_2 \quad (3.2\text{-}25)$$

where the subscripts A and B refer to the fact that the derivatives are taken at the points A and B, respectively. If we use Eqs. (3.2-9) and (3.2-10) in Eq. (3.2-25) we obtain

$$V(x_1, y_1, z_1; x_2, y_2, z_2) - V_{AB} = -(p_1 x_1 + q_1 y_1 + r_1 z_1) + (p_2 x_2 + q_2 y_2 + r_2 z_2) \quad (3.2\text{-}26)$$

where (p_1, q_1, r_1) and (p_2, q_2, r_2) represent the optical direction cosines of the rays at the points A and B. Thus using Eq. (3.2-24) we obtain

$$p_2 x_2 + q_2 y_2 + r_2 z_2 - (p_1 x_1 + q_1 y_1 + r_1 z_1) = C(x_1, y_1, z_1; x_2, y_2, z_2) \quad (3.2\text{-}27)$$

Let us first consider the case when P lies vertically above A. Then Q would also lie vertically above or below B (depending on whether the magnification is positive or negative). If we choose the x axis along AP and the z axis along AB, we have $y_1 = z_1 = 0$ and $y_2 = z_2 = 0$. Thus Eq. (3.2-27) reduces to

$$p_2 x_2 - p_1 x_1 = C(x_1, 0, 0; x_2, 0, 0) \quad (3.2\text{-}28)$$

Since the system is assumed to be rotationally symmetric, a ray incident along the z axis (i.e., with $p_1 = 0$) emerges from the system along the z axis,* i.e., with $p_2 = 0$. Imposing this condition on Eq. (3.2-28), we find that $C(x_1, 0, 0; x_2, 0, 0)$ must be zero. Thus we obtain

$$x_1 p_1 = x_2 p_2 \quad (3.2\text{-}29)$$

* Since the derivatives are evaluated at the axial points [see Eq. (3.2-25)], then p_1, p_2, etc., correspond to the direction cosines of rays through the axial points A and B (see Fig. 3.3a).

Since p_1 and p_2 are the optical direction cosines of the rays along the x axis, if n_1 and n_2 are the refractive indices of media I and II (see Fig. 3.3a; media I and II are assumed to be homogeneous), then

$$p_1 = n_1 \sin \theta_1, \qquad p_2 = n_2 \sin \theta_2 \tag{3.2-30}$$

where θ_1 and θ_2 are the angles made by the rays with the z axis (see Fig. 3.3a). Thus Eq. (3.2-29) reduces to

$$n_1 x_1 \sin \theta_1 = n_2 x_2 \sin \theta_2 \tag{3.2-31}$$

which is called the *Abbe sine condition*. Notice that we have not made any restrictions on θ_1 and θ_2. Thus if the sine condition is satisfied, then nearby off-axis points (in the object plane) get sharply imaged by rays making arbitrary angles with the axis.

Problem 3.3. Let SS' be a spherical surface separating two media of refractive indices n_1 and n_2 (see Fig. 3.3b). Let C be the center of curvature of the spherical surface. From geometrical considerations and use of Snell's law at the point T, derive Eq. (3.2-31). This result implies that if the imaging is perfect then the sine condition has to be satisfied.

Problem 3.4. Show that if a line element along the z axis has to be imaged perfectly, then the condition to be satisfied is

$$\frac{z_2}{z_1} = \frac{n_1 \sin^2 \theta_1/2}{n_2 \sin^2 \theta_2/2} \tag{3.2-32}$$

which is known as Herschel's condition.

3.3. Mixed Characteristic Function

3.3.1. Definition and Properties

In addition to the point characteristic function, one can define two more characteristic functions, namely, the mixed characteristic function (which will be discussed in this section) and the angle characteristic function (which will be discussed in Section 3.4). The mixed characteristic function of a system is defined by

$$W = V - (x_2 p_2 + y_2 q_2 + z_2 r_2) = V - \sum x_2 p_2 \tag{3.3-1}$$

where

$$\sum x_2 p_2 \equiv x_2 p_2 + y_2 q_2 + z_2 r_2 \tag{3.3-2}$$

W is called the mixed characteristic function because, as will be shown, it depends on the coordinates of the object point and the optical direction cosines of the final ray. The physical significance of W will be discussed later. We find from Eq. (3.3-1) that

$$dW = dV - (x_2\,dp_2 + p_2\,dx_2 + y_2\,dq_2 + q_2\,dy_2 + z_2\,dr_2 + r_2\,dz_2) \quad (3.3\text{-}3)$$

Substituting for dV from Eq. (3.2-14) we find that

$$dW = -x_2\,dp_2 - y_2\,dq_2 - z_2\,dr_2 - p_1\,dx_1 - q_1\,dy_1 - r_1\,dz_1 \quad (3.3\text{-}4)$$

Thus we obtain

$$p_1 = -\frac{\partial W}{\partial x_1}, \qquad q_1 = -\frac{\partial W}{\partial y_1}, \qquad r_1 = -\frac{\partial W}{\partial z_1} \quad (3.3\text{-}5)$$

and

$$x_2 = -\frac{\partial W}{\partial p_2}, \qquad y_2 = -\frac{\partial W}{\partial q_2}, \qquad z_2 = -\frac{\partial W}{\partial r_2} \quad (3.3\text{-}6)$$

Thus the mixed characteristic function is found to be a function of the initial position coordinates of the ray (x_1, y_1, z_1) and the final optical direction cosines of the ray (p_2, q_2, r_2), i.e., $W \equiv W(x_1, y_1, z_1; p_2, q_2, r_2)$. Since $p_1^2 + q_1^2 + r_1^2 = n_1^2$, we obtain

$$\left(\frac{\partial W}{\partial x_1}\right)^2 + \left(\frac{\partial W}{\partial y_1}\right)^2 + \left(\frac{\partial W}{\partial z_1}\right)^2 = n_1^2 \quad (3.3\text{-}7)$$

From Eqs. (3.3-5) and (3.3-6) it is clear that if the mixed characteristic function is known, then one can obtain the initial direction cosines (p_1, q_1, r_1) and the final position coordinates (x_2, y_2, z_2) of the ray by simple differentiation.

For the physical interpretation of W, we refer to Fig. 3.4, where for simplicity we have assumed the points (x_1, y_1, z_1) and (x_2, y_2, z_2) to lie in homogeneous media. Let (x_1, y_1, z_1) be the coordinates of the point A with respect to the origin O_1 and let (x_2, y_2, z_2) be the coordinates of the point

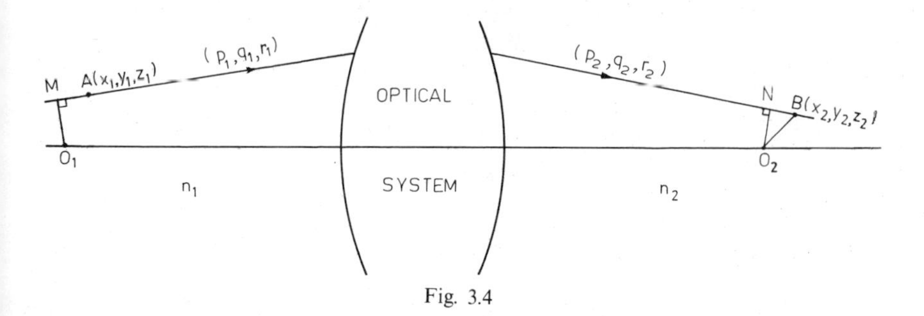

Fig. 3.4

B with respect to the origin O_2. Let O_2N represent the normal from O_2 to the ray. Now,

$$\overrightarrow{O_2B} = \mathbf{r}_2 = x_2\hat{\mathbf{i}} + y_2\hat{\mathbf{j}} + z_2\hat{\mathbf{k}} \tag{3.3-8}$$

Since (p_2, q_2, r_2) represent the optical direction cosines of NB, the unit vector along NB is given by

$$\hat{\mathbf{s}}_2 = \frac{1}{n_2}(p_2\hat{\mathbf{i}} + q_2\hat{\mathbf{j}} + r_2\hat{\mathbf{k}}) \tag{3.3-9}$$

Thus the optical pathlength of NB is given by

$$n_2 \cdot NB = n_2\mathbf{r}_2 \cdot \hat{\mathbf{s}}_2 = (x_2p_2 + y_2q_2 + z_2r_2) \tag{3.3-10}$$

Substituting the above value of $(x_2p_2 + y_2q_2 + z_2r_2)$ into Eq. (3.3-1), we obtain

$$\begin{aligned} W &= V - n_2 \cdot NB \\ &= \text{optical pathlength from the point } A \text{ to the point } N \end{aligned} \tag{3.3-11}$$

In case $B(x_2, y_2, z_2)$ lies in an inhomogeneous medium, W represents the optical pathlength from A to the foot of the perpendicular drawn from O_2 to the tangent to the ray at B, n_2 being the refractive index at B.

It may be noted that if the point B is in a homogeneous medium, n_2 is constant around B. Thus differentiating the equation $p_2^2 + q_2^2 + r_2^2 = n_2^2$, we obtain

$$p_2\,dp_2 + q_2\,dq_2 = -r_2\,dr_2 \tag{3.3-12}$$

Using the value of dr_2 (as given above) in Eq. (3.3-4), we obtain

$$dW = -\left(x_2 - \frac{z_2p_2}{r_2}\right)dp_2 - \left(y_2 - \frac{z_2q_2}{r_2}\right)dq_2 - p_1\,dx_1 - q_1\,dy_1 - r_1\,dz_1 \tag{3.3-13}$$

i.e., the mixed characteristic function can be expressed as a function of five variables. We also obtain

$$x_2 - \frac{z_2p_2}{r_2} = -\frac{\partial W}{\partial p_2}, \qquad y_2 - \frac{z_2q_2}{r_2} = -\frac{\partial W}{\partial q_2} \tag{3.3-14}$$

In addition to the mixed characteristic function defined above, one can define another mixed characteristic function W' by the relation

$$W' = V + (x_1p_1 + y_1q_1 + z_1r_1) = V + \sum x_1p_1 \tag{3.3-15}$$

where $\sum x_1p_1 = x_1p_1 + y_1q_1 + z_1r_1$. One can proceed in an exactly similar manner as above and obtain

$$dW' = x_1\,dp_1 + y_1\,dq_1 + z_1\,dr_1 + p_2\,dx_2 + q_2\,dy_2 + r_2\,dz_2 \tag{3.3-16}$$

When $A(x_1, y_1, z_1)$ lies in a homogeneous medium, then we obtain

$$dW' = \left(x_1 - \frac{z_1 p_1}{r_1}\right) dp_1 + \left(y_1 - \frac{z_1 q_1}{r_1}\right) dq_1 + p_2\, dx_2 + q_2\, dy_2 + r_2\, dz_2 \tag{3.3-17}$$

Furthermore, W' satisfies the differential equation

$$\left(\frac{\partial W'}{\partial x_2}\right)^2 + \left(\frac{\partial W'}{\partial y_2}\right)^2 + \left(\frac{\partial W'}{\partial z_2}\right)^2 = n_2^2 \tag{3.3-18}$$

Problem 3.5. Show that W' represents the optical pathlength from the point M to the point B, where M is the foot of the perpendicular drawn from O_1 to the initial ray (see Fig. 3.4).

Problem 3.6. Determine the equation of a surface that renders parallel, after reflection, all rays emanating from a point.

Solution. It is clear from the problem that the surface must be a surface of revolution and the parallel bundle of rays must be parallel to the axis of revolution. Also, the point from which the rays are emanating must lie on the axis of revolution. Let the axis of revolution be chosen as the z axis. Let the vertex of the surface be the origin of coordinates. Let $A(0, 0, z_1)$ be the point from which rays are emanating (see Fig. 3.5). The mixed characteristic function W is a function of the initial coordinates and the final optical direction cosines of the ray. Hence, for a given initial point, if all final rays have to be parallel, the mixed characteristic function must be a constant. Let a ray start from the point $A(0, 0, z_1)$, intersect the surface SS' at $P(x, y, z)$, and become parallel to the z axis, as shown in Fig. 3.5. The ray will hit any plane $z = z_2$ at $Q(x, y, z_2)$. Thus from Eq. (3.3-1) the mixed characteristic function W is given by

$$W = [x^2 + y^2 + (z - z_1)^2]^{1/2} + (z_2 - z) - z_2 = [x^2 + y^2 + (z - z_1)^2]^{1/2} - z \tag{3.3-19}$$

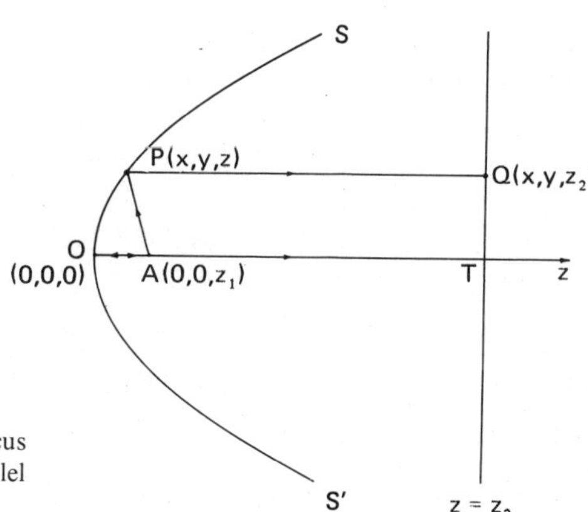

Fig. 3.5. All rays emanating from the focus of a paraboloid are, after reflection, parallel to its axis.

where we have used the fact that $p_2 = q_2 = 0$ and $r_2 = 1$. Thus the equation of the surface is given by

$$[x^2 + y^2 + (z - z_1)^2]^{1/2} - z = C \tag{3.3-20}$$

Corresponding to the ray traveling in the direction AO, $x = 0$, $y = 0$, $z = 0$, and therefore $C = z_1$. Hence after simplification we get

$$x^2 + y^2 = 4z_1 z \tag{3.3-21}$$

which is a surface of revolution obtained by revolving the curve

$$x^2 = 4z_1 z \tag{3.3-22}$$

about the z axis. Equation (3.3-22) is the equation of a parabola with focus at $(0, z_1)$, and Eq. (3.3-21) is the equation of a paraboloid. Thus a paraboloid renders all rays emanating from its focus parallel to its axis of revolution.

3.3.2. *Third-Order Aberration of Rotationally Symmetric Systems*

In Section 2.2 we obtained explicit expressions for the third-order aberration coefficients of rotationally symmetric systems. Now, we use the theory of characteristic functions to obtain the form of the third-order aberrations in rotationally symmetric systems. In order to study the aberrations, we will make use of the mixed characteristic function W, which is a function of the object point coordinates (x_1, y_1, z_1) and the final optical direction cosines of the ray (p_2, q_2, r_2). The final coordinates of the ray are determined by Eq. (3.3-14), where the final medium is assumed to be homogeneous. If the plane $z_2 = 0$ is chosen as the paraxial image plane on which the aberrations have to be determined, then from Eq. (3.3-14) one obtains

$$x_2 = -\frac{\partial W}{\partial p_2}, \qquad y_2 = -\frac{\partial W}{\partial q_2} \tag{3.3-23}$$

Thus for a given object plane $z_1 = \text{const}$, W is a function of x_1, y_1, p_2, and q_2. In order to determine what combinations of x_1, y_1, p_2, and q_2 appear in W for a rotationally symmetric system, we form the following three rotationally invariant quantities:

$$\xi = x_1^2 + y_1^2, \qquad \eta = x_1 p_2 + y_1 q_2, \qquad \zeta = p_2^2 + q_2^2 \tag{3.3-24}$$

The rotational invariance of these quantities can easily be seen by transforming to cylindrical coordinates:

$$\begin{aligned} x_1 &= r_1 \cos\theta, \qquad & p_2 &= \kappa \cos\varphi \\ y_1 &= r_1 \sin\theta, \qquad & q_2 &= \kappa \sin\varphi \end{aligned} \tag{3.3-25}$$

Then

$$\xi = r_1^2, \qquad \eta = r_1\kappa\cos(\theta - \varphi), \qquad \zeta = \kappa^2 \tag{3.3-26}$$

Thus when the system is rotated about the z axis, ξ, η, and ζ do not change and are therefore rotationally invariant. It can be seen easily that there cannot be more than three independent rotationally invariant quantities because, if there were a fourth independent quantity σ (which is also rotationally invariant) then it would be possible to express x_1, y_1, p_2, and q_2 in terms of ξ, η, ζ, and σ; then x_1, y_1, p_2, and q_2 would also have been rotationally invariant, which is not true. Thus the characteristic function W depends on x_1, y_1, p_2, and q_2 only through ξ, η, and ζ.

We expand $W(\xi, \eta, \zeta)$ in ascending powers of its arguments to obtain

$$W(\xi, \eta, \zeta) = W_0 + W_1(\xi, \eta, \zeta) + W_2(\xi^2, \eta^2, \zeta^2, \xi\eta, \xi\zeta, \eta\zeta) + \cdots \tag{3.2-27}$$

where W_n is a homogeneous polynomial of degree n in ξ, η, and ζ. Paraxial rays are represented by infinitesimal values of ξ, η, and ζ, and as such, for paraxial rays, one need retain terms only up to W_1 in the expansion (3.3-27). The paraxial values of the image point (x_2, y_2) are given by

$$\begin{aligned} x_{2p} &= -\frac{\partial W_1}{\partial p_2} = -\left(\frac{\partial W_1}{\partial \eta}\frac{\partial \eta}{\partial p_2} + \frac{\partial W_1}{\partial \zeta}\frac{\partial \zeta}{\partial p_2}\right) \\ &= -2\left(\frac{\partial W_1}{\partial \eta}x_1 + \frac{\partial W_1}{\partial \zeta}p_2\right) \end{aligned} \tag{3.3-28}$$

$$y_{2p} = -\frac{\partial W_1}{\partial q_2} = -2\left(\frac{\partial W_1}{\partial \eta}y_1 + \frac{\partial W_1}{\partial \zeta}q_2\right) \tag{3.3-29}$$

where we have used Eq. (3.3-23). Since W_1 is linear in ξ, η, and ζ, $(\partial W_1/\partial \eta)$ and $(\partial W_1/\partial \zeta)$ are constants. Thus Eqs. (3.3-28) and (3.3-29) give a linear relationship between the image coordinates and the object coordinates. This leads to the formation of perfect images, as expected from paraxial optics.

Thus it is clear that the higher-order terms in the expansion of W, namely, W_2, W_3, etc., represent aberrations present in the image (W_2 represents the third-order aberration, etc.). If W_2 is also considered in Eq. (3.3-27), then we obtain

$$x_2 = -\frac{\partial W_1}{\partial p_2} - \frac{\partial W_2}{\partial p_2} = x_{2p} - \frac{\partial W_2}{\partial p_2} \tag{3.3-30}$$

Thus the aberration along the x direction is given by*

$$\Delta x_2 = x_2 - x_{2p} = -2\left(\frac{\partial W_2}{\partial \eta} x_1 + \frac{\partial W_2}{\partial \zeta} p_2\right) \quad (3.3\text{-}31)$$

Similarly,

$$\Delta y_2 = y_2 - y_{2p} = -2\left(\frac{\partial W_2}{\partial \eta} y_1 + \frac{\partial W_2}{\partial \zeta} q_2\right) \quad (3.3\text{-}32)$$

Since W_2 is a polynomial of second degree in ξ, η, and ζ, we write

$$W_2 = -\tfrac{1}{2}(\tfrac{1}{2}F'\xi^2 + E'\xi\eta + C'\xi\zeta + \tfrac{1}{2}D'\eta^2 + B'\eta\zeta + \tfrac{1}{2}A'\zeta^2) \quad (3.3\text{-}33)$$

where A', B', C', D', E', and F' are constants and the factors have been introduced for convenience. Thus Δx_2 becomes

$$\Delta x_2 = (E'\xi + D'\eta + B'\zeta)\, x_1 + (C'\xi + B'\eta + A'\zeta)\, p_2 \quad (3.3\text{-}34)$$

Similarly,

$$\Delta y_2 = (E'\xi + D'\eta + B'\zeta)\, y_1 + (C'\xi + B'\eta + A'\zeta)\, q_2 \quad (3.3\text{-}35)$$

Thus in the third-order aberrations only five constants appear, implying that the complete third-order aberration of a rotationally symmetric system can indeed be described by five constants. Using the same analysis as used in Section 2.3, one can show that A' represents spherical aberration, B' represents coma, C' and D' together represent astigmatism and curvature of field, and E' represents distortion.

Problem 3.7. Show that, to determine the fifth-order aberration of a rotationally symmetric system, one requires nine constants. Obtain the image shapes of a point object in the presence of each of these terms (see, e.g., Buchdahl, 1970).

3.4. Angle Characteristic Function

The angle characteristic function T is defined through the relation

$$T = V + \sum x_1 p_1 - \sum x_2 p_2 \quad (3.4\text{-}1)$$

Thus we get

$$dT = dV + \sum x_1\, dp_1 + \sum p_1\, dx_1 - \sum x_2\, dp_2 - \sum p_2\, dx_2 \quad (3.4\text{-}2)$$

which, using Eq. (3.2-14), reduces to

$$dT = x_1\, dp_1 + y_1\, dq_1 + z_1\, dr_1 - x_2\, dp_2 - y_2\, dq_2 - z_2\, dr_2 \quad (3.4\text{-}3)$$

* For convenience, we are using here a notation that is slightly different from that used in Chapter 2.

Thus we obtain

$$x_1 = \frac{\partial T}{\partial p_1}, \qquad y_1 = \frac{\partial T}{\partial q_1}, \qquad z_1 = \frac{\partial T}{\partial r_1} \tag{3.4-4}$$

$$x_2 = -\frac{\partial T}{\partial p_2}, \qquad y_2 = -\frac{\partial T}{\partial q_2}, \qquad z_2 = -\frac{\partial T}{\partial r_2} \tag{3.4-5}$$

Hence T is a function of six variables $p_1, q_1, r_1, p_2, q_2, r_2$. For (x_1, y_1, z_1) and (x_2, y_2, z_2) lying in homogeneous media, using Eq. (3.2-11), we can write

$$dT = \left(x_1 - \frac{z_1 p_1}{r_1}\right) dp_1 + \left(y_1 - \frac{z_1 q_1}{r_1}\right) dq_1 - \left(x_2 - \frac{z_2 p_2}{r_2}\right) dp_2 - \left(y_2 - \frac{z_2 q_2}{r_2}\right) dq_2 \tag{3.4-6}$$

Thus, if the initial and final rays lie in homogeneous media, then T is a function of the four variables p_1, q_1, p_2, and q_2 only. It also follows from Eq. (3.4-6) that

$$x_1 - \frac{z_1 p_1}{r_1} = \frac{\partial T}{\partial p_1}, \qquad y_1 - \frac{z_1 q_1}{r_1} = \frac{\partial T}{\partial q_1} \tag{3.4-7}$$

$$x_2 - \frac{z_2 p_2}{r_2} = -\frac{\partial T}{\partial p_2}, \qquad y_2 - \frac{z_2 q_2}{r_2} = -\frac{\partial T}{\partial q_2} \tag{3.4-8}$$

i.e., if the angle characteristic is known, then one can at once determine (x_1, y_1, z_1) and (x_2, y_2, z_2). T is called the angle characteristic function because it depends on the direction cosines of the initial and final rays.

Problem 3.8. Show that the angle characteristic function is the optical pathlength between the points M and N (see Fig. 3.4).

3.5. Explicit Evaluation of Characteristic Functions

As already stated, the exact determination of characteristic functions is in general quite difficult. In this section we will give two specific examples, one in which exact determination is possible and one where only an approximate form is determined.

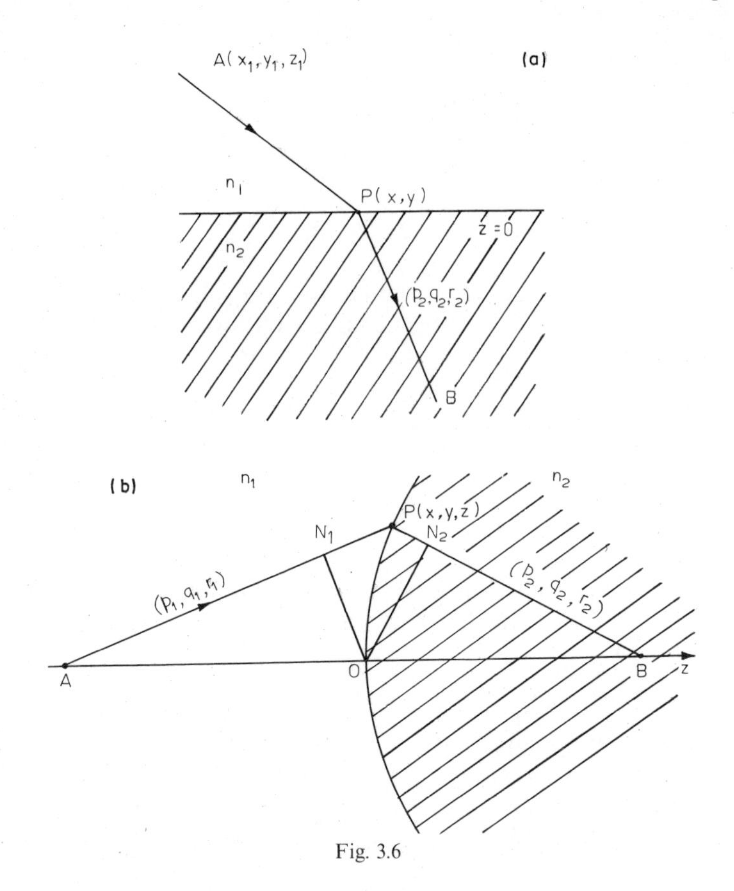

Fig. 3.6

3.5.1. Mixed Characteristic Function for a Plane Surface of Refraction

Let $z = 0$ be a plane surface of refraction separating two media of refractive indices n_1 and n_2, as shown in Fig. 3.6a. A ray starts from the point $A(x_1, y_1, z_1)$, intersects the surface at the point $P(x, y)$, and proceeds along PB with optical direction cosines (p_2, q_2, r_2). Thus the mixed characteristic function is

$$W = n_1[(x_1 - x)^2 + (y_1 - y)^2 + z_1^2]^{1/2} - xp_2 - yq_2 \tag{3.5-1}$$

We must now eliminate x and y from Eq. (3.5-1) using the condition that APB is a ray. For APB to be a ray

$$\frac{\partial W}{\partial x} = 0, \qquad \frac{\partial W}{\partial y} = 0$$

i.e.,

$$n_1 \frac{(x_1 - x)}{d_1} + p_2 = 0, \qquad n_1 \frac{(y_1 - y)}{d_1} + q_2 = 0 \tag{3.5-2}$$

where $d_1 = [(x_1 - x)^2 + (y_1 - y)^2 + z_1^2]^{1/2}$. From Eq. (3.5-2) we obtain

$$(x_1 - x)^2 + (y_1 - y)^2 = d_1^2 - z_1^2 = (p_2^2 + q_2^2)\, d_1^2/n_1^2$$

Thus

$$d_1 = n_1 z_1/(n_1^2 - p_2^2 - q_2^2)^{1/2} \tag{3.5-3}$$

from which one readily obtains

$$W = z_1(n_1^2 - p_2^2 - q_2^2)^{1/2} - x_1 p_2 - y_1 q_2 \tag{3.5-4}$$

which is the mixed characteristic function of a surface of refraction.

Problem 3.9. Show that the point characteristic function of a plane mirror is

$$V = [(x_2 - x_1)^2 + (y_2 - y_1)^2 + (z_2 + z_1)^2]^{1/2}$$

where $z = 0$ represents the plane of the mirror.

3.5.2. *Angle Characteristic of a Spherical Surface of Refraction*

We will calculate an approximate value for the angle characteristic T for a spherical surface separating two media of refractive indices n_1 and n_2. Let O represent the origin of our coordinate system chosen on the surface, let Oz be along the line joining O and the center of curvature C, and let the incident and refracted rays be AP and PB, as shown in Fig. 3.6b. Let N_1 and N_2 be the feet of the normals from O to AP and PB, respectively. If the coordinates of P are (x, y, z), and (p_1, q_1, r_1) and (p_2, q_2, r_2) represent the optical direction cosines of the incident and refracted rays, then T (defined as the optical length of $N_1P + PN_2$) is

$$T = (p_2 - p_1)\,x + (q_2 - q_1)\,y + (r_2 - r_1)\,z \tag{3.5-5}$$

We eliminate x, y, and z from the above equation by making use of Snell's law, according to which the vector with components $[(p_2 - p_1), (q_2 - q_1), (r_2 - r_1)]$ must be along the normal to the surface (see Problem 1.2). The approximate equation of the surface near the origin can be written

$$z = (x^2 + y^2)/2R \tag{3.5-6}$$

The normal to this surface has components $(x/R,\ y/R,\ -1)$. Thus we have

$$\frac{p_2 - p_1}{r_2 - r_1} = -\frac{x}{R}, \qquad \frac{q_2 - q_1}{r_2 - r_1} = -\frac{y}{R} \tag{3.5-7}$$

Substituting for x and y from Eq. (3.5-7) into Eq. (3.5-6) and eliminating x, y, and z from Eq. (3.5-5), we get

$$T = -\frac{R}{2(r_2 - r_1)}[(p_2 - p_1)^2 + (q_2 - q_1)^2] \tag{3.5-8}$$

which is an approximate form for T for a spherical surface, because Eq. (3.5-6) is an approximate form of the equation for a spherical surface. Since Eq. (3.5-6) is an exact equation for a paraboloid of revolution, Eq. (3.5-8) is an exact form for T for a paraboloid of revolution. Since

$$r_1 = (n_1^2 - p_1^2 - q_1^2)^{1/2}, \qquad r_2 = (n_2^2 - p_2^2 - q_2^2)^{1/2} \tag{3.5-9}$$

for infinitesimal values of p_1, q_1, p_2, and q_2 (paraxial rays) we expand the square root in Eq. (3.5-9) to get

$$r_1 = n_1 - \frac{p_1^2 + q_1^2}{2n_1}, \qquad r_2 = n_2 - \frac{p_2^2 + q_2^2}{2n_2} \tag{3.5-10}$$

Thus

$$\frac{1}{r_2 - r_1} \simeq \frac{1}{n_2 - n_1}\left[1 - \frac{p_1^2 + q_1^2}{2n_1(n_2 - n_1)} + \frac{p_2^2 + q_2^2}{2n_2(n_2 - n_1)} + \cdots\right] \tag{3.5-11}$$

Thus to first order* we get

$$T = \frac{R}{2(n_2 - n_1)}[(p_2 - p_1)^2 + (q_2 - q_1)^2] \tag{3.5-12}$$

Problem 3.10. Using the value of T as given by Eq. (3.5-12), obtain the law of refraction at a spherical surface of refraction:

$$\frac{n_2}{z_2} - \frac{n_1}{z_1} = \frac{n_2 - n_1}{R} \tag{3.5-13}$$

Solution. We consider an axial object point A with coordinates $(0, z_1)$. The x coordinate of the image point has to be automatically zero because a ray with $p_1 = 0$ from $(0, z_1)$ passes undeviated. Thus substituting the value of T obtained in Eq. (3.5-12) into Eqs. (3.4-7) and (3.4-8) we obtain

$$z_1\frac{p_1}{n_1} = \frac{R}{n_2 - n_1}(p_2 - p_1) = z_2\frac{p_2}{n_2} \tag{3.5-14}$$

where we have approximated r_1 and r_2 by n_1 and n_2, respectively [see Eq. (3.5-10)]. Eliminating p_1 and p_2 we obtain Eq. (3.5-13).

* For an expression for T for a refracting surface, accurate up to the next order of approximation, see Born and Wolf (1975).

4

Diffraction

4.1. Introduction

Let us consider a point source P_0 in front of a rectangular slit AB as shown in Fig. 4.1. According to geometrical optics (i.e., under the assumption of the validity of the rectilinear propagation of light), the source will cast a sharp shadow on the screen SS'; thus, the region $A'B'$ will be uniformly illuminated and there will be complete darkness in the remaining portion of the screen. However, if one carefully observes the region near A' (or B') then he will find either a gradual variation of intensity or even a fringe pattern. This is due to the phenomenon of diffraction, and because of the smallness of λ (the wavelength of light), these effects cannot be observed unless one makes very careful observations. Indeed, in the limit of $\lambda \to 0$, the gradual variation of the intensity (or the fringe pattern) will tend to disappear.

In this chapter we will examine the phenomenon of diffraction in

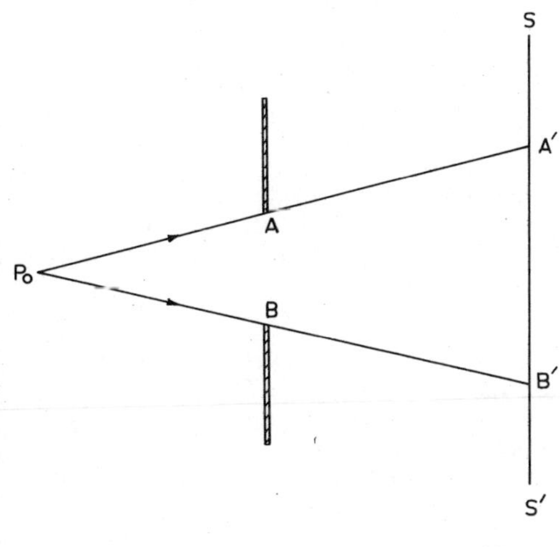

Fig. 4.1. According to the laws of geometrical optics, the point source P_0 will produce a sharp shadow at A' and B' on the screen SS'. However, due to diffraction one will observe a fringe pattern (or a gradual variation of intensity) around A' and B'.

considerable detail. Our starting point will be the scalar wave equation*

$$\nabla^2 U = \frac{1}{c^2}\frac{\partial^2 U}{\partial t^2} \tag{4.1-1}$$

which is satisfied by each component of the electric or magnetic field vectors in a uniform homogeneous medium (see Problem 10.2). In Eq. (4.1-1), c represents the speed of propagation of the wave. The intensity of the wave will be proportional to $|U|^2$.

4.2. The Spherical Wave

We consider a monochromatic wave, i.e., we assume the time dependence to be of the form $\exp(i\omega t)$:

$$U(\mathbf{r}, t) = u(\mathbf{r})\exp(i\omega t) \tag{4.2-1}$$

If we substitute the above form of $U(\mathbf{r}, t)$ into Eq. (4.1-1), we obtain

$$\nabla^2 u + k^2 u = 0 \tag{4.2-2}$$

where k $(=\omega/c)$ represents the wave vector. Equation (4.2-2) is known as the Helmholtz equation. For a spherical wave, u is a function of the radial coordinate r only, and consequently Eq. (4.2-2) becomes

$$\frac{1}{r^2}\frac{d}{dr}\left(r^2\frac{du}{dr}\right) + k^2 u = 0 \tag{4.2-3}$$

The function $\chi(r) = ru(r)$ satisfies

$$\frac{d^2\chi}{dr^2} + k^2\chi(r) = 0 \tag{4.2-4}$$

or

$$\chi(r) = e^{\mp ikr} \tag{4.2-5}$$

Thus, for monochromatic spherical waves,

$$U(\mathbf{r}, t) = \frac{A}{r}e^{i(\omega t \mp kr)} \tag{4.2-6}$$

* It is of interest to point out that the treatment of diffraction phenomena using the scalar wave equation was known much before Maxwell expounded his famous electromagnetic theory of light. The results obtained by using the scalar wave equation agree quite well with most experimental observations using optical beams. For a more rigorous diffraction theory in which one incorporates the effects due to polarization, the reader is referred to Chapter XI of Born and Wolf (1975).

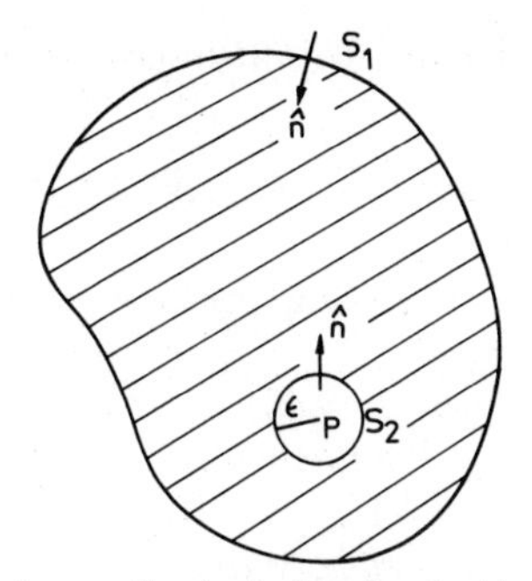

Fig. 4.2. S_2 is a small spherical surface inside S_1; $\hat{\mathbf{n}}$ represents the unit inward normal.

where A represents the amplitude of the wave; the upper and lower signs correspond to the outgoing and incoming spherical waves, respectively. The factor $1/r$ in front of the exponential implies that the intensity associated with spherical waves emanating from a point source will decrease as $1/r^2$, which is the inverse square law.

4.3. Integral Theorem of Helmholtz and Kirchhoff

Let us consider a closed surface S surrounding a volume V. If the functions u and u' along with their first and second derivatives are continuous inside and on the surface S, then Green's theorem* tells us that

$$\iiint_V (u\nabla^2 u' - u'\nabla^2 u)\,dV = -\iint_S (u\nabla u' - u'\nabla u)\cdot\hat{\mathbf{n}}\,dS \qquad (4.3\text{-}1)$$

where the integrals appearing on the left- and right-hand side represent a volume integral (over the volume V) and a surface integral (over the surface S), respectively; $\hat{\mathbf{n}}$ represents a unit inward normal on the surface (see Fig. 4.2).

If we assume that u and u' represent solutions of the Helmholtz equation, then

$$\nabla^2 u = -k^2 u, \qquad \nabla^2 u' = -k^2 u'$$

* The proof of Eq. (4.3-1) is very simple. We consider a vector $\mathbf{F} = (u\nabla u' - u'\nabla u)$. Thus

$$\nabla\cdot\mathbf{F} = (u\nabla^2 u' - u'\nabla^2 u)$$

If we now use Gauss' theorem

$$\iiint_V \nabla\cdot\mathbf{F}\,dV = -\iint_S \mathbf{F}\cdot\hat{\mathbf{n}}\,dS$$

(the negative sign appearing because $\hat{\mathbf{n}}$ represents an inward normal) we should obtain Eq. (4.3-1).

and the integrand appearing on the left-hand side of Eq. (4.3-1) vanishes everywhere. Thus

$$\iint_S (u\nabla u' - u'\nabla u)\cdot\hat{\mathbf{n}}\, dS = 0 \tag{4.3-2}$$

We now assume u' to represent a spherical wave, i.e.,

$$u'(\mathbf{r}) = e^{-ikr}/r \tag{4.3-3}$$

where r $[=(x^2 + y^2 + z^2)^{1/2}]$ represents the distance from a chosen origin. Clearly, since the function u' and its derivative are unbounded at $r = 0$, the volume V cannot include the origin, which has been denoted by point P in Fig. 4.2. We surround the point P by a small sphere of radius ε and assume the domain of integration to be over the shaded region shown in Fig. 4.2.* Thus the surface integral will now consist of two parts: one over the surface S_1 and the other over the surface S_2. Thus Eq. (4.3-2) becomes

$$\iint_{S_1} (u\nabla u' - u'\nabla u)\cdot\hat{\mathbf{n}} dS + \iint_{S_2} (u\nabla u' - u'\nabla u)\cdot\hat{\mathbf{n}} dS = 0 \tag{4.3-4}$$

Clearly, the inward normal on the surface S_2 is along the radius vector (see Fig. 4.2), and therefore

$$\begin{aligned}\iint_{S_2} (u\nabla u' - u'\nabla u)\cdot\hat{\mathbf{n}}\, dS &= \iint_{S_2} (u\nabla u' - u'\nabla u)\cdot\hat{\mathbf{r}}\, dS \\ &= \iint \left(u\frac{\partial u'}{\partial r} - u'\frac{\partial u}{\partial r}\right)\varepsilon^2\, d\Omega \end{aligned}\tag{4.3-5}$$

where $d\Omega$ represents the solid angle.† Substituting for u' from Eq. (4.3-3), we obtain

$$\begin{aligned}\iint_{S_2} (u\nabla u' - u'\nabla u)\cdot\hat{\mathbf{n}}\, dS &= \iint_{\Omega}\left[u\,\frac{\partial}{\partial r}\left(\frac{e^{-ikr}}{r}\right) - \frac{e^{-ikr}}{r}\frac{\partial u}{\partial r}\right]\varepsilon^2\, d\Omega \\ &= \iint\left[-iku - \frac{u}{r} - \frac{\partial u}{\partial r}\right]\frac{e^{-ikr}}{r}\,\varepsilon^2\, d\Omega \\ &= -e^{-ik\varepsilon}\left[\iint u\, d\Omega + \varepsilon\iint\left(\frac{\partial u}{\partial r} + iku\right)d\Omega\right]\end{aligned}\tag{4.3-6}$$

* Clearly in the shaded region the function u' is well behaved; we assume that the function u is also well behaved in this region.

† We have used the relation

$$dS = r^2 \sin\theta\, d\theta\, d\varphi = \varepsilon^2\, d\Omega$$

where $d\Omega = \sin\theta\, d\theta\, d\varphi$ and use has been made of the fact that on the surface S_2, r has the constant value ε.

where we have used the fact that $r = \varepsilon$ on the surface S_2. We assume ε to be infinitesimal and since the function u is regular around the point P, the second integral on the right-hand side of Eq. (4.3-6) tends to zero and the first integral simply becomes $4\pi u(P)$. Thus

$$\lim_{\varepsilon \to 0} \iint_{S_2} (u\nabla u' - u'\nabla u) \cdot \hat{\mathbf{n}}\, dS = -4\pi u(P) \tag{4.3-7}$$

Substituting this expression into Eq. (4.3-4), we get

$$u(P) = \frac{1}{4\pi} \iint_{S_1} \left[u\nabla \left(\frac{e^{-ikr}}{r} \right) - \frac{e^{-ikr}}{r} \nabla u \right] \cdot \hat{\mathbf{n}}\, dS$$

or

$$u(P) = \frac{1}{4\pi} \iint_{S_1} \left[u \frac{\partial}{\partial r} \left(\frac{e^{-ikr}}{r} \right) \hat{\mathbf{r}} \cdot \hat{\mathbf{n}} - \frac{e^{-ikr}}{r} \nabla u \cdot \hat{\mathbf{n}} \right] dS \tag{4.3-8}$$

Equation (4.3-8) is known as the integral theorem of Helmholtz and Kirchhoff and gives us the remarkable result that if a function u (which satisfies the Helmholtz equation) along with its normal derivative is known on a closed surface, then the value of the function at any point inside the surface can also be determined.

4.4. The Fresnel–Kirchhoff Diffraction Formula

Let us consider an aperture A in an infinite opaque screen SS' illuminated by spherical waves emanating from a point source P_0. We wish to calculate the field at the point P on the other side of the aperture (see Fig. 4.3a); the aperture is assumed to be on the x-y plane. With the point P as the center, we draw a sphere of radius R, which intersects the screen in a circular area; this circular area comprises two parts: (a) the aperture A and (b) a portion of the opaque screen denoted by B. In order to calculate the field at the point P we use the integral theorem of Helmholtz and Kirchhoff [i.e., Eq. (4.3-8)] and choose the surface S to comprise the following parts: (1) the spherical surface C, (2) the illuminated aperture A, and (3) the nonilluminated portion of the screen B. Thus Eq. (4.3-8) takes the form

$$u(P) = \frac{1}{4\pi} \left[\iint_A + \iint_B + \iint_C \right] \left[u \frac{\partial}{\partial r} \left(\frac{e^{-ikr}}{r} \right) \hat{\mathbf{r}} \cdot \hat{\mathbf{n}} - \frac{e^{-ikr}}{r} \nabla u \cdot \hat{\mathbf{n}} \right] dS \tag{4.4-1}$$

We will now show that the contribution from the spherical surface C goes

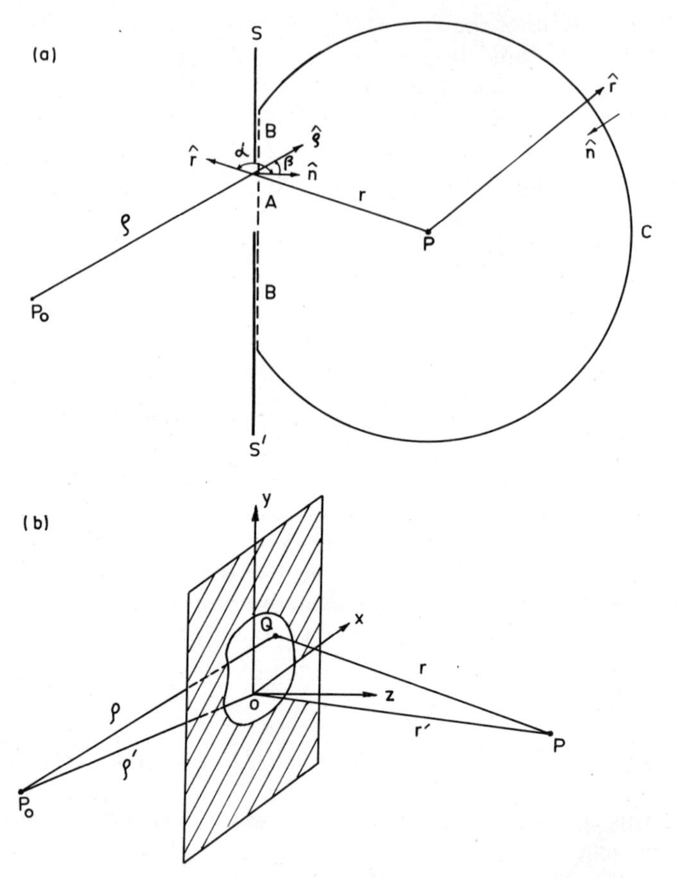

Fig. 4.3. Spherical waves emanating from the point source P_0 illuminate an aperture that lies in the x-y plane. We wish to calculate the field at the point P.

to zero as $R \to \infty$.* On the surface C, $\hat{\mathbf{n}} = -\hat{\mathbf{r}}$ and $r = R$ (see Fig. 4.3a); thus,

$$\iint_C \left[u \frac{\partial}{\partial r} \left(\frac{e^{-ikr}}{r} \right) \hat{\mathbf{r}} \cdot \hat{\mathbf{n}} - \frac{e^{-ikr}}{r} \nabla u \cdot \hat{\mathbf{n}} \right] dS$$

$$= \iint \left[-u \left(-ik - \frac{1}{r} \right) \frac{e^{-ikr}}{r} + \frac{e^{-ikr}}{r} \frac{\partial u}{\partial r} \right]_{r=R} R^2 \, d\Omega$$

$$= e^{-ikR} \iint \left[\left(\frac{\partial u}{\partial R} + iku \right) R + u \right] d\Omega \tag{4.4-2}$$

* One may be tempted to reason that since the functions $u(R)$ and e^{-ikR}/R go to zero as $R \to \infty$, the contribution from the surface C must go to zero; however, this is not obvious because as $R \to \infty$, the surface area also increases as R^2.

Consequently, the integral over the spherical surface C will vanish if u decreases in such a way that

$$\lim_{R\to\infty}\left(\frac{\partial u}{\partial R}+iku\right)R=0 \tag{4.4-3}$$

Equation (4.4-3) is known as the Sommerfeld radiation condition and is satisfied by a spherical wave. Thus, if u decreases at least as fast as a diverging spherical wave (which is indeed true for all practical cases), the contribution from the surface C vanishes identically.* Thus Eq. (4.4-1) becomes

$$u(P)=\frac{1}{4\pi}\left[\iint_A+\iint_B\right]\left[u\frac{\partial}{\partial r}\left(\frac{e^{-ikr}}{r}\right)\hat{\mathbf{r}}\cdot\hat{\mathbf{n}}-\frac{e^{-ikr}}{r}\nabla u\cdot\hat{\mathbf{n}}\right]dS \tag{4.4-4}$$

In order to evaluate the above integrals one makes the following assumptions:

(1) Over the surface B (which lies just behind the opaque portion of the screen) the field u and its derivative are zero.

(2) On the surface A, the field along with its derivative has exactly the same value as it would have had if the screen had been absent. Thus if the screen is illuminated by a point source, then on the aperture A the field is given by

$$u=(A/\rho)\,e^{-ik\rho} \tag{4.4-5}$$

where ρ represents the distance of an arbitrary point on the aperture from the point source (see Fig. 4.3a) and A represents the amplitude of the wave.

The above two assumptions are known as Kirchhoff's boundary conditions and are valid for most optical experiments.† If we use these conditions we obtain

$$u(P)=\frac{1}{4\pi}\iint_A\left[u\frac{\partial}{\partial r}\left(\frac{e^{-ikr}}{r}\right)\hat{\mathbf{r}}\cdot\hat{\mathbf{n}}-\frac{e^{-ikr}}{r}\nabla u\cdot\hat{\mathbf{n}}\right]dS \tag{4.4-6}$$

A further simplification is obtained if we assume that the distance of the

* It is often argued that if one assumes a source that is switched on at $t=0$, then at any instant t one can choose a large enough value of R such that the disturbance has not reached the spherical surface C, i.e., $R > ct$. Consequently, the integrand is zero everywhere on the surface C, which makes the integral vanish. However, this argument is not rigorously correct because it is inconsistent with our assumption of a strictly monochromatic wave, which (by definition) is assumed to exist at all times.

† A more rigorous analysis based on the electromagnetic character of light shows that the fields are appreciable even in the nonilluminated portion of the screen at distances of order λ. Similarly, the fields in the aperture at distances of order λ from the edges also differ considerably from the expression given by Eq. (4.4-5). However, as long as the dimensions of the apertures are large in comparison to wavelength, the use of Kirchhoff's boundary conditions gives fairly accurate results.

point P from the aperture A is many times the wavelength. Under this approximation

$$\frac{\partial}{\partial r}\left(\frac{e^{-ikr}}{r}\right) = \left(-ik - \frac{1}{r}\right)\frac{e^{-ikr}}{r} \simeq -ik\frac{e^{-ikr}}{r} \tag{4.4-7}$$

Thus

$$u(P) = -\frac{1}{4\pi}\iint_A [iku\cos\alpha + \nabla u\cdot\hat{\mathbf{n}}]\frac{e^{-ikr}}{r}\,dS \tag{4.4-8}$$

where α is the angle between the vectors $\hat{\mathbf{r}}$ and $\hat{\mathbf{n}}$ (see Fig. 4.3a). Next, assuming the aperture A to be illuminated by spherical waves emanating from the point P_0 and assuming the validity of Kirchhoff's boundary conditions, we obtain $u = (A/\rho)\,e^{-ik\rho}$ on the aperture A. Thus

$$\nabla u\cdot\hat{\mathbf{n}} = -A\left(ik + \frac{1}{\rho}\right)\frac{e^{-ik\rho}}{\rho}\hat{\boldsymbol{\rho}}\cdot\hat{\mathbf{n}} \simeq -iAk\frac{e^{-ik\rho}}{\rho}\cos\beta \tag{4.4-9}$$

where we have assumed $k\rho \gg 1$, i.e., the distance between the source and the aperture is large compared to the wavelength; β represents the angle between $\hat{\boldsymbol{\rho}}$ and $\hat{\mathbf{n}}$ (see Fig. 4.3a). Substituting from Eq. (4.4-9) into Eq. (4.4-8), we finally get

$$u(P) = ik\frac{A}{4\pi}\iint_A \frac{e^{-ik(r+\rho)}}{r\rho}(\cos\beta - \cos\alpha)\,dS \tag{4.4-10}$$

The above result, which corresponds to the illumination of the aperture by a single point source, is known as the Fresnel–Kirchhoff diffraction formula. It is of interest to mention that since the above expression is symmetrical with respect to an interchange of r and ρ, the field at the point P due to a point source at the point P_0 will be the same as the field at the point P_0 due to a point source at the point P. This is known as the reciprocity theorem of Helmholtz.

The factor $(\cos\beta - \cos\alpha)$ is known as the obliquity factor and for most practical problems is approximately equal to 2. This follows from the fact that the apertures are usually of small dimensions, so that the angles α and β are very close to π and zero, respectively.* In this approxi-

* For a plane wave incident normally, the obliquity factor becomes $(1 + \cos\alpha)$. This is the familiar factor that one uses in solving diffraction problems using Huygens' principle (see, e.g., Jenkins and White, 1957).

mation Eq. (4.4-10) becomes*

$$u(P) = i\frac{Ak}{2\pi}\iint_A \frac{e^{-ik(r+\rho)}}{r\rho}\,dS \tag{4.4-11}$$

In most problems of interest, since the dimensions of the aperture are small, the quantities r and ρ do not vary appreciably over the aperture A and as such, very little error will be involved if we replace r and ρ in the denominator of the integrand by r' and ρ', where r' and ρ' are the distances of the points P and P_0 from a conveniently chosen origin in the aperture. Thus we may write†

$$u(P) = \frac{iA}{\lambda}\frac{1}{r'\rho'}\iint_A e^{-ik(r+\rho)}\,dS \tag{4.4-12}$$

4.5. Fraunhofer and Fresnel Diffraction

For a given point source, in order to calculate the diffraction pattern, one has to evaluate the integral appearing on the right-hand side of Eq. (4.4-12). This, in general, is quite difficult and one has to resort to certain approximations that correspond to specific experimental conditions. We assume the x-y plane to coincide with the aperture and we choose a suitable origin O of the coordinate system, which lies inside the aperture (see Fig. 4.3b). Let the coordinates of the points P_0 and P be (x_0, y_0, z_0) and (x, y, z), respectively. Let $Q(\xi, \eta, 0)$ represent an arbitrary point inside the aperture. Clearly,

$$\begin{aligned} r = PQ &= [(x-\xi)^2 + (y-\eta)^2 + z^2]^{1/2} \\ &= r'\left[1 - \frac{2(x\xi + y\eta)}{r'^2} + \frac{\xi^2+\eta^2}{r'^2}\right]^{1/2} \end{aligned} \tag{4.5-1}$$

* Equation (4.4-11) can be easily understood from Huygens' principle. If we consider an elemental area ΔS on the aperture A and if we use the concept of secondary wavelets, then the field at the point P due to the secondary wavelets emanating from the area ΔS is proportional to $(1/r)\,\Delta S\exp(-ikr)$. But the amplitude at the area ΔS due to a point source at P_0 is proportional to $(1/\rho)\exp(-ik\rho)$. Consequently, the field at the point P is proportional to

$$\sum (1/r\rho)\exp[-ik(r+\rho)]\,\Delta S$$

where the summation is carried over all the elemental areas of the aperture A. The above expression is the same as given by Eq. (4.4-11).

† It may be worthwhile to mention that although the quantities r and ρ do not vary appreciably over the aperture, the factor $\exp[-ik(r+\rho)]$ does not have a very rapid variation. For example, for $\lambda = 6\times 10^{-5}$ cm, the factor $\cos kr$ becomes $\cos(\frac{1}{3}\pi \times 10^5 r)$. As the value of r is changed from, say, 30 to 30.00002 cm, the cosine factor will change from $+1$ to -0.5. This shows the rapidity with which the exponential factor will vary in the domain of integration.

where $r' = (x^2 + y^2 + z^2)^{1/2}$ represents the distance of the point P from the origin. Since the dimension of the aperture is very small compared to the distances r' and ρ', we may make a binomial expansion to obtain

$$r = r' - \frac{x\xi + y\eta}{r'} + \left[-\frac{(x\xi + y\eta)^2}{2r'^3} + \frac{\xi^2 + \eta^2}{2r'} \right] + \cdots \tag{4.5-2}$$

Similarly,

$$\rho = \rho' - \frac{x_0\xi + y_0\eta}{\rho'} + \left[-\frac{(x_0\xi + y_0\eta)^2}{2\rho'^3} + \frac{\xi^2 + \eta^2}{2\rho'} \right] + \cdots \tag{4.5-3}$$

where $\rho' = (x_0^2 + y_0^2 + z_0^2)^{1/2}$ is the distance of the point P_0 from the origin O. Substituting the above expressions for r and ρ into Eq. (4.4-12), we obtain

$$u(x, y, z) = \frac{iA}{\lambda} \frac{\exp[-ik(r' + \rho')]}{r'\rho'} \iint_A \exp[ikf(\xi, \eta)]\, d\xi\, d\eta \tag{4.5-4}$$

where

$$f(\xi, \eta) = (l - l_0)\,\xi + (m - m_0)\,\eta - \frac{1}{2}\left[\left(\frac{1}{r'} + \frac{1}{\rho'} \right)(\xi^2 + \eta^2) - \frac{(l\xi + m\eta)^2}{r'} - \frac{(l_0\xi + m_0\eta)^2}{\rho'} \right] + \cdots \tag{4.5-5}$$

with

$$l = x/r', \qquad m = y/r', \qquad l_0 = -x_0/\rho', \qquad m_0 = -y_0/\rho' \tag{4.5-6}$$

represent the direction cosines of OP and P_0O along the x and y directions.

If r' and ρ' are so large that the quadratic (and higher-order terms) in ξ and η are negligible, then we have what is known as Fraunhofer diffraction. On the other hand, if it is necessary to retain terms that are quadratic in ξ and η, then the corresponding diffraction pattern is said to be of Fresnel type. It is obvious that the Fraunhofer diffraction pattern is much easier to calculate and fortunately it is of greater importance in optics. We will first make a study of the Fraunhofer diffraction pattern produced by different kinds of apertures, which will be followed by a discussion on the Fresnel diffraction pattern.

If we neglect the terms quadratic in ξ and η in the expression for $f(\xi, \eta)$, then Eq. (4.5-4) assumes the form

$$u(x, y, z) = C \iint_A \exp[ik(p\xi + q\eta)]\, d\xi\, d\eta \tag{4.5-7}$$

where

$$p = l - l_0, \qquad q = m - m_0, \qquad C = \frac{iA}{\lambda r' \rho'} \exp[-ik(r' + \rho')] \qquad (4.5\text{-}8)$$

It should be mentioned that neglecting terms quadratic in ξ and η essentially implies that the source and the screen are at infinite distances from the aperture. However, if r' and ρ' are very large, then the factor C appearing in Eq. (4.5-7) will become exceedingly small [see Eq. (4.5-8)]. This is obvious because if the source is at a large distance, the correspond-

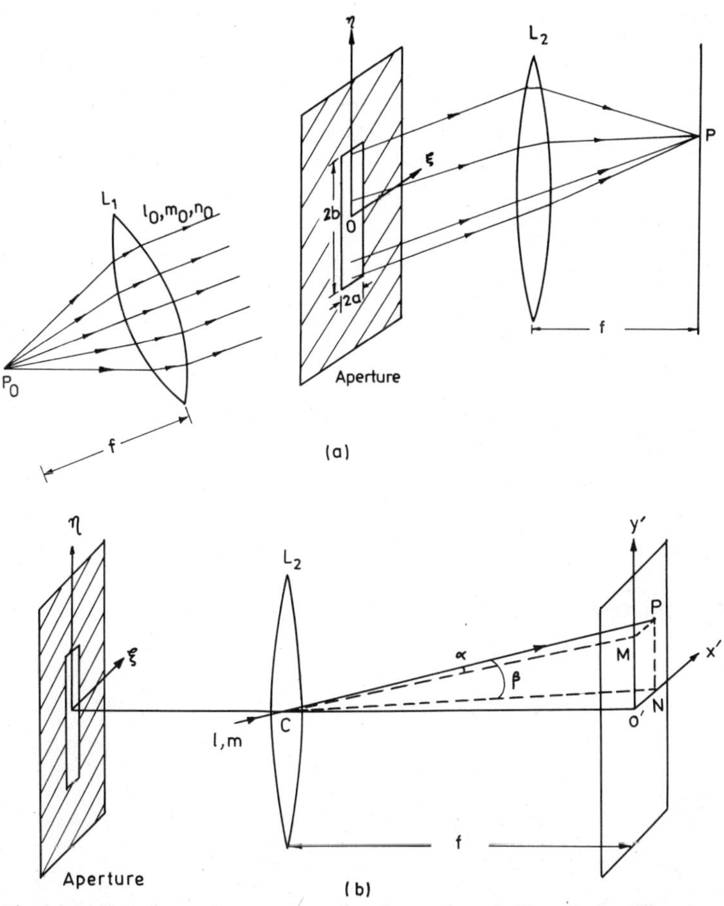

Fig. 4.4. (a) Experimental arrangement for observation of a Fraunhofer diffraction pattern. The point source P_0 (placed at the focal plane of a convex lens) produces a parallel beam of light that is incident on the aperture. The Fraunhofer diffraction pattern is observed at the back focal plane of the lens L_2. (b) Fraunhofer diffraction by a rectangular aperture. The intensity at the point P corresponds to the direction $l = \sin\alpha$ and $m = \sin\beta$.

ing amplitude at the aperture will be very small; similarly, if the observation screen is kept at a very large distance, then the intensity distribution will be very small. The first difficulty is removed by placing a point source on the focal plane of a corrected convex lens, so that one has a plane wave of finite amplitude incident on the aperture. In a similar manner, instead of removing the observation screen to a large distance, we may place a convex lens in front of the aperture. On its focal plane we will observe the Fraunhofer diffraction pattern. The arrangement is shown in Fig. 4.4a. Using such a procedure the amplitude distribution on the focal plane will be essentially the same as given by Eq. (4.5-7), where C represents a constant factor.

Equation (4.5-7) implies that there is no amplitude distribution along the wavefront at the plane of the aperture. If instead of an aperture we have an amplitude distribution along the wavefront,* then Eq. (4.5-7) modifies to

$$u(x, y, z) = C \iint A(\xi, \eta)\, e^{ik(p\xi + q\eta)}\, d\xi\, d\eta \tag{4.5-9}$$

where $A(\xi, \eta)$ represents the field distribution along the wavefront. Equation (4.5-9) shows that the Fraunhofer diffraction pattern is essentially a Fourier transform of the function $A(\xi, \eta)$ (see Chapter 6).

4.6. Fraunhofer Diffraction by a Rectangular Aperture

Let us consider a rectangular aperture of width $2a$ and length $2b$ (see Fig. 4.4). Let the direction cosines of the incident plane wave be represented by l_0, m_0, n_0. Assuming that the origin of our coordinate system is at the center of the rectangle, Eq. (4.5-7) assumes the form

$$u = C \int_{-a}^{a} \int_{-b}^{b} \exp[ik(p\xi + q\eta)]\, d\xi\, d\eta \tag{4.6-1}$$

Carrying out the integrations we obtain

$$u = 4Cab\left(\frac{\sin kpa}{kpa}\right)\left(\frac{\sin kqb}{kqb}\right) \tag{4.6-2}$$

* Indeed, for most laser beams the intensity distribution is Gaussian along the wavefront. The diffraction of a Gaussian beam is discussed in Problem 4.8.

a

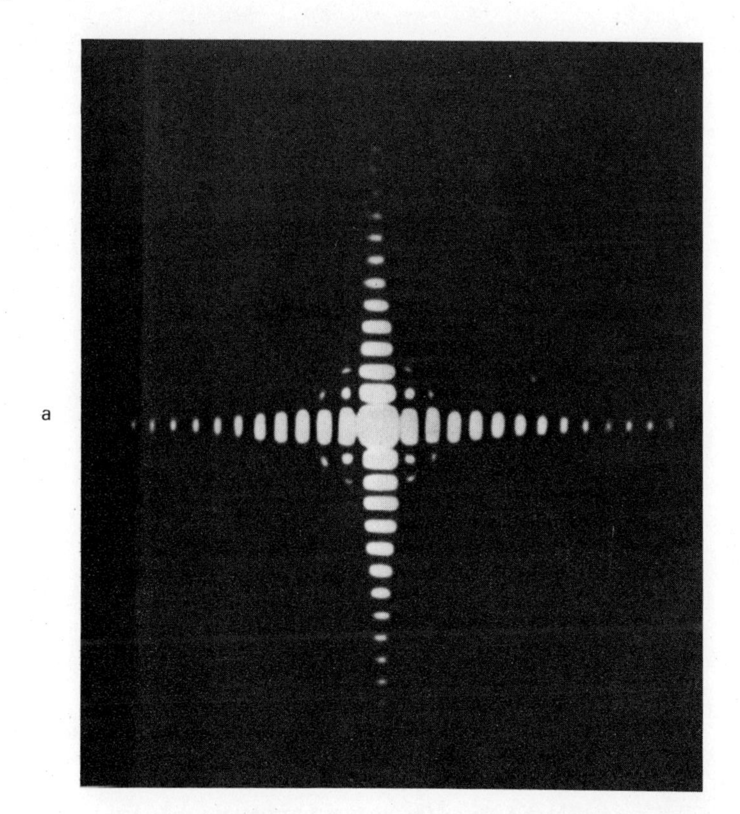

b

0·0007	0·00014	0·00039	0·0083	0·00039	0·00014	0·00007
0·00014	0·00027	0·00077	0·0165	0·00077	0·00027	0·00014
0·00039	0·00077	0·00219	0·0468	0·00219	0·00077	0·00039
0·0083	0·0165	0·0468	1·0	0·0468	0·0165	0·0083
0·00039	0·00077	0·00219	0·0468	0·00219	0·00077	0·00039
0·00014	0·00027	0·00077	0·0165	0·00077	0·00027	0·00014
0·00007	0·00014	0·00039	0·0083	0·00039	0·00014	0·00007

Fig. 4.5. (a) Fraunhofer diffraction pattern produced by a square aperture. Photograph adapted from Preston (1972). The points denote the positions of maxima and the numbers denote their relative intensity.

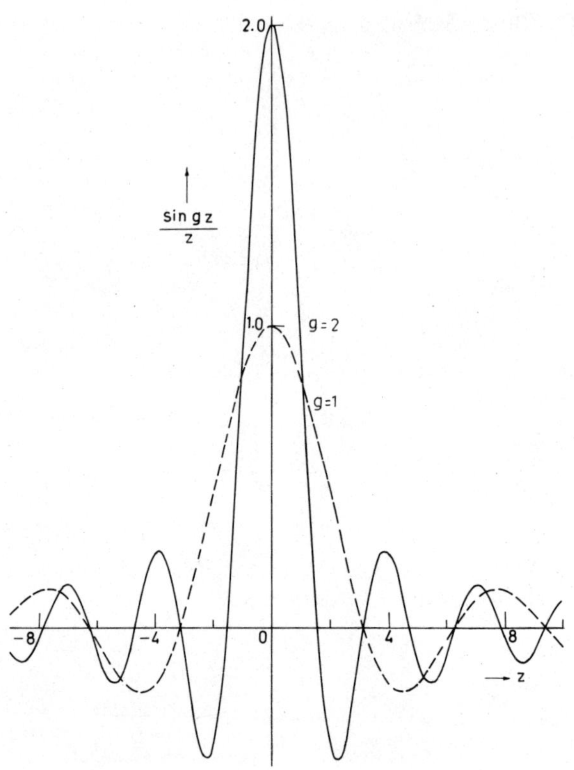

Fig. 4.6a. The variation of the function $(\sin gz)/z$ for $g = 1.0$ and $g = 2.0$. Notice that as the value of g becomes larger, the function becomes more and more sharply peaked at $z = 0$.

Thus, the intensity distribution will be of the form

$$I = I_0 \left(\frac{\sin kpa}{kpa} \right)^2 \left(\frac{\sin kqb}{kqb} \right)^2 \tag{4.6-3}$$

where I_0 represents the intensity at the point P_0 (see Fig. 4.4a), which corresponds to $l = l_0$ and $m = m_0$. We now consider the particular case in which the beam is incident normally so that $l_0 = m_0 = 0$. Let α be the angle between CP and CM and β the angle between CP and CN (see Fig. 4.4b); P represents an arbitrary point on the back focal plane of the lens L_2, PM and PN being the normals to the axes from the point P. Thus

$$p = l = \frac{x'}{r'} = \frac{PM}{CP} = \sin \alpha, \qquad q = m = \frac{y'}{r'} = \frac{PN}{CP} = \sin \beta \tag{4.6-4}$$

Thus for a plane wave incident normally, we obtain

$$I = I_0 \left[\frac{\sin(ka \sin \alpha)}{ka \sin \alpha} \right]^2 \left[\frac{\sin(kb \sin \beta)}{kb \sin \beta} \right]^2 \tag{4.6-5}$$

The minima occur when either $ka \sin \alpha$ or $kb \sin \beta$ is an integral multiple of π, which implies

$$2a \sin \alpha = n\lambda, \qquad 2b \sin \beta = m\lambda \tag{4.6-6}$$

where $n, m = 1, 2, 3, \ldots$. Thus when

$$\alpha = \sin^{-1}\left(\frac{\lambda}{2a}\right), \quad \sin^{-1}\left(\frac{2\lambda}{2a}\right), \quad \sin^{-1}\left(\frac{3\lambda}{2a}\right) \tag{4.6-7}$$

etc., the intensity will be zero for all values of β. The minima will therefore occur along straight lines parallel to the $O'y'$ axis. Similarly, the condition $2b \sin \beta = m\lambda$ will give rise to minima parallel to the $O'x'$ axis. Figure 4.5a shows the diffraction pattern of a square aperture for which $a = b$. In Fig. 4.5b, the points denote the positions of the maxima and the numbers denote the relative intensity of the maxima with respect to the intensity at O'. The straight lines represent the loci of the points where the intensity is zero.

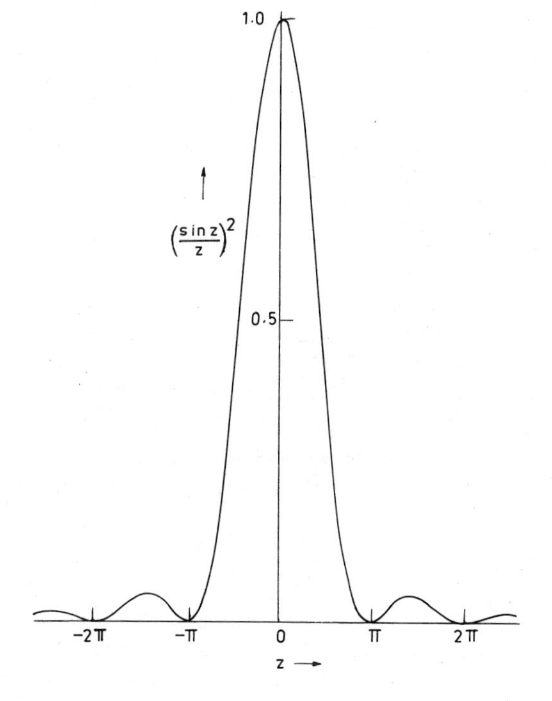

Fig. 4.6b. The variation of intensity on the x axis, corresponding to the diffraction pattern produced by a long narrow slit ($b \gg a$).

Let us next consider the case of a long narrow slit for which b is very large. We know that the function $(\sin gz)/z$ in the limit of a large value of g is a very sharply peaked function around $z = 0$ (see Fig. 4.6a). Indeed,

$$\lim_{g \to \infty} \frac{\sin gz}{z} = \pi\,\delta(z) \tag{4.6-8}$$

where $\delta(z)$ represents the Dirac delta function (see Appendix A). Thus for large values of b, the function $[(\sin kqb)/kq]^2$ will be negligible except around $q = 0$. Consequently, the intensity in the focal plane will be negligible at all points except those lying on the x axis (which correspond to $q = 0$). The corresponding intensity distribution is shown in Fig. 4.6b. Thus for a long narrow slit the intensity distribution will be approximately described by the relation

$$I = \begin{cases} I_0 \left[\dfrac{\sin(ka \sin\alpha)}{ka \sin\alpha}\right]^2 & \text{for } \beta = 0 \text{ (i.e., on the } x \text{ axis)} \\ 0 & \text{everywhere else} \end{cases} \tag{4.6-9}$$

Problem 4.1. Obtain the positions and relative intensities of maxima and minima when $b = 1$ cm and $a = 0.01$ cm for $\lambda = 5 \times 10^{-5}$ cm. Show that appreciable intensity appears only along the x axis.

Let us next consider the familiar experimental arrangement in which an incoherent line source is placed at the focal plane of a convex lens L_1 as shown in Fig. 4.7a. The point P_0 produces a plane wave that is incident normally on the slit, which produces a pattern along the $O'x'$ axis. A different point P_0' will produce a pattern along KK' that is parallel to the x axis (see Fig. 4.7a). Thus a line source will produce a pattern as shown in Fig. 4.7b; the extent of the fringes along the y' axis will be proportional to the length of the line source and will be independent of the length of the slit.

4.7. Fraunhofer Diffraction by a Circular Aperture

We next consider the diffraction of a plane wave by a circular aperture of radius a. We assume that the plane wave is incident normally so that $l_0 = 0$ and $m_0 = 0$. Thus Eq. (4.5-7) becomes

$$u = C \iint \exp[ik(l\xi + m\eta)]\, d\xi\, d\eta \tag{4.7-1}$$

We choose the origin of our coordinate system to be at the center of the

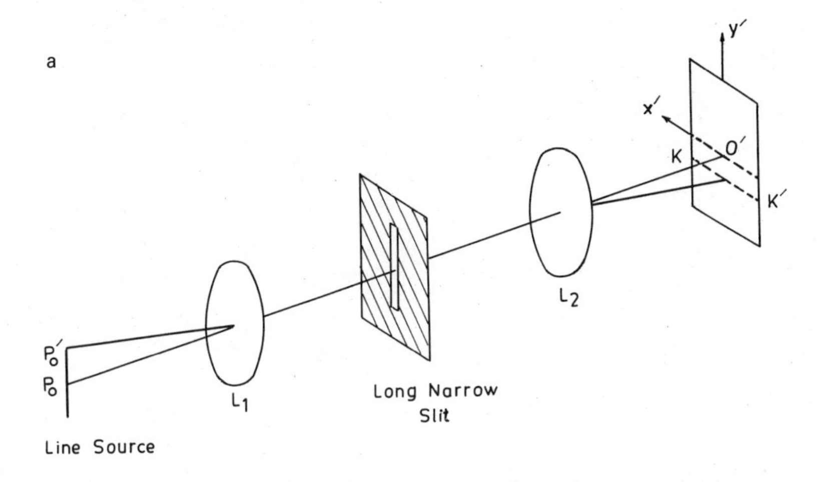

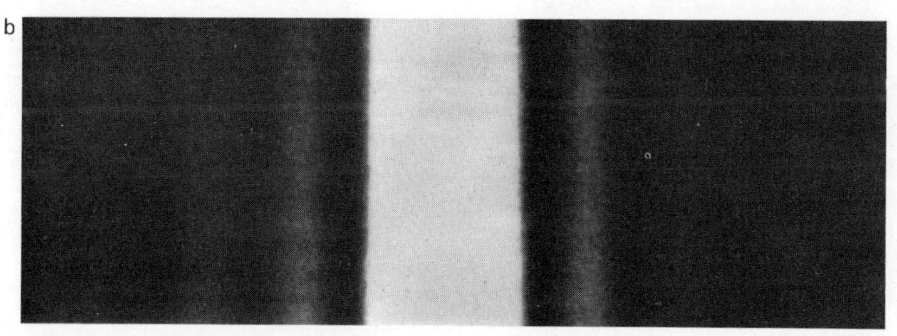

Fig. 4.7. (a) Single-slit diffraction pattern produced by two points of a line source. (b) The single-slit diffraction pattern produced by a line source.

circular aperture. Thus for any point Q on the circular aperture, we may write

$$\xi = \sigma \cos \varphi, \qquad \eta = \sigma \sin \varphi \tag{4.7-2}$$

where $\sigma = (\xi^2 + \eta^2)^{1/2}$ represents the distance of the point Q from the center and φ is the angle that the line OQ makes with the ξ axis (see Fig. 4.8b). Using this transformation, we obtain

$$u = C \int_0^a \int_0^{2\pi} \exp[ik\sigma(l \cos \varphi + m \sin \varphi)]\, \sigma \, d\sigma \, d\varphi \tag{4.7-3}$$

We wish to calculate the intensity distribution at the focal plane of the lens L_2, and in particular we consider an arbitrary point P whose x and y co-

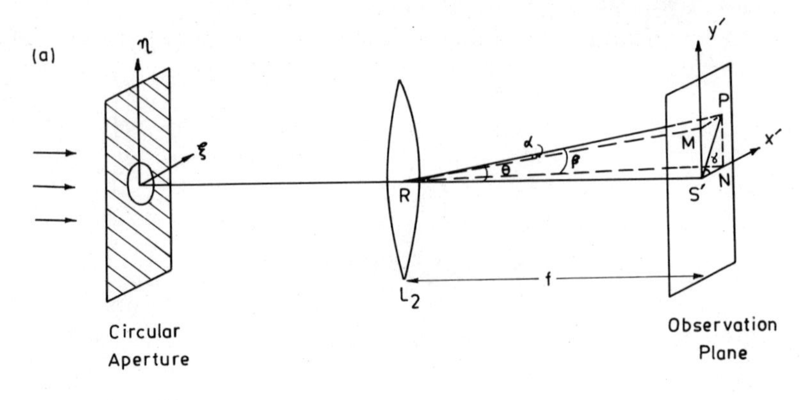

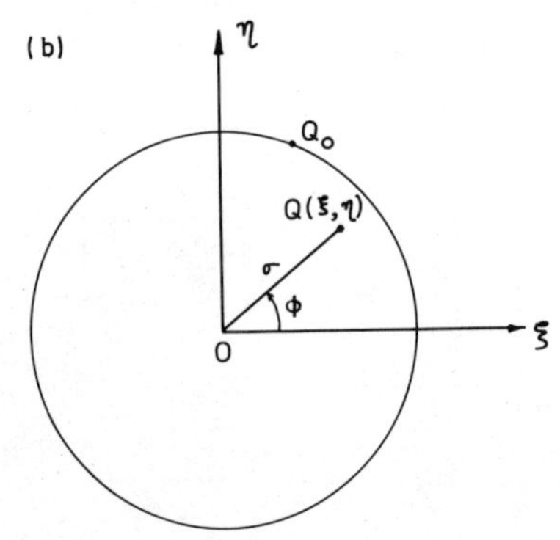

Fig. 4.8. (a) The Fraunhofer diffraction pattern of the circular aperture is obtained on the focal plane of the lens L_2. PN and PM are perpendiculars on the Ox' and Oy' axes respectively; $\angle PS'N = \gamma$, $\angle PRM = \alpha$, $\angle PRN = \beta$, and $\angle PRS' = \theta$. (b) The coordinate system.

ordinates are x' and y', respectively. Now the direction cosine l is given by

$$l = \sin\alpha = \frac{x'}{r} = \frac{PM}{PR} = \frac{S'N}{PR} = \frac{PS'}{PR}\cos\gamma = \sin\theta\cos\gamma \tag{4.7-4}$$

where θ is the angle that the line RP makes with the axial line RS' and γ is the angle made by PS' with the x axis. Similarly,

$$m = \sin\beta = \frac{y'}{r} = \frac{PN}{PR} = \frac{PS'}{PR}\sin\gamma = \sin\theta\sin\gamma \tag{4.7-5}$$

Substituting the above values for l and m into Eq. (4.7-3), we obtain

$$\begin{aligned} u &= C\int_0^a\int_0^{2\pi} \exp[ik\sigma\sin\theta(\cos\varphi\cos\gamma + \sin\varphi\sin\gamma)]\,\sigma\,d\sigma\,d\varphi \\ &= C\int_0^a \sigma\,d\sigma\int_0^{2\pi}\exp[ik\sigma\sin\theta\cos(\varphi-\gamma)]\,d\varphi \end{aligned} \tag{4.7-6}$$

Now for any periodic function $F(\varphi)$ with periodicity 2π [i.e., $F(\varphi + 2\pi) = F(\varphi)$], we have

$$\int_0^{2\pi} F(\varphi - \gamma)\, d\varphi = \int_0^{2\pi} F(\varphi)\, d\varphi \tag{4.7-7}$$

Thus*

$$\begin{aligned} u(P) &= C \int_0^a \sigma\, d\sigma \int_0^{2\pi} \exp[ik\sigma \sin\theta \cos\varphi]\, d\varphi \\ &= 2\pi C \int_0^a J_0(k\sigma \sin\theta)\, \sigma\, d\sigma \\ &= \frac{2\pi C}{(k \sin\theta)^2} \int_0^{ka\sin\theta} J_0(x)\, x\, dx \end{aligned} \tag{4.7-8}$$

where we have used the well-known integral representation of $J_0(x)$:

$$J_0(x) = \frac{1}{2\pi} \int_0^{2\pi} e^{ix\cos\alpha}\, d\alpha \tag{4.7-9}$$

Since

$$\frac{d}{dx}[x^{n+1} J_{n+1}(x)] = x^{n+1} J_n(x) \tag{4.7-10}$$

we have for $n = 0$,

$$\frac{d}{dx}[x J_1(x)] = x J_0(x) \tag{4.7-11}$$

Thus Eq. (4.7-8) becomes

$$\begin{aligned} u &= \frac{2\pi C}{(k\sin\theta)^2} \int_0^{ka\sin\theta} \frac{d}{dx}[x J_1(x)]\, dx \\ &= 2\pi C a \frac{J_1(ka\sin\theta)}{k\sin\theta} \end{aligned} \tag{4.7-12}$$

Thus

$$I = I_0 \left[\frac{2J_1(ka\sin\theta)}{ka\sin\theta}\right]^2 = I_0 \left[\frac{2J_1(w)}{w}\right]^2 \tag{4.7-13}$$

* Equation (4.7-8) is also obvious from the fact that because of the cylindrical symmetry of the problem, the result should be independent of the value of γ, i.e., the amplitude at all points on the circumference of a circle whose center is at S' must be the same.

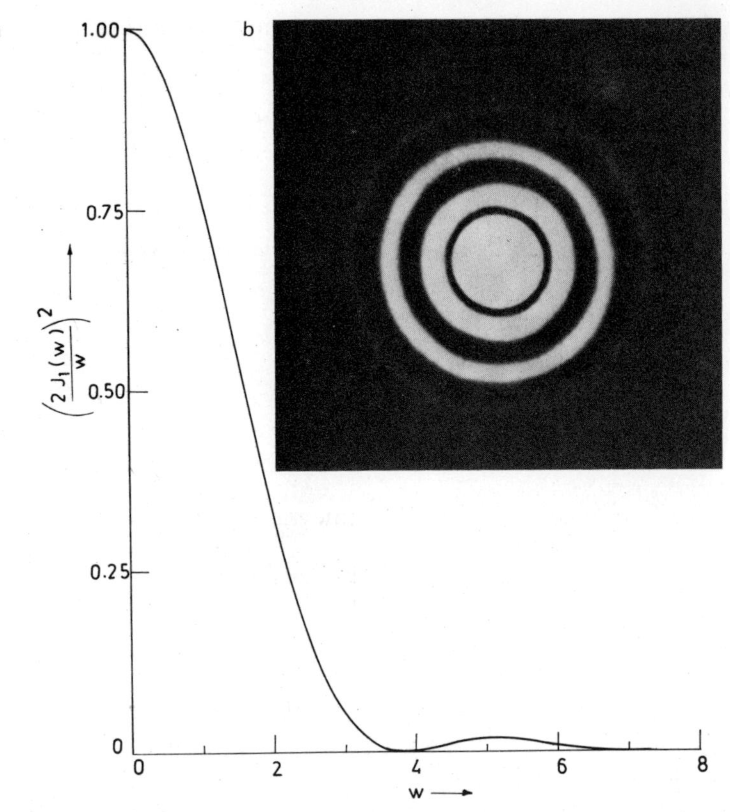

Fig. 4.9. (a) The intensity distribution in the Fraunhofer diffraction pattern produced by a circular aperture. (b) The observed pattern. Photograph adapted from Preston (1972).

where $w = ka \sin\theta$ and I_0 represents the intensity at the central spot, i.e., corresponding to $\theta = 0$.* Equation (4.7-13) represents the well-known Airy pattern, which was first derived by Airy by a somewhat different method. A plot of the intensity variation as a function of w is shown in Fig. 4.9a. The minima in the intensity pattern occur at values of θ given by

$$w = ka \sin\theta = 3.83, 7.02, 10.17, \text{etc.} \tag{4.7-14}$$

and the maxima at

$$w = 0, 5.14, 8.46, 11.62, \text{etc.} \tag{4.7-15}$$

* $\frac{2J_1(x)}{x} = \frac{2}{x}\sum_{r=0}^{\infty}(-1)^r\left(\frac{x}{2}\right)^{2r+1}\frac{1}{(r+1)!}$

Thus $\lim_{x\to 0}[2J_1(x)/x] = 1$.

Thus the intensity pattern consists of a bright central disk surrounded by alternate dark and bright rings. Such a pattern is known as the Airy pattern (see Fig. 4.9b). The values of $[2J_1(w)/w]^2$ at the maxima [see Eq. (4.7-15)] are 1.00, 0.0175, 0.0042, 0.0016, etc. Thus, the central spot is very bright and the intensity of the rings decreases as one moves away from the center.

Problem 4.2. Equation (4.7-13) represents the Fraunhofer diffraction pattern of a circular opening of radius a. If instead of a circular opening one has an annular aperture, i.e., the aperture is bounded by two circles of radii a_1 and a_2 (see Fig. 4.10a), then calculate the diffraction pattern of such an aperture.

Solution. For an aperture of the form shown in Fig. 4.10a, the Fraunhofer diffraction pattern is given by

$$u(P) = \frac{2\pi C}{(k \sin \theta)^2} \int_{ka_1 \sin \theta}^{ka_2 \sin \theta} J_0(x)\, x\, dx \tag{4.7-16}$$

where $x = k\sigma \sin \theta$. Using Eq. (4.7-11) the integration can easily be carried out. The corresponding intensity distribution can be written in the form

$$I(P) = \frac{I_0}{(1-\kappa^2)^2} \left[\frac{2J_1(ka_2 \sin \theta)}{ka_2 \sin \theta} - \kappa^2 \frac{2J_1(k\kappa a_2 \sin \theta)}{k\kappa a_2 \sin \theta} \right]^2 \tag{4.7-17}$$

where $I_0 = \pi^2 C^2 (a_2^2 - a_1^2)^2$ is the intensity at the point $\theta = 0$ and $\kappa = a_1/a_2$ $(0 < \kappa < 1)$. The intensity distribution is plotted in Fig. 4.10b for various values

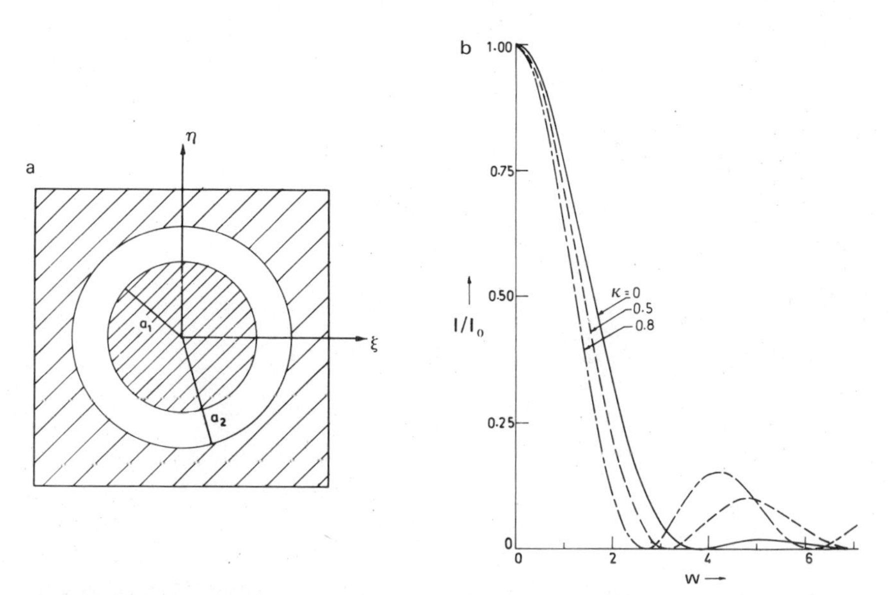

Fig. 4.10. (a) An annular aperture bounded by circles of radii a_1 and a_2. (b) Fraunhofer diffraction pattern of an annular aperture. The solid curve corresponds to $\kappa = 0$ (i.e., a circular opening) and the dashed curve corresponds to $\kappa = \frac{1}{2}$.

of κ. Notice that $\kappa = 0$ corresponds to a circular aperture. The minima of the diffraction pattern are the solutions of

$$J_1(ka_2 \sin \theta) - \kappa J_1(k\kappa a_2 \sin \theta) = 0 \tag{4.7-18}$$

For nonzero value of κ, the first zero of Eq. (4.7-18) lies closer to the central maximum (which occurs at $\theta = 0$) compared to the case $\kappa = 0$. Thus, the effect of putting a central obstruction has the effect of increasing the resolution. But as κ increases, i.e., as the central obstruction increases in diameter, I_0 decreases and hence the brightness of the image decreases. It can also be seen from Fig. 4.10b that as κ increases, the energy in the side lobes also increases.

Problem 4.3. Show that in the limit $\kappa \to 1$, the first dark ring occurs at the value of θ given by $ka_2 \sin \theta = 2.4$.

4.8. Distribution of Intensity in the Airy Pattern

Let $F(r)$ represent the fractional energy contained in a circle of radius r in the focal plane, i.e.,

$$F(r) = \int_0^r I(r')\, 2\pi r'\, dr' \Big/ \int_0^\infty I(r')\, 2\pi r'\, dr' \tag{4.8-1}$$

where $2\pi r' I(r')\, dr'$ represents the flux in the annular aperture whose radii lie between r' and $r' + dr'$. Now,

$$w' = ka \cdot \sin \theta' \simeq kar'/f \tag{4.8-2}$$

where f is the focal length of the lens L_2. Thus,

$$F(r) = \int_0^w \left[\frac{J_1(w')}{w'}\right]^2 w'\, dw' \Big/ \int_0^\infty \left[\frac{J_1(w')}{w'}\right]^2 w'\, dw' \tag{4.8-3}$$

Using Eq. (4.7-11), we obtain

$$\begin{aligned} \frac{J_1^2(w')}{w'} &= J_1(w')\left[J_0(w') - \frac{dJ_1(w')}{dw'}\right] \\ &= -\left[J_0(w')\frac{dJ_0(w')}{dw'} + J_1(w')\frac{dJ_1(w')}{dw'}\right] \\ &= -\frac{1}{2}\frac{d}{dw'}\left[J_0^2(w') + J_1^2(w')\right] \end{aligned} \tag{4.8-4}$$

Consequently,

$$F(r) = [J_0^2(w') + J_1^2(w')]_{w'=0}^{w} / [J_0^2(w') + J_1^2(w')]_{w'=0}^{\infty} = 1 - J_0^2(w) - J_1^2(w) \tag{4.8-5}$$

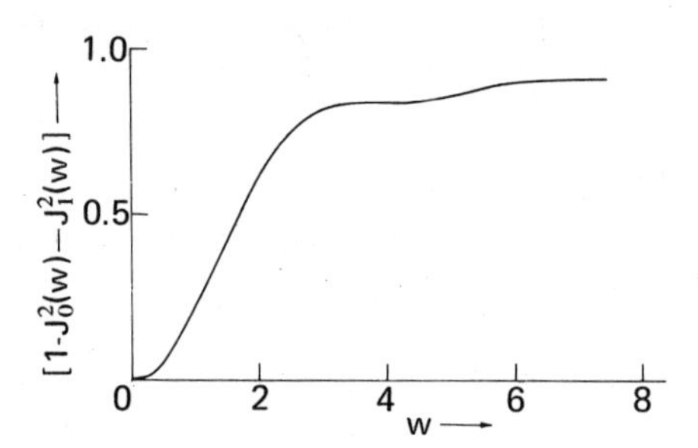

Fig. 4.11. Variation of $F(r)$ with w ($\simeq kar'/f$).

This function is plotted in Fig. 4.11. Corresponding to the first, second, and third dark rings (i.e., for $w = 3.832$, 7.015, and 10.173) the values of $F(r)$ are 0.839, 0.910, and 0.938, respectively. Thus, about 84% of the light is contained within the circle bounded by the first dark ring, about 7% is contained within the first and the second dark rings, etc.

4.9. Fresnel Diffraction by a Circular Aperture

Let us consider a circular aperture of radius a and let us suppose that a point source is located on the axis at the point P_0 (see Fig. 4.12a). We will restrict our discussion to the case when the source P_0 as well as the observation point P lie on the axis of the system.* It is obvious that it will be more convenient to use the cylindrical system of coordinates and in this system the coordinates of an arbitrary point Q (on the circular aperture) will be (σ, φ) where σ ($= OQ$) represents the distance of Q from the origin O and φ is the angle that OQ makes with the x axis. Thus, using Eq. (4.4-11), the field at the point P will be given by

$$u(P) = \frac{iA}{\lambda} \int_0^a \int_0^{2\pi} \frac{1}{r\rho} \exp[-ik(r+\rho)]\, \sigma\, d\sigma\, d\varphi \tag{4.9-1}$$

The distances r and ρ have been shown in Fig. 4.12a and since the points P_0 and P lie on the axis, r and ρ will be independent of φ. Consequently

$$u(P) = \frac{2\pi i A}{\lambda} \int_0^a \frac{e^{-ik(r+\rho)}}{r\rho}\, \sigma\, d\sigma \tag{4.9-2}$$

Since the angles QOP_0 and QOP are right angles, we will have $\sigma^2 = \rho^2 -$

* The axis is defined to be the line that passes through the center of the circle and is normal to the plane of the circle.

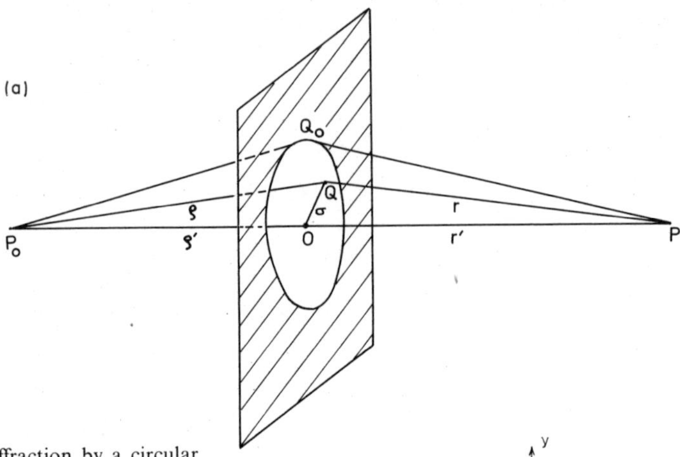

Fig. 4.12. (a) Diffraction by a circular aperture of radius a. The point source P_0, the center of the circle O, and the observation point P lie on the same straight line, which is normal to the plane of the aperture. Q represents an arbitrary point inside the aperture and Q_0 lies on the rim of the aperture. (b) Construction of Fresnel half-period zones.

$\rho'^2 = r^2 - r'^2$. Thus

$$\sigma\, d\sigma = \rho\, d\rho = r\, dr \tag{4.9-3}$$

Therefore,

$$d(r + \rho) = \left(\frac{1}{r} + \frac{1}{\rho}\right)\sigma\, d\sigma \tag{4.9-4}$$

On substituting the value of $\sigma\, d\sigma$ from Eq. (4.9-4) into Eq. (4.9-2) we obtain

$$u(P) = \frac{2\pi i A}{\lambda}\int \frac{e^{-ik(r+\rho)}}{(r+\rho)}\, d(r+\rho) \simeq \frac{2\pi i A}{\lambda(r' + \rho')}\int_{\xi_0}^{\xi_1} e^{-ik\xi}\, d\xi \tag{4.9-5}$$

where $\xi = r + \rho$; the quantity $r + \rho$ has been taken out of the integral because for a small aperture it has negligible variation over the area of the aperture. The quantities ξ_0 and ξ_1 correspond to the minimum and maximum values of ξ. Clearly $\xi_0 = r' + \rho'$, and

$$\xi_1 = P_0Q_0 + Q_0P = (\rho'^2 + a^2)^{1/2} + (r'^2 + a^2)^{1/2} \tag{4.9-6}$$

where Q_0 lies on the rim of the aperture. We write

$$\xi_1 = \xi_0 + p(\lambda/2) \tag{4.9-7}$$

where

$$p(\lambda/2) \equiv (P_0Q_0 + Q_0P) - (P_0O + OP)$$
$$= [(\rho'^2 + a^2)^{1/2} + (r'^2 + a^2)^{1/2}] - (r' + \rho') \tag{4.9-8}$$

Carrying out the integration in Eq. (4.9-5), we obtain

$$u(P) \simeq u_0(P)(1 - e^{-i\pi p}) \tag{4.9-9}$$

where

$$u_0(P) = A\exp[-ik(r' + \rho')]/(r' + \rho') \tag{4.9-10}$$

represents the field at the point P in the absence of the aperture. Thus,

$$I(P) = 4I_0 \sin^2(p\pi/2) \tag{4.9-11}$$

which shows that the intensity is zero or maximum when p is an even or odd integer, i.e., $(P_0Q_0 + Q_0P) - (P_0O + OP)$ is an even or odd multiple of $\lambda/2$. This can be understood physically by using the concept of Fresnel half-period zones, according to which the circular aperture is divided in a large number of concentric circular zones; the radius of the nth circle is defined by the following relation (see Fig. 4.12b):

$$(P_0Q_n + Q_nP) - (P_0O + OP) = n\lambda/2 \tag{4.9-12}$$

The field produced at the point P by the nth zone will be out of phase with the field produced by the $(n - 1)$th zone. Consequently, if the aperture contains an even number of half-period zones, the intensity at the point P will be negligibly small, and conversely, if the circular aperture contains an odd number of zones, the intensity at the point P will be maximum.

4.10. *Fresnel Diffraction by a Single Slit*

Let us consider a rectangular slit of dimensions $2a$ and $2b$. We consider a point source at P_0 and let the point of observation be the point P (see Fig. 4.13). We consider the special case when the line joining the points P_0 and P intersects the aperture normally. The point of intersection O is chosen as the origin of the coordinate system. Returning to Eq. (4.4-12), the field at the point P is given by the expression

$$u(P) = \frac{iA}{\lambda r'\rho'} \int_{\xi_1}^{\xi_2} \int_{\eta_1}^{\eta_2} \exp[-ik(r + \rho)]\, d\xi\, d\eta \tag{4.10-1}$$

where $\xi_2 - \xi_1 = 2a$ and $\eta_2 - \eta_1 = 2b$. (If the origin O lies inside the

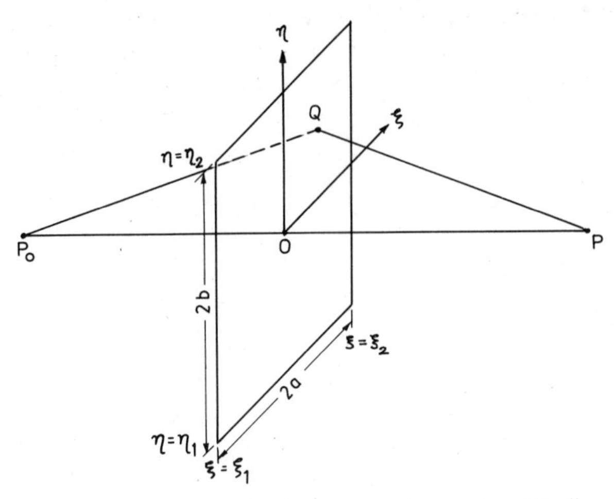

Fig. 4.13. Fresnel diffraction by a rectangular aperture. The line joining the point source P_0 and the observation point P is normal to the plane of the aperture.

rectangle, then ξ_1 and η_1 will be negative quantities.) Now

$$\rho = (OP_0^2 + OQ^2)^{1/2} = (\rho'^2 + \xi^2 + \eta^2)^{1/2} = \rho'[1 + (\xi^2 + \eta^2)/\rho'^2]^{1/2}$$
$$\simeq \rho' + (\xi^2 + \eta^2)/2\rho' \quad (4.10\text{-}2)$$

where we have retained terms quadratic in ξ and η. Similarly,

$$r \simeq r' + (\xi^2 + \eta^2)/2r' \quad (4.10\text{-}3)$$

Thus,

$$r + \rho \simeq (r' + \rho') + (\xi^2 + \eta^2)(r' + \rho')/2r'\rho' \quad (4.10\text{-}4)$$

We introduce two dimensionless variables σ and τ:

$$\sigma = \xi\left[\frac{2(r' + \rho')}{\lambda r'\rho'}\right]^{1/2}, \qquad \tau = \eta\left[\frac{2(r' + \rho')}{\lambda r'\rho'}\right]^{1/2} \quad (4.10\text{-}5)$$

Hence,

$$ik(r + \rho) \simeq ik(r' + \rho') + \tfrac{1}{2}i\pi(\sigma^2 + \tau^2) \quad (4.10\text{-}6)$$

Substituting the above relations in Eq. (4.10-1), we obtain

$$u(P) = \frac{i}{2}u_0(P)\int_{\sigma_1}^{\sigma_2}\exp\left(-i\frac{\pi}{2}\sigma^2\right)d\sigma\int_{\tau_1}^{\tau_2}\exp\left(-i\frac{\pi}{2}\tau^2\right)d\tau \quad (4.10\text{-}7)$$

where $u_0(P)$ is the same as given by Eq. (4.9-10) and represents the field at the point P in the absence of the aperture. In order to discuss the diffraction pattern, it is necessary to introduce the Fresnel integrals. We briefly digress here to define these integrals and study some of their salient features.

The Fresnel Integrals

The Fresnel integrals are defined by the following equations:

$$C(w) \equiv \int_0^w \cos\left(\frac{\pi}{2}\sigma^2\right) d\sigma, \qquad S(w) \equiv \int_0^w \sin\left(\frac{\pi}{2}\sigma^2\right) d\sigma \quad (4.10\text{-}8)$$

These integrals have the following properties:

$$C(0) = S(0) = 0, \qquad C(\infty) = S(\infty) = \tfrac{1}{2} \quad (4.10\text{-}9)$$

$$C(w) = -C(-w), \qquad S(w) = -S(-w) \quad (4.10\text{-}10)$$

The variations of $C(w)$ and $S(w)$ with w are shown in Fig. 4.14a. If we expand the cosine and the sine functions inside the integral [see Eq. (4.10-8)] and integrate term by term, we can show that

$$C(w) = w\left[1 - \frac{1}{2!\,5}\left(\frac{\pi}{2}w^2\right)^2 + \frac{1}{4!\,9}\left(\frac{\pi}{2}w^2\right)^4 - \cdots\right] \quad (4.10\text{-}11)$$

$$S(w) = w\left[\frac{1}{1!\,3}\left(\frac{\pi}{2}w^2\right) - \frac{1}{3!\,7}\left(\frac{\pi}{2}w^2\right)^3 + \frac{1}{5!\,11}\left(\frac{\pi}{2}w^2\right)^5 - \cdots\right] \quad (4.10\text{-}12)$$

For large values of w, one can integrate by parts and obtain the following asymptotic expansions:

$$C(w) = \frac{1}{2} - \frac{1}{\pi w}\left[P(w)\cos\left(\frac{\pi}{2}w^2\right) - Q(w)\sin\left(\frac{\pi}{2}w^2\right)\right] \quad (4.10\text{-}13)$$

$$S(w) = \frac{1}{2} - \frac{1}{\pi w}\left[P(w)\sin\left(\frac{\pi}{2}w^2\right) + Q(w)\cos\left(\frac{\pi}{2}w^2\right)\right] \quad (4.10\text{-}14)$$

where

$$Q(w) = 1 - \frac{1\cdot 3}{(\pi w^2)^2} + \frac{1\cdot 3\cdot 5\cdot 7}{(\pi w^2)^4} - \cdots \quad (4.10\text{-}15)$$

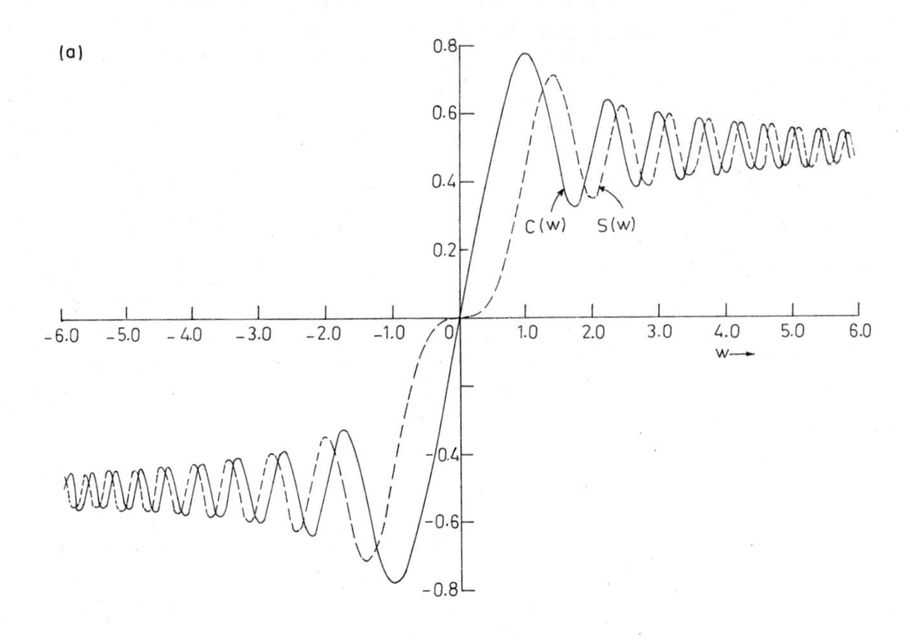

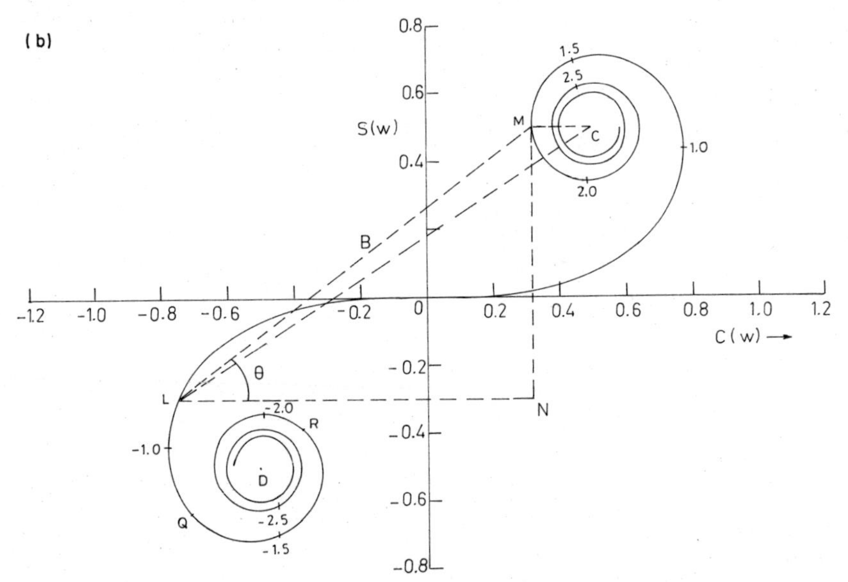

Fig. 4.14. (a) The variation of the Fresnel integrals $C(w)$ and $S(w)$ with w. (b) The Cornu spiral.

$$P(w) = \frac{1}{\pi w^2} - \frac{1 \cdot 3 \cdot 5}{(\pi w^2)^3} + \frac{1 \cdot 3 \cdot 5 \cdot 7 \cdot 9}{(\pi w^2)^5} + \cdots \tag{4.10-16}$$

The numerical values of $C(w)$ and $S(w)$ are given in Table 4.1.

Figure 4.14b gives a geometrical representation of the Fresnel integrals and is known as Cornu's spiral. The numbers written on the spiral are the values of w. For example, for $w = 1.2$, $C(w) = 0.7154$ and $S(w) = 0.6234$, as we can see from the figure. The main advantage of Cornu's spiral lies in the following interesting property: Let us write

$$[C(\tau_2) - C(\tau_1)] - i[S(\tau_2) - S(\tau_1)] \equiv Be^{-i\theta} \tag{4.10-17}$$

Thus

$$C(\tau_2) - C(\tau_1) = B\cos\theta, \qquad S(\tau_2) - S(\tau_1) = B\sin\theta$$

Let the points L and M on Cornu's spiral (see Fig. 4.14b) correspond to $w = \tau_1$ and $w = \tau_2$, respectively. It is obvious that

$$LN = C(\tau_2) - C(\tau_1) = B\cos\theta \tag{4.10-18}$$

$$MN = S(\tau_2) - S(\tau_1) = B\sin\theta \tag{4.10-19}$$

Thus the length of the line joining the points L and M will be B and the angle that the line makes with the abscissa will be θ.

Table 4.1. The Fresnel Integrals

$$C(w) \equiv \int_0^w \cos\left(\frac{\pi}{2}\sigma^2\right) d\sigma \qquad S(w) \equiv \int_0^w \sin\left(\frac{\pi}{2}\sigma^2\right) d\sigma$$

w	$C(w)$	$S(w)$	w	$C(w)$	$S(w)$
0.0	0.00000	0.00000	2.6	0.38894	0.54999
0.2	0.19992	0.00419	2.8	0.46749	0.39153
0.4	0.39748	0.03336	3.0	0.60572	0.49631
0.6	0.58110	0.11054	3.2	0.46632	0.59335
0.8	0.72284	0.24934	3.4	0.43849	0.42965
1.0	0.77989	0.43826	3.6	0.58795	0.49231
1.2	0.71544	0.62340	3.8	0.44809	0.56562
1.4	0.54310	0.71353	4.0	0.49843	0.42052
1.6	0.36546	0.63889	4.2	0.54172	0.56320
1.8	0.33363	0.45094	4.4	0.43833	0.46227
2.0	0.48825	0.34342	4.6	0.56724	0.51619
2.2	0.63629	0.45570	4.8	0.43380	0.49675
2.4	0.55496	0.61969	5.0	0.56363	0.49919

Returning to the diffraction problem, we note that

$$\int_{\sigma_1}^{\sigma_2} \exp\left(-i\frac{\pi}{2}\sigma^2\right) d\sigma = \left[\int_0^{\sigma_2} \cos\left(\frac{\pi}{2}\sigma^2\right) d\sigma - \int_0^{\sigma_1} \cos\left(\frac{\pi}{2}\sigma^2\right) d\sigma\right]$$

$$- i\left[\int_0^{\sigma_2} \sin\left(\frac{\pi}{2}\sigma^2\right) d\sigma - \int_0^{\sigma_1} \sin\left(\frac{\pi}{2}\sigma^2\right) d\sigma\right]$$

$$= [C(\sigma_2) - C(\sigma_1)] - i[S(\sigma_2) - S(\sigma_1)] \quad (4.10\text{-}20)$$

Thus, Eq. (4.10-7) becomes

$$u(P) = \tfrac{1}{2} i u_0(P) \{[C(\sigma_2) - C(\sigma_1)] - i[S(\sigma_2) - S(\sigma_1)]\}$$

$$\times \{[C(\tau_2) - C(\tau_1)] - i[S(\tau_2) - S(\tau_1)]\} \quad (4.10\text{-}21)$$

It is interesting to note that for an infinitely extended aperture, i.e., for $\sigma_1 = \tau_1 = -\infty$ and $\sigma_2 = \tau_2 = +\infty$ one obtains

$$u(P) = \tfrac{1}{2} i u_0(P)(1-i)(1-i) = u_0(P) \quad (4.10\text{-}22)$$

Thus, in spite of all the approximations that have been made, one does obtain the correct result for an unobstructed wavefront. In general, in order to evaluate the field at the point P (see Fig. 4.13), one determines the values of ξ_1, ξ_2, η_1, and η_2; knowing these values one obtains the field from Eq. (4.10-21). We would like to recall that the present analysis assumes that for a given position of the point source P_0, the point P should be such that P_0P is normal to the aperture. However, if we have a plane wavefront incident normally (i.e., if $\rho' = \infty$), then we can vary the point P; for each case we drop a perpendicular from the point P on the aperture and calculate ξ_1, ξ_2, η_1, and η_2 and then determine $u(P)$ from Eq. (4.10-21).* We now consider the diffraction produced by a straight edge and by a long narrow slit.

Case I: Fresnel Diffraction of a Plane Wave by a Straight Edge. For a plane wave incident normally, $\rho' = \infty$ and

$$\sigma = \xi(2/\lambda r')^{1/2}, \qquad \tau = \eta(2/\lambda r')^{1/2} \quad (4.10\text{-}23)$$

* It should be pointed out that even when the source is not at an infinite distance, the diffraction pattern around the point P can be determined by the following approximate procedure: the field at a point that is at a distance ε from the point P—say along the x axis (see Fig. 4.13)—is the same as the field at the point P provided we shift the aperture in the downward direction by a distance ε. This approximate procedure is valid if (1) $a, b \ll \rho'$, (2) a is small but b is large and the diffraction pattern is obtained along the y axis, or (3) a and b are large but the diffraction pattern is obtained close to the edges of the geometrical shadow.

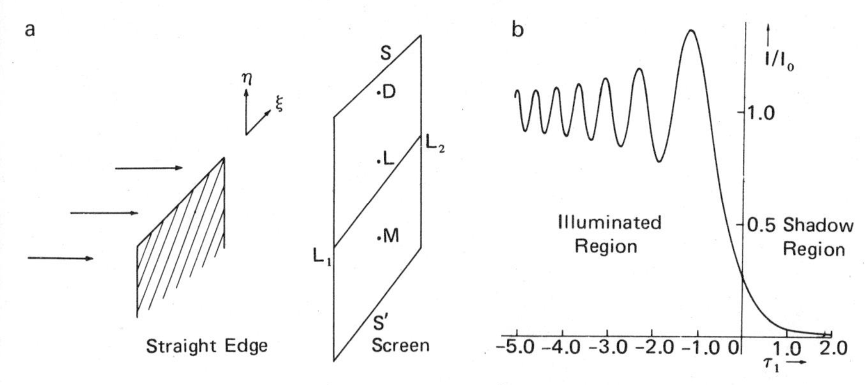

Fig. 4.15. (a) A plane wave incident on a straight edge. The diffraction pattern is observed on the screen SS'. LL' denotes the edge of the geometrical shadow and P represents an arbitrary point on the observation screen. (b) The variation of intensity corresponding to a straight-edge diffraction pattern. The negative and positive values of τ_1 correspond to the illuminated and the shadow regions, respectively.

Further, for a straight edge parallel to the x axis (see Fig. 4.15a),

$$\sigma_1 = -\infty, \qquad \sigma_2 = +\infty, \qquad \tau_2 = +\infty \tag{4.10-24}$$

and the value of τ_1 will depend on the position of the observation point. Thus, using Eqs. (4.10-9) and (4.10-21), the field distribution is given by

$$u(P) = \tfrac{1}{2}i(1-i)\,u_0(P)\{[\tfrac{1}{2} - C(\tau_1)] - i[\tfrac{1}{2} - S(\tau_1)]\} \tag{4.10-25}$$

If the point P is such that it lies on the edge of the geometrical shadow (i.e., on the line L_1L_2, see Fig. 4.15a), then $\tau_1 = 0$ and

$$u(P) = \tfrac{1}{4}i(1-i)^2\,u_0(P) = \tfrac{1}{2}u_0(P)$$

or

$$I(P) = \tfrac{1}{4}I_0(P) \tag{4.10-26}$$

Thus, the intensity on the edge of the geometrical shadow is one-fourth the intensity that it would have been in the absence of the edge. On the other hand, when the point P is deep inside the geometrical shadow (i.e., when $\tau_1 \to \infty$), we obtain $u(P) \to 0$, as it indeed should be.

In order to determine the field at an arbitrary point on the screen, we use Cornu's spiral. Let us first consider a point of observation M in the geometrical shadow region. Consequently the corresponding value of τ_1 will be positive. Let the point M on the spiral (see Fig. 4.14b) correspond to $w = \tau_1$. Since the point C in the curve corresponds to $w = \infty$, we have

$$[\tfrac{1}{2} - C(\tau_1)] - i[\tfrac{1}{2} - S(\tau_1)] = MCe^{-i\psi} \tag{4.10-27}$$

$\frac{1}{2}i(1-i)\,u_0(P)\,MCe^{-i(kz-\omega t)}$

where ψ is the angle that MC makes with the abscissa [see Eq. (4.10-18)]. Substituting in Eq. (4.10-25) one obtains for the intensity distribution,

$$I(M) = \tfrac{1}{2}I_0(MC)^2 \tag{4.10-28}$$

We can easily see that as the point of observation moves into the shadow region, the value of τ_1 increases. Thus the point M keeps moving on the spiral toward the point C and the length MC decreases uniformly. Hence in the shadow region the intensity uniformly decreases to zero (see Fig. 4.15b).

As we move away from the edge of the geometrical shadow to the illuminated region, the value of τ_1 becomes negative and the corresponding point L (on Cornu's spiral) lies in the third quadrant, as shown in Fig. 4.14b. The intensity is again given by

$$I(L) = \tfrac{1}{2}I_0(LC)^2 \tag{4.10-29}$$

As the value of τ_1 becomes more and more negative, the length LC keeps increasing until the point L reaches Q, which corresponds to $\tau_1 \simeq -1.22$. The intensity at this point is maximum and is $\simeq 1.37\ I_0$. As the value of τ_1 becomes further negative, the length CL starts decreasing until it reaches the point R, which corresponds to $\tau_1 \simeq -1.87$. Thus the intensity keeps oscillating with decreasing amplitude about I_0 as we move further into the illuminated region (see Fig. 4.15b). For $\tau_1 \to -\infty$ (which corresponds to a point like D in Fig. 4.15a), the intensity approaches the value

$$\tfrac{1}{2}I_0(DC)^2 = I_0$$

The variation of intensity with τ_1 is plotted in Fig. 4.15b and, as is obvious, this represents a universal curve, i.e., for given values of λ and r', one simply has to calculate τ_1 as the observation point moves along the y axis. For example, for $\lambda = 5 \times 10^{-5}$ cm, $r' = 4$ cm, one obtains $\eta = 0.01\tau$ cm [see Eq. (4.10-23)]. Since the first three maxima correspond to $\tau_1 \simeq -1.22$, -2.34, and -3.08, the positions of the maxima in the actual diffraction pattern occur at -0.0122, -0.0234, and -0.0308 cm. As the value of r' increases, the distances between the consecutive maxima will also increase.

Case II: Fresnel Diffraction of a Plane Wave by a Long Narrow Slit and Transition to the Fraunhofer Region. As in the previous case, for a plane wave incident normally, $\rho' = \infty$ and σ and τ are again given by Eq. (4.10-23). Further, for a long narrow slit, as shown in Fig. 4.16, $a \to \infty$ and one obtains $\sigma_1 = -\infty$ and $\sigma_2 = +\infty$. Thus, Eq. (4.10-21) becomes

$$u(P) = \tfrac{1}{2}i(1 - i)\,u_0(P)\{[C(\tau_2) - C(\tau_1)] - i[S(\tau_2) - S(\tau_1)]\} \tag{4.10-30}$$

If $2b$ represents the width of the slit, then $\eta_2 - \eta_1 = 2b$; consequently

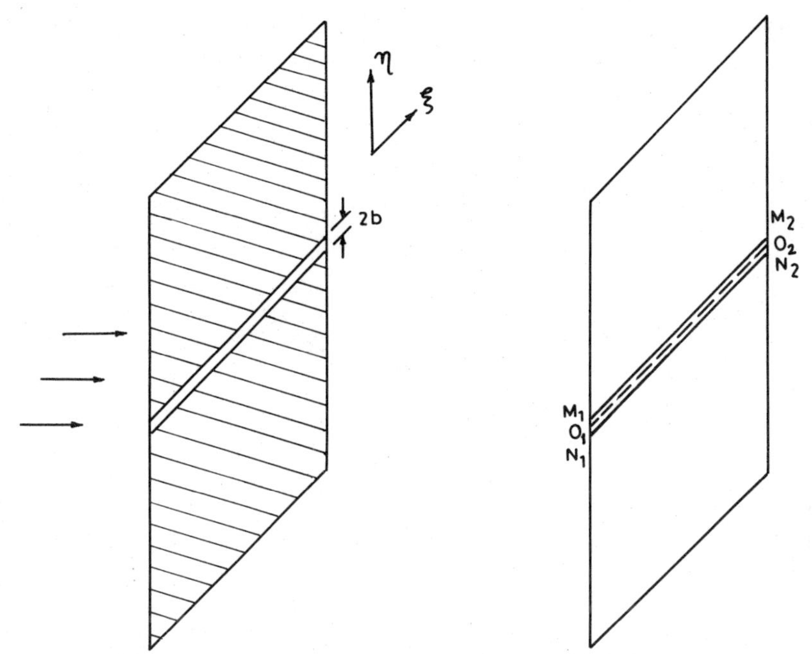

Fig. 4.16. A plane wave incident normally on a long narrow slit of width $2b$. M_1M_2 and N_1N_2 are the edges of the geometrical shadow. The diffraction pattern is symmetrical about the line O_1O_2.

$\tau_2 = \tau_1 + \tau_0$, where

$$\tau_0 = 2b(2/\lambda r')^{1/2} \tag{4.10-31}$$

which, for given values of b, λ, and r', is a fixed quantity. Thus, using Eq. (4.10-30) the intensity distribution will be given by

$$I(P) = \tfrac{1}{2}I_0\{[C(\tau_1 + \tau_0) - C(\tau_1)]^2 + [S(\tau_1 + \tau_0) - S(\tau_1)]^2\} \tag{4.10-32}$$

It may be noted that as $\tau_0 \to \infty$, the intensity distribution will be identical to the one obtained for a straight edge. Consequently, for large b one will obtain straight-edge diffraction patterns around the lines M_1M_2 and N_1N_2, which represent the edges of the geometrical shadow (see Fig. 4.16). In general, the diffraction pattern will be symmetrical about the line O_1O_2 and therefore it will be sufficient to calculate the intensity distribution only in the region below O_1O_2; the intensity in the upper region could be obtained by symmetry.

In Fig. 4.17 we have plotted the variation of intensity with τ_1 for $\tau_0 = 0.7$ and 8.0. One can see that for large values of τ_0 (i.e., when the observation screen is very close to the aperture) the diffraction pattern

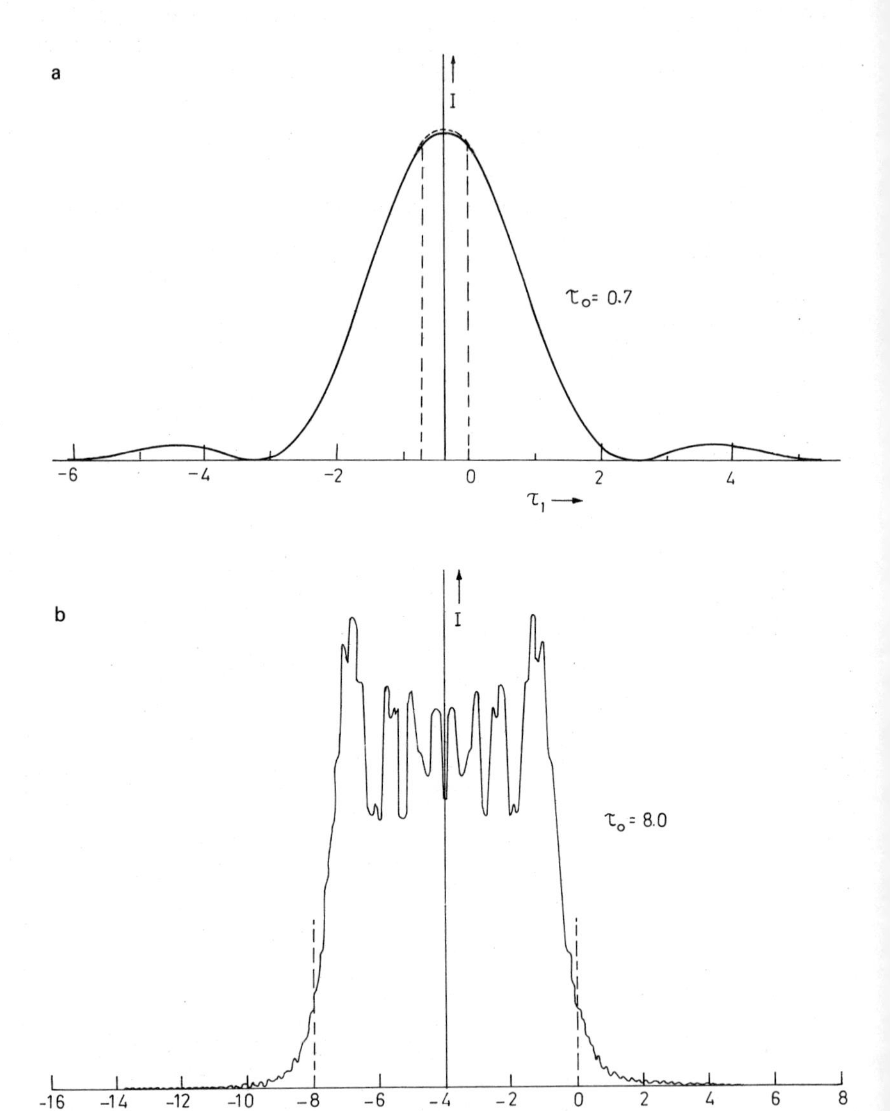

Fig. 4.17. The diffraction pattern produced by a slit for $\tau_0 = 0.7$ (a) and 8.0 (b). The dashed curve represents the Fraunhofer pattern, the vertical lines represent the edges of the geometrical shadow.

essentially consists of two straight-edge diffraction patterns. On the other hand, for small values of τ_0 (i.e., when the observation screen is far from the aperture) the diffraction becomes essentially of the Fraunhofer type.

Problem 4.4. In order to study the diffraction pattern at large distances from the slit, it is convenient to rewrite Eq. (4.10-32) in the form

$$I(P) = \tfrac{1}{2}I_0\{[C(\tau' + \tau_0/2) - C(\tau' - \tau_0/2)]^2 + [S(\tau' + \tau_0/2) - S(\tau' - \tau_0/2)]^2\} \tag{4.10-33}$$

where $\tau' = \tau_1 + \tau_0/2$. Assuming τ_0 to be small, make a Taylor series expansion of $C(\tau' + \tau_0/2)$, $C(\tau' - \tau_0/2)$, etc., about τ' and show that

$$I(P) \simeq \tfrac{1}{2}I_0\tau_0^2\left[1 - \frac{\pi^2\tau'^2\tau_0^2}{12} + \frac{\pi^2\tau_0^4}{24 \times 120}(2\pi^2\tau'^4 - 4) + \cdots\right] \tag{4.10-34}$$

Notice that because of the introduction of the new variable τ', the above intensity distribution is automatically symmetric about the point $\tau' = 0$.

Problem 4.5. In the above problem we have introduced the variable

$$\tau' = \tau_1 + \frac{\tau_0}{2} = x\left(\frac{2}{\lambda r'}\right)^{1/2} = \left(\frac{2r'}{\lambda}\right)^{1/2}\frac{x}{r'} \simeq \left(\frac{2r'}{\lambda}\right)^{1/2}\theta$$

where θ is the angle of diffraction (see Fig. 4.18). For a finite value of θ, as r' increases, τ' also increases. Consequently, in the Fraunhofer region we must assume τ' to be large and τ_0 to be small. Use the asymptotic forms of the Fresnel integrals to study the transition to the Fraunhofer region.

Solution. Using the asymptotic forms [see Eqs. (4.10-13) and (4.10-14)] we

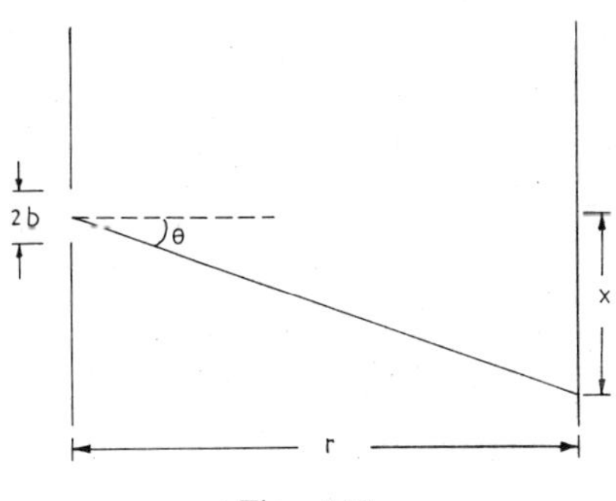

Fig. 4.18

obtain

$$[C(\tau' + \tau_0/2) - C(\tau' - \tau_0/2)] \simeq \frac{1}{\pi\tau'}\left\{\sin\left[\frac{\pi}{2}\left(\tau' + \frac{\tau_0}{2}\right)^2\right] - \sin\left[\frac{\pi}{2}\left(\tau' - \frac{\tau_0}{2}\right)^2\right]\right\}$$

$$= \frac{2}{\pi\tau'}\cos\left[\frac{\pi}{2}\left(\tau'^2 + \frac{\tau_0^2}{4}\right)\right]\sin\frac{\pi\tau'\tau_0}{2}$$

$$S\left(\tau' + \frac{\tau_0}{2}\right) - S\left(\tau' - \frac{\tau_0}{2}\right) = \frac{2}{\pi\tau'}\sin\left[\frac{\pi}{2}\left(\tau'^2 + \frac{\tau_0^2}{4}\right)\right]\sin\frac{\pi\tau'\tau_0}{2}$$

Thus

$$I(P) = \frac{2I_0}{\pi^2\tau'^2}\sin^2\frac{\pi\tau'\tau_0}{2} = I_{00}\frac{\sin^2\beta}{\beta^2} \tag{4.10-35}$$

where

$$I_{00} = \frac{1}{2}I_0\tau_0^2, \quad \beta = \frac{1}{2}\pi\tau'\tau_0 = \frac{1}{2}\pi\left(\frac{2r'}{\lambda}\right)^{1/2}\tan\theta \times 2b\left(\frac{2}{\lambda r'}\right)^{1/2} \simeq \frac{2\pi b\theta}{\lambda}$$

Hence we obtain the Fraunhofer diffraction pattern. Notice that even for small values of τ', Eq. (4.10-35) gives the same result as Eq. (4.10-34). In Fig. 4.17a, we have also plotted the intensity distribution given by Eq. (4.10-35) and, as can be seen, the pattern is almost of the Fraunhofer type.

4.11. Diffraction of Waves Having Amplitude Distribution along the Wavefront

When an aperture is illuminated by a point source, then the field at any point P on the other side of the aperture is given by Eq. (4.4-11). In writing this equation, it is implicitly assumed that the aperture consists of either transparent or opaque regions. In general, the different points in the aperture could have different transmittances. If $T(\xi, \eta)$ represents the amplitude transmittance of the aperture (see, for example, Fig. 4.3b), and if we assume that the aperture is illuminated by a plane wavefront of amplitude C, then the amplitude at any point $P(x, y, z)$ is given by

$$u(x, y, z) = \frac{i}{\lambda}C\iint T(\xi, \eta)\frac{e^{-ikr}}{r}\,d\xi\,d\eta \tag{4.11-1}$$

where $r = [(x - \xi)^2 + (y - \eta)^2 + z^2]^{1/2}$ and we have assumed the aperture to lie in the plane $z = 0$.* If we assume that $x, y, \xi, \eta \ll z$, then we have

$$r \simeq z + \frac{(x-\xi)^2 + (y-\eta)^2}{2z} \tag{4.11-2}$$

As already discussed in Section 4.4, we replace the factor r in the denominator of Eq. (4.11-1) by z to obtain†

$$u(x, y, z) \simeq i\frac{C}{\lambda z}e^{-ikz} \iint T(\xi, \eta) \exp\left\{-i\frac{k}{2z}[(x-\xi)^2 + (y-\eta)^2]\right\} d\xi\, d\eta \tag{4.11-3}$$

If we restrict ourselves to the case when the amplitude in the $z = 0$ plane depends only on ξ [i.e., if $T(\xi, \eta)$ is independent of η], then Eq. (4.11-3) becomes

$$u(x, y, z) = i\frac{C}{\lambda z}e^{-ikz} \int_{-\infty}^{\infty} T(\xi) \exp\left[-ik\frac{(x-\xi)^2}{2z}\right] d\xi \times \int_{-\infty}^{\infty} \exp\left[-i\frac{k}{2z}(y-\eta)^2\right] d\eta \tag{4.11-4}$$

Since

$$\int_{-\infty}^{\infty} \exp\left[-i\frac{k}{2z}(y-\eta)^2\right] d\eta = \left(\frac{\lambda z}{i}\right)^{1/2} \tag{4.11-5}$$

Eq. (4.11-4) reduces to

$$u(x, y, z) = C\left(\frac{i}{\lambda z}\right)^{1/2} e^{-ikz} \int_{-\infty}^{\infty} T(\xi) \exp\left[-i\frac{k}{2z}(x-\xi)^2\right] d\xi \tag{4.11-6}$$

In general, it is not possible to evaluate the integral appearing in Eq. (4.11-6) analytically. But there exists an interesting case of a periodic object with a transmittance of the form

$$T(\xi) = \cos(2\pi\xi/\xi_0) \tag{4.11-7}$$

for which the integral can be evaluated exactly. Notice that if a plane wave is incident on an object with a transmittance of the form given by Eq. (4.11-7), then the emerging wave has not only an amplitude dependence but also a phase variation, e.g., the wavefront emerging between $\xi = -\xi_0/2$

* Alternatively, one could consider the diffraction of a beam having intensity distribution along its wavefront, as is indeed the case for most laser beams (see Problems 4.8 and 4.9).

† Equation (4.11-3) represents the convolution between $T(\xi, \eta)$ and $\exp[-i(k/2z)(\xi^2 + \eta^2)]$ (see Chapter 6).

and $\xi = \xi_0/2$ differs in phase by π from that emerging between $\xi = \xi_0/2$ and $\xi = 3\xi_0/2$. To produce such an object, one can make use of half-wave plates, which change the phase by π. If we substitute Eq. (4.11-7) into Eq. (4.11-6) and carry out the integration, we obtain

$$u(x, z) = Ce^{-ikz} \exp\left(i\frac{\pi\lambda z}{\xi_0^2}\right) \cos\left(\frac{2\pi x}{\xi_0}\right) \tag{4.11-8}$$

and hence the intensity distribution at any value of z is given by

$$I(x, z) = C^2 \cos^2\left(\frac{2\pi x}{\xi_0}\right) \tag{4.11-9}$$

which is the same as the intensity distribution across the object plane (i.e., $z = 0$). Since $I(x, z)$ is independent of z, we obtain the interesting result that the intensity distribution at all planes normal to the z axis is the same as that in the object plane.

Problem 4.6. Show that for an object with a transmittance $\sin(2\pi\xi/\xi_0)$ the amplitude at any plane z is given by

$$u(x, z) = Ce^{-ikz} \exp\left(i\frac{\pi\lambda z}{\xi_0^2}\right) \sin\left(\frac{2\pi x}{\xi_0}\right) \tag{4.11-10}$$

Problem 4.7. (a) Let a plane object with a periodicity x_0 be illuminated by plane-parallel light. (Assume the object plane to be $z = 0$.) By making a Fourier series expansion, show that there exist values of z at which the original object pattern is reproduced. (b) Show also that in planes given by $z = (2m + 1)\, x_0^2/\lambda$ the original object distribution is reproduced but is displaced from the axis by one-half the periodicity of the object.

Solution. (a) Let $F(x)$ be the amplitude transmittance of the object with periodicity x_0. Then we can write

$$F(x) = \sum_n A_n \cos\frac{2\pi nx}{x_0} + \sum_n B_n \sin\frac{2\pi nx}{x_0} \tag{4.11-11}$$

where A_n and B_n are the Fourier cosine and sine coefficients. Each term of the sum on the right-hand side of Eq. (4.11-11) represents a cosine or a sine object. Making use of Eqs. (4.11-8) and (4.11-10), we see that the field at any value of z is given by

$$u(x, z) = Ce^{-ikz} \sum_n \exp\left(i\frac{\pi\lambda z n^2}{x_0^2}\right)\left[A_n \cos\left(\frac{2\pi nx}{x_0}\right) + B_n \sin\left(\frac{2\pi nx}{x_0}\right)\right] \tag{4.11-12}$$

Thus at any value of z, the Fourier components add with different relative phases as compared to that in the plane $z = 0$. The original object distribution will be reproduced whenever the exponential term in Eq. (4.11-12) becomes unity, i.e., whenever the Fourier components again add in the same relative phase as they did in the plane $z = 0$. This occurs whenever $z = (2x_0^2/\lambda)\, m$, where $m = 1, 2, 3, \ldots$. For these values of z, the intensity distribution is the same as the object intensity distribution, or one has the formation of what are known as self-images or "Fourier images."

(b) *Hint:* It will be helpful to use the relation

$$\sum_n A_n e^{in\pi} \cos\frac{2\pi n x}{x_0} = \sum_n A_n \cos\left[\frac{2\pi n}{x_0}\left(x + \frac{x_0}{2}\right)\right]$$

Problem 4.8. If in the plane $z = 0$, instead of an aperture we have an amplitude distribution given by $\psi(x, y, z = 0)$, then the field at any arbitrary point $P(x, y, z)$ would be given by

$$u(x, y, z) = \frac{i}{\lambda z} e^{-ikz} \iint \psi(\xi, \eta, z = 0) \exp\left\{ -\frac{ik}{2z}[(x-\xi)^2 + (y-\eta)^2] \right\} d\xi\, d\eta \tag{4.11-13}$$

where, as before, we have assumed $x, y, \xi, \eta \ll z$. Assuming*

$$\psi(\xi, \eta, z = 0) = A \exp\left[-\frac{(\xi^2 + \eta^2)}{w_0^2} \right] \tag{4.11-14}$$

calculate the field distribution in any plane z.

Solution. Substituting Eq. (4.11-14) into Eq. (4.11-13), we obtain

$$\begin{aligned} u(x, y, z) &= \frac{iA}{\lambda z} e^{-ikz} \iint_{-\infty}^{\infty} \exp\left[-\frac{\xi^2 + \eta^2}{w_0^2} - ik\frac{(x-\xi)^2 + (y-\eta)^2}{2z} \right] d\xi\, d\eta \\ &= \frac{iA}{\lambda z} \exp\left[-ikz - \frac{ik}{2z}(x^2 + y^2) - \frac{k^2\alpha^2}{4z^2}(x^2 + y^2) \right] \\ &\quad \times \int_{-\infty}^{\infty} d\xi \exp\left[-\frac{(\xi - \xi_0)^2}{\alpha^2} \right] \int_{-\infty}^{\infty} d\eta \exp\left[-\frac{(\eta - \eta_0)^2}{\alpha^2} \right] \end{aligned} \tag{4.11-15}$$

where

$$\frac{1}{\alpha^2} = \frac{1}{w_0^2} + \frac{ik}{2z}, \qquad \xi_0 = \frac{ik\alpha^2}{2z} x, \qquad \eta_0 = \frac{ik\alpha^2}{2z} y \tag{4.11-16}$$

Since

$$\int_{-\infty}^{\infty} \exp\left(-\frac{\xi^2}{\alpha^2}\right) d\xi = 2\int_0^{\infty} \exp\left(-\frac{\xi^2}{\alpha^2}\right) d\xi = \alpha \int_0^{\infty} \exp(-x)\, x^{-1/2}\, dx = \alpha\pi^{1/2}$$

and

$$\alpha^2 = \frac{2w_0^2 z}{2z + ikw_0^2}\,\frac{2z - ikw_0^2}{2z - ikw_0^2} = \frac{4z^2 w_0^2}{4z^2 + k^2 w_0^4} - i\frac{2w_0^4 kz}{4z^2 + k^2 w_0^4}$$

* Equation (4.11-14) corresponds to a Gaussian intensity distribution, and indeed one does obtain such a distribution for most laser beams (see Section 4.15).

we obtain

$$u(x, y, z) = \frac{iA\pi}{\lambda}\frac{2w_0^2}{2z + ikw_0^2}\exp\left\{-ik\left[z + \frac{x^2 + y^2}{2z(1 + k^2w_0^4/4z^2)}\right]\right\}\exp\left[-\frac{x^2 + y^2}{w^2(z)}\right] \tag{4.11-17}$$

where

$$w^2(z) = w_0^2\left[1 + \frac{\lambda^2 z^2}{\pi^2 w_0^4}\right] \tag{4.11-18}$$

Thus, according to Eq. (4.11-17), a beam that has an initial Gaussian intensity distribution remains Gaussian as it propagates through space. The width of the Gaussian beam [represented by $w(z)$] increases as the beam progresses (see Fig. 10.4b). Notice that since in Eq. (4.11-18), z occurs in an even power, the beam width increases along both the positive and negative z direction. This is because we had assumed the phase front to be plane at $z = 0$. Thus the Gaussian beam has minimum width at $z = 0$, which is also referred to as the waist of the Gaussian beam. From Eq. (4.11-17) it is clear that as the beam progresses, the phase front does not remain plane. In order to determine the radius of curvature of the phase front, we note that e^{-ikr}/r represents a spherical wave of radius of curvature r. At large distances from the origin and for points near the z axis, the above function can be approximated by

$$\frac{1}{r}\exp\left[-ikz\left(1 + \frac{x^2 + y^2}{z^2}\right)^{1/2}\right] \simeq \frac{1}{r}\exp\left[-ik\left(z + \frac{x^2 + y^2}{2z}\right)\right]$$

If we compare the above expression with Eq. (4.11-17), we obtain the following approximate expression for the radius of curvature of the phase front:

$$R(z) = z\left[1 + \frac{k^2 w_0^4}{4z^2}\right] \tag{4.11-19}$$

Problem 4.9. The above problem can also be solved by noting that the field must satisfy the scalar-wave equation [see Eq. (4.2-2)]. Since the solution of Eq. (4.2-2) if of the form

$$\exp[-i(k_x x + k_y y + k_z z)]$$

with $k_x^2 + k_y^2 + k_z^2 = k^2$, its general solution can be written in the form

$$u(x, y, z) = \iint U(k_x, k_y)\exp\{-i[k_x x + k_y y + (k^2 - k_x^2 - k_y^2)^{1/2} z]\}\, dk_x\, dk_y \tag{4.11-20}$$

If we use Fourier's theorem, we can invert the above equation to obtain

$$U(k_x, k_y) = \frac{1}{4\pi^2}\iint u(x, y, z = 0)\exp[i(k_x x + k_y y)]\, dx\, dy \tag{4.11-21}$$

Use Eqs. (4.11-20) and (4.11-21) to study the diffraction of a Gaussian beam.

Solution. For a Gaussian beam with its waist in the plane $z = 0$,

$$u(x, y, z = 0) = A \exp\left[-\frac{x^2 + y^2}{w_0^2}\right] \tag{4.11-22}$$

and from Eq. (4.11-21) we obtain

$$U(k_x, k_y) = \frac{A}{4\pi^2}\int_{-\infty}^{\infty} \exp\left[-\left(\frac{x^2}{w_0^2} - ik_x x\right)\right]dx \int_{-\infty}^{\infty} \exp\left[-\left(\frac{y^2}{w_0^2} - ik_y y\right)\right]dy$$

$$= \frac{Aw_0^2}{4\pi}\exp\left[-\frac{w_0^2}{4}(k_x^2 + k_y^2)\right] \tag{4.11-23}$$

For points near the z axis, we may write

$$\exp\{-i[k_x x + k_y y + (k^2 - k_x^2 - k_y^2)^{1/2} z]\}$$
$$\simeq \exp\left\{-i\left[k_x x + k_y y + kz - \frac{k_x^2 + k_y^2}{2k} z\right]\right\} \tag{4.11-24}$$

Substituting Eqs. (4.11-23) and (4.11-24) into Eq. (4.11-20) and carrying out the integration, for the diffraction pattern we get an expression identical to the one obtained in the previous problem.

4.12. Babinet's Principle

Two screens are said to be complementary to each other if the open regions in one correspond to opaque regions in the other and conversely. Babinet's principle gives us a simple method of determining the diffraction pattern due to one of the screens when the diffraction pattern due to the other is known. Let $u_1(P)$ and $u_2(P)$ represent the amplitudes at a point P due to the two complementary screens. If $u(P)$ is the amplitude at the point P in the absence of any screen, then since the diffraction pattern is determined by integrating over the open portions of the screen and since the sum of the open portions of the two screens is the whole space, we obtain

$$u(P) = u_1(P) + u_2(P) \tag{4.12-1}$$

This is called the Babinet's principle.

Problem 4.10. Using Babinet's principle, determine the intensity at a point on the axis of an opaque circular disk.

Solution. Referring to Eq. (4.9-9), the amplitude at a point on the axis due to a circular *opening* is given by

$$u_1(P) = u_0(P)(1 - e^{-i\pi p}) \tag{4.12-2}$$

where $u_0(P)$ represents the amplitude at the point P in the absence of any aperture.

Thus if $u_2(P)$ represents the amplitude at the point P due to an opaque circular disk of the same radius, then from Babinet's principle, we have

$$u_2(P) = u_0(P) - u_1(P) = u_0(P)\, e^{-i\pi p} \tag{4.12-3}$$

Thus the intensity at the point P on the axis of a circular disk is

$$I_2(P) = |u_2(P)|^2 = I_0(P) \tag{4.12-4}$$

which gives us the remarkable result that the intensity at a point on the axis of an opaque disk is equal to the intensity at the point in the absence of the disk! Notice that this result is true only as long as Eq. (4.12-2) is true, i.e., only as long as the disk is so small that the variation in $(r + \rho)$ is negligibly small over the disk.

4.13. Periodic Apertures

Let us consider an array of a large number of identical apertures as shown in Fig. 4.19. Clearly, the Fraunhofer diffraction pattern is a sum of the fields produced by the individual apertures and is given by the expression

$$u = C\left[\iint_{S_1} + \iint_{S_2} + \cdots\right]\{e^{-ik(p\xi + q\eta)}\, d\xi\, d\eta\} \tag{4.13-1}$$

where the integrals represent contributions from individual apertures. Let $Q_1, Q_2, Q_3, \ldots$ represent points that are identically situated inside the apertures. For example, if the apertures are cicular in nature, then the points $Q_1, Q_2, Q_3, \ldots$ could represent the centers of the circles. Let the coordinates of the points $Q_1, Q_2, Q_3, \ldots$ be $(\xi_1, \eta_1), (\xi_2, \eta_2), (\xi_3, \eta_3), \ldots$.

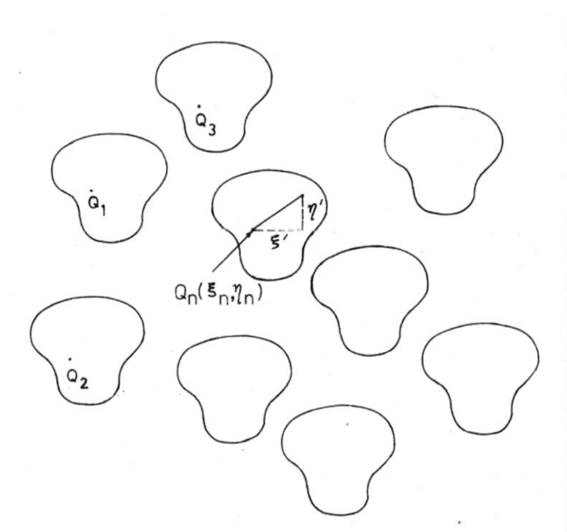

Fig. 4.19. An array of identical apertures. Q_n represents a conveniently chosen origin in the nth aperture.

Obviously

$$u = C\sum_n \iint \exp\{-ik[p(\xi_n + \xi') + q(\eta_n + \eta')]\}\, d\xi'\, d\eta' \quad (4.13\text{-}2)$$

where (ξ', η') are the coordinates of an arbitrary point in a given aperture with respect to the point (ξ_n, η_n) (see Fig. 4.19). Thus

$$u = Cu_0 \sum_n e^{-ik(p\xi_n + q\eta_n)} \quad (4.13\text{-}3)$$

where

$$u_0 = \iint e^{-ik(p\xi' + q\eta')}\, d\xi'\, d\eta' \quad (4.13\text{-}4)$$

represents the Fraunhofer diffraction pattern due to any one of the apertures. The corresponding intensity distribution can be written

$$I = I_0 S \quad (4.13\text{-}5)$$

where

$$S = \left| \sum_n e^{-ik(p\xi_n + q\eta_n)} \right|^2 \quad (4.13\text{-}6)$$

In Eq. (4.13-5), I_0 represents the intensity produced by a single aperture and S represents the interference pattern produced by n point sources situated at $(\xi_1, \eta_1), (\xi_2, \eta_2), \ldots$. For example, if we have two apertures, then

$$\begin{aligned} S &= \left| e^{-ik(p\xi_1 + q\eta_1)} + e^{-ik(p\xi_2 + q\eta_2)} \right|^2 \\ &= 2 + 2\cos\{k[p(\xi_1 - \xi_2) + q(\eta_1 - \eta_2)]\} \end{aligned}$$

or

$$S = 4\cos^2 \tfrac{1}{2}\delta \quad (4.13\text{-}7)$$

where $\delta = k[p(\xi_1 - \xi_2) + q(\eta_1 - \eta_2)]$. Thus

$$I = 4I_0 \cos^2 \tfrac{1}{2}\delta \quad (4.13\text{-}8)$$

The factor $\cos^2 \frac{1}{2}\delta$ is characteristic of the interference pattern produced by two point sources. Thus, if we consider long parallel slits, each of width $2a$, then

$$I_0 = \begin{cases} I_{00}(\sin^2\beta)/\beta^2 & \text{for } q = 0 \\ 0 & \text{for } q \neq 0 \end{cases} \quad (4.13\text{-}9)$$

where $\beta = kpa = (2\pi a \sin\theta)/\lambda$ and I_{00} represents the intensity at $\theta = 0$ produced by either of the slits, θ being the angle of diffraction. We have assumed normal incidence so that $p = \sin\theta$. Further, since we may assume

$q = 0$, we may write

$$\delta = kp(\xi_1 - \xi_2) = (2\pi/\lambda)\, d \sin\theta \tag{4.13-10}$$

where d represents the distance between the two slits. Thus,

$$I = I_{00} \frac{\sin^2\beta}{\beta^2} \left[4\cos^2\left(\frac{2\pi}{\lambda} d \sin\theta \right) \right] \tag{4.13-11}$$

which is the diffraction pattern produced by two parallel slits.

For a diffraction grating, we will have N equally spaced parallel slits, and thus

$$\begin{aligned} S &= |\exp(-ikp\xi_1) + \exp(-ikp\xi_2) + \cdots + \exp(-ikp\xi_N)|^2 \\ &= |1 + \exp(-ikpd) + \exp(-2ikpd) + \cdots + \exp(-i(N-1)\,kpd)|^2 \\ &= \left| \frac{1 - \exp(-iNkpd)}{1 - \exp(-ikpd)} \right|^2 = \frac{\sin^2(\frac{1}{2}Nkpd)}{\sin^2(\frac{1}{2}kpd)} \end{aligned} \tag{4.13-12}$$

which represents the interference pattern produced by N equidistant point sources that lie on a straight line. Thus the resultant diffraction pattern is given by (see Fig. 6.6)

$$I = I_{00} \frac{\sin^2\beta}{\beta^2} \frac{\sin^2 N\gamma}{\sin^2\gamma} \tag{4.13-13}$$

where $\gamma = \frac{1}{2}kpd = \pi d \sin\theta/\lambda$. When $\gamma = \pi, 2\pi, \ldots$, i.e., when

$$d \sin\theta = m\lambda, \qquad m = 1, 2, \ldots \tag{4.13-14}$$

the intensity is

$$N^2 I_{00} \frac{\sin^2(m\pi a/d)}{(m\pi a/d)^2}$$

Thus, one obtains intense maxima in those directions that satisfy Eq. (4.13-14) unless the factor

$$\sin^2(m\pi a/d)/(m\pi a/d)^2$$

is very small; $m = 1, 2, 3, \ldots$ correspond to what are known as the first-order, second-order, third-order, ... grating spectra.

Problem 4.11. Show that when $a = \frac{1}{2}d$, even-order grating spectra are absent.

Problem 4.12. The angular dispersion of the grating spectrum is given by

$$\frac{d\theta}{d\lambda} = \frac{1}{\cos\theta} \frac{m}{d} \tag{4.13-15}$$

Compare this with a typical prism spectrum.

Problem 4.13. For an oblique incidence of a plane wave, show that the grating condition modifies to

$$d(\cos\theta + \cos\theta') = m\lambda$$

4.14. *Intensity Distribution near the Focal Plane*

If a point source is placed on the axis of a lens (which is free from spherical aberration), then the wave emerging from the lens would be a converging spherical wave. In this section, we will calculate the intensity pattern produced by such a converging spherical wave near the focus. We will use Eq. (4.11-3), which gives us the field at any point (x, y, z) given the field $T(x, y)$ on a plane specified by $z = 0$. Thus, we will first have to determine the field distribution $T(x, y)$ of the converging spherical wave on a plane $z = 0$ (see Fig. 4.20a).

We know that the amplitude distribution of a *diverging* spherical wave is of the form $(A/r)\exp(-ikr)$ (see Section 4.2). Thus, for a *converging* spherical wave, the amplitude distribution would be given by $(A/r)\exp(+ikr)$, where the distance r is measured from the focal point O of the spherical wave.

If d represents the distance between the plane $z = 0$ and the focus O of the converging spherical wave, then for any point $P(x, y)$ lying on the plane $z = 0$,

$$r = [x^2 + y^2 + d^2]^{1/2} = d\left[1 + \frac{x^2 + y^2}{d^2}\right]^{1/2}$$

$$\simeq d + \frac{x^2 + y^2}{2d} \tag{4.14-1}$$

where we have assumed that $d \gg a$, a being the radius of the aperture at $z = 0$. Thus, the field distribution on the plane $z = 0$ of the converging spherical wave is

$$T(x, y) \simeq \frac{A}{d}\exp(ikd)\exp[ik(x^2 + y^2)/2d] \tag{4.14-2}$$

where in the denominator we have replaced r by its approximate value d. Thus, the field at any point (x, y, z) is given by

$$u(x, y, z) = \frac{i}{\lambda z}\frac{A}{d}e^{ikd}e^{-ikz}\iint_{\mathscr{A}} \exp[ik(\xi^2 + \eta^2)/2d]$$

$$\times \exp\left\{-\frac{ik}{2z}[(x - \xi)^2 + (y - \eta)^2]\right\} d\xi\, d\eta \tag{4.14-3}$$

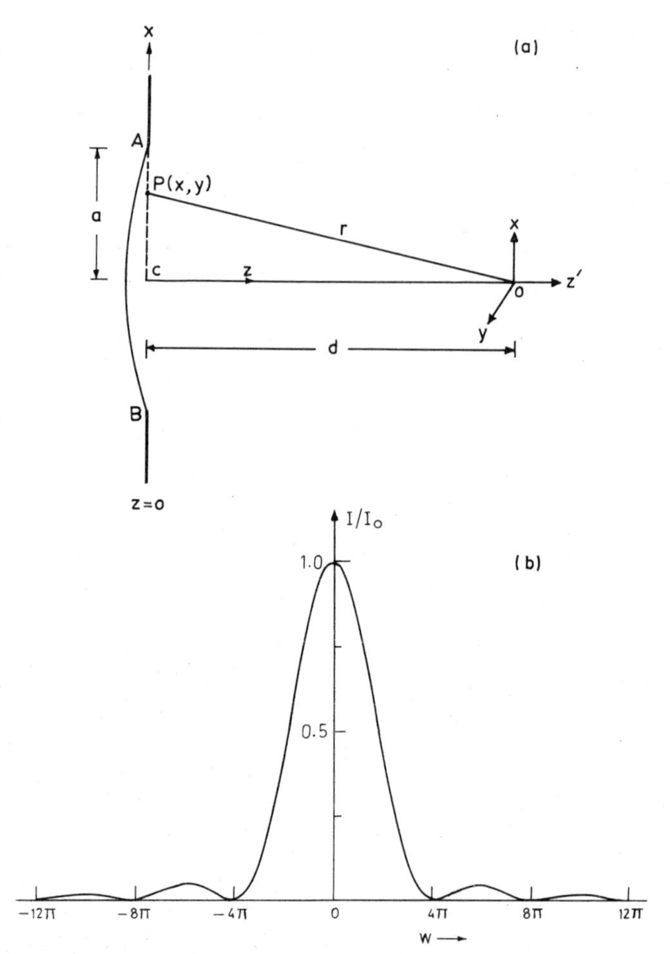

Fig. 4.20. (a) A converging spherical wave with its focus at the point O. (b) Variation of intensity along the axis near the focus of a converging spherical wave of finite cross section.

where we have used Eqs. (4.11-3) and (4.14-2). Here $\mathscr{A}$ represents the area of the aperture.

As we are interested in calculating the field distribution around the focus O, we choose a coordinate system (x, y, z') centered at the focus and restrict ourselves to a region close to the focus so that we may neglect all terms quadratic in x, y, and z'. Thus, Eq. (4.14-3) reduces to

$$u(x, y, z) \simeq \frac{iA}{\lambda d^2} e^{-ikz'} \iint_{\mathscr{A}} \exp\left[\frac{ik}{2d^2} z'(\xi^2 + \eta^2) + \frac{ik}{d}(x\xi + y\eta)\right] d\xi\, d\eta \tag{4.14-4}$$

where we have used

$$z' = z - d \tag{4.14-5}$$

with $z' \ll d$. We now introduce the coordinates

$$\begin{aligned} \xi &= a\tau \cos \theta, \qquad \eta = a\tau \sin \theta \\ x &= \sigma \cos \psi, \qquad y = \sigma \sin \psi \end{aligned} \tag{4.14-6}$$

(Notice that τ is a dimensionless quantity.) Clearly, the limits of integration on τ and θ are 0 to 1 and 0 to 2π, respectively. Thus, Eq. (4.14-4) reduces to

$$u(x, y, z) \simeq \frac{iAa^2}{\lambda d^2} e^{-ikz'} \int_0^1 \int_0^{2\pi} \exp\left[\frac{ika^2}{2d^2} z'\tau^2 + \frac{ik}{d} a\tau\sigma \cos(\theta - \psi)\right] \tau \, d\tau \, d\theta \tag{4.14-7}$$

Following Born and Wolf (1975), we introduce the following dimensionless variables:

$$w = k\left(\frac{a}{d}\right)^2 z', \qquad v = k\left(\frac{a}{d}\right)\sigma = k\left(\frac{a}{d}\right)(x^2 + y^2)^{1/2} \tag{4.14-8}$$

Thus Eq. (4.14-7) may be written in the form

$$\begin{aligned} u(x, y, z) \simeq \frac{iAa^2}{\lambda d^2} \exp\left[-i\left(\frac{d}{a}\right)^2 w\right] \int_0^1 \exp\left(\frac{1}{2} iw\tau^2\right) \\ \times \left[\int_0^{2\pi} e^{iv\tau \cos(\theta - \psi)} \, d\theta\right] \tau \, d\tau \end{aligned} \tag{4.14-9}$$

Since

$$\frac{1}{2\pi} \int_0^{2\pi} \exp[iv\tau \cos(\theta - \psi)] \, d\theta = J_0(v\tau) \tag{4.14-10}$$

where $J_0(v\tau)$ is the Bessel function of zero order [see Eq. (4.7-9)] Eq. (4.14-9) reduces to

$$u(x, y, z) = \frac{2\pi i A a^2}{\lambda d^2} \exp\left[-i\left(\frac{d}{a}\right)^2 w\right] \int_0^1 J_0(v\tau) \exp\left(\frac{1}{2} iw\tau^2\right) \tau \, d\tau \tag{4.14-11}$$

We may now consider two special cases:

(1) Along the z axis, $x = y = 0$ and therefore $v = 0$. Consequently, Eq. (4.14-11) becomes

$$\begin{aligned} u(0, 0, z) &= \frac{2\pi i A a^2}{\lambda d^2} \exp\left[-i\left(\frac{d}{a}\right)^2 w\right] \int_0^1 \exp\left[\frac{1}{2} iw\tau^2\right] \tau \, d\tau \\ &= \frac{\pi i A a^2}{\lambda d^2} \exp\left[-i\left(\frac{d}{a}\right)^2 w\right] e^{iw/4} \frac{\sin(w/4)}{w/4} \end{aligned} \tag{4.14-12}$$

Thus, the intensity distribution will be given by

$$I(0, 0, z) = I_0 \left[\frac{\sin(w/4)}{w/4} \right]^2 \tag{4.14-13}$$

where I_0 represents the intensity at the focal point O. Figure 4.20b gives the variation of I with w. From this figure one can see that the first minimum occurs at $w = \pm 4\pi$ or

$$z' = \pm 2\lambda (d/a)^2 \tag{4.14-14}$$

Further, the intensity drops by about 20% at $u \simeq 3.2$; the corresponding value of z' is

$$z' \simeq \pm 3.2 \frac{\lambda}{2\pi} \left(\frac{d}{a} \right)^2 \tag{4.14-15}$$

which is usually referred to as the focal tolerance. For $\lambda = 6 \times 10^{-5}$ cm, $d = 20$ cm, and $a \doteq 2$ cm, the focal tolerance is approximately 3×10^{-3} cm.

(2) At the focal plane, $z' = 0$, and therefore $w = 0$. Consequently, Eq. (4.14-11) becomes

$$u(x, y, 0) = \frac{2\pi i A a^2}{\lambda d^2} \int_0^1 J_0(v\tau)\, \tau \, d\tau = \frac{2\pi i A a^2}{\lambda d^2} \frac{J_1(v)}{v} \tag{4.14-16}$$

which is the familiar Airy pattern, as it indeed should be.

It is interesting to study the behavior of the phase of the field at the focal plane. Equation (4.14-16) can also be written

$$u(x, y, 0) = \frac{2\pi A a^2}{\lambda d^2} e^{i\pi/2} \frac{J_1(v)}{v} \tag{4.14-17}$$

The dark rings occur at $v = 3.833$, 7.016, 10.174, etc. For $0 < v < 3.833$, $J_1(v)$ is positive and the phase will be $\pi/2$. For $3.833 < v < 7.016$, $J_1(v)$ is negative, and therefore at the first dark ring, the phase undergoes an abrupt change of π. Thus we get the remarkable result that inside the first dark ring, the phase remains constant and has the value $\pi/2$; between the first and the second dark ring, the phase again remains constant but differs from the phase of the central bright spot by π. Notice that Huygens' method of construction of secondary wavelets completely breaks down around the focal point.

Problem 4.14. When $|w/v| < 1$, the point of observation lies inside the geometrical shadow; indeed $|w/v| = 1$ corresponds to the edge of the geometrical shadow. Show that when $|w/v| < 1$, the intensity distribution can be written in the form

$$I(w, v) = \left(\frac{2}{w} \right)^2 [U_1^2(w, v) + U_2^2(w, v)]\, I_0 \tag{4.14-18}$$

where

$$U_n(w, v) = \sum_{s=0}^{\infty} (-1)^s \left(\frac{w}{v}\right)^{n+2s} J_{n+2s}(v) \tag{4.14-19}$$

Reference: Born and Wolf, 1975.

Problem 4.15. Using the identities

$$\cos(w \cos\theta) = J_0(w) + 2 \sum_{s=1}^{\infty} (-1)^s J_{2s}(w) \cos 2s\theta \tag{4.14-20}$$

$$\sin(w \cos\theta) = 2 \sum_{s=0}^{\infty} (-1)^s J_{2s+1}(w) \cos[(2s+1)\theta] \tag{4.14-21}$$

show that along the edge of the geometrical shadow, the intensity distribution is of the form

$$\frac{I_0}{w^2}[1 - 2J_0(w) \cos w + J_0^2(w)]$$

Reference: Born and Wolf, 1975.

Problem 4.16. (a) For $|w/v| > 1$, i.e., for points lying in the illuminated region, show that the intensity distribution can be written in the form

$$I(w, v) = \left(\frac{2}{w}\right)^2 \left\{1 + V_0^2(w, v) + V_1^2(w, v) - 2V_0(w, v) \cos\left[\frac{1}{2}\left(w + \frac{v^2}{w}\right)\right]\right.$$
$$\left. - 2V_1(w, v) \sin\left[\frac{1}{2}\left(w + \frac{v^2}{w}\right)\right]\right\} I_0 \tag{4.14-22}$$

where

$$V_n(w, v) = \sum_{s=0}^{\infty} (-1)^s \left(\frac{v}{w}\right)^{n+2s} J_{n+2s}(v) \tag{4.14-23}$$

(The functions $U_n(w, v)$ and $V_n(w, v)$ are known as the Lommel functions.)

(b) Show that

$$\frac{1}{2I_0} \int_0^{v_0} I(w = 0, v)\, v\, dv = 1 - J_0^2(v_0) - J_1^2(v_0) \tag{4.14-24}$$

The left-hand side represents the fraction of energy contained in the region $0 < v < v_0$ at the focal plane. The above equation agrees with the one derived in Section 4.8.
Reference: Born and Wolf, 1975.

Figure 4.21 shows lines of equal intensity (also called isophotes) in the meridional plane near the focus ($w = 0, v = 0$) of the converging spherical wave. The intensity distribution is symmetric about $v = 0$. The dashed lines in the figure correspond to the boundary of the geometrical shadow.

4.15. Optical Resonators

We conclude this chapter with a brief discussion of optical resonators, which form an integral part of a laser system. In a laser, if the active medium is to act as an oscillator, part of the output energy must be fed back. The feedback is brought about by the use of mirrors placed at the two ends of the active medium. Such a system forms a resonator. One of the mirrors is completely reflecting while the other is only partially reflecting, so as to couple out the laser beam.

The resonators used at optical wavelengths are termed open resonators, because the resonator system is bound by just two mirrors and the sides of the cavity are open. In such resonators only those waves that are approximately propagating along the axis of the resonator can have sufficient gain. Waves that make large angles with the axis escape from the resonator without much amplification. Also, because the mirrors have to be of finite size, every time a wave suffers a reflection at the mirror, a part of its energy spills over, leading to what is known as the diffraction loss.

In this section, we will make a wave analysis of the resonator and ob-

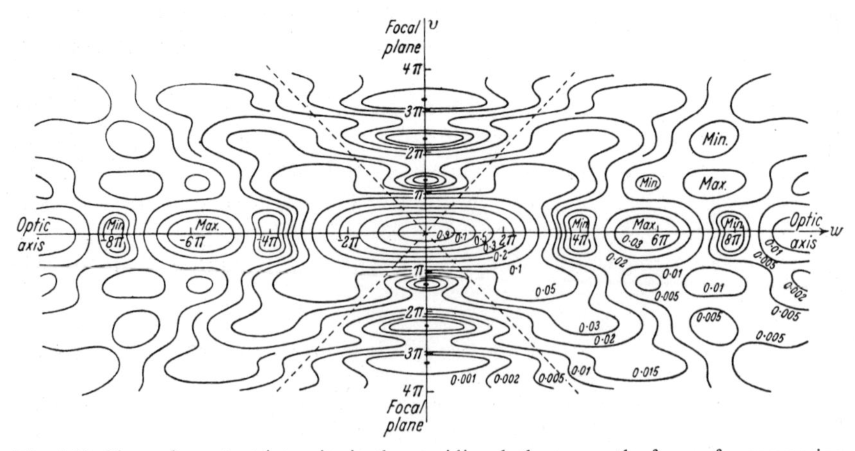

Fig. 4.21. Lines of constant intensity in the meridional plane near the focus of a converging spherical wave. The dashed lines correspond to the boundary of the geometrical shadow. The minima along the v axis correspond to the dark rings of the Airy pattern. (Adapted from Linfoot and Wolf, 1956. Photograph courtesy of Professor E. Wolf.)

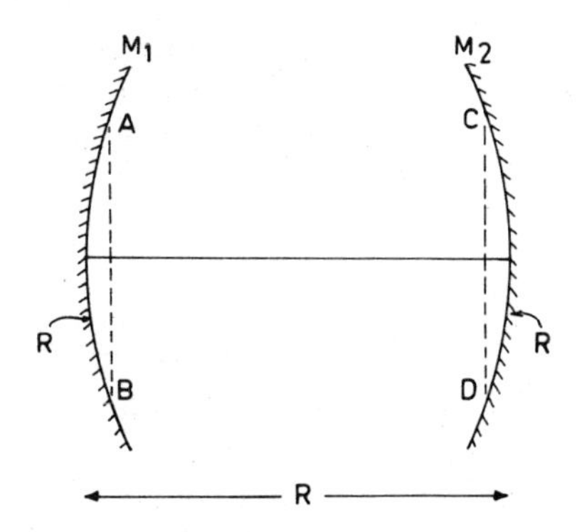

Fig. 4.22. A confocal resonator system consisting of two mirrors of equal radii of curvature and separated by a distance equal to the radius. In such a configuration, the foci of both the mirrors coincide at the center.

tain the approximate field patterns associated with various modes. We will consider a resonator* consisting of two concave mirrors of equal radii of curvature R, which are separated by a distance equal to R (see Fig. 4.22). In such a system, the foci of the mirrors coincide (and lie at the center of the resonator system) and hence such a resonator configuration is termed confocal. In the following analysis, we will assume that $R \gg \lambda$, λ being the wavelength of the electromagnetic radiation in the medium enclosed by the resonator mirrors. We will also restrict ourselves to a region close to the axis of the resonator. The problem is to find the field configuration that can oscillate back and forth between the two mirrors, reproducing itself after every traversal. Such a field configuration is known as a mode† of the resonator. In addition, one must also satisfy the condition that the total phase shift suffered by the field as it completes one round trip must be an even multiple of π. This is because of the fact that standing waves be formed in the space between the resonator mirrors. We will see that this condition restricts the frequencies to a set of discrete values.

In order to find the modal patterns of the resonator, we must first determine the effect of a mirror of radius of curvature R on an incident field.‡ We will assume the z axis to lie along the forward direction of propagation of the wave. Consider spherical waves emanating from a point P at a distance u from the mirror (see Fig. 4.23). The spherical waves after reflection from the mirror will converge toward a point Q at a distance

* In Problem 4.19 we will discuss a more general resonator system, consisting of two mirrors of radii of curvature R_1 and R_2 separated by a distance d.

† Modes also appear in problems dealing with waveguides (see Chapter 10). The amplitude of the modes in open resonators decays exponentially with time; this is due to the diffraction losses that the field suffers at every reflection. Hence these modes are also referred to as quasimodes.

‡ The procedure is similar to that given in Section 6.3, where we obtained the effect of a thin lens on an incident field distribution.

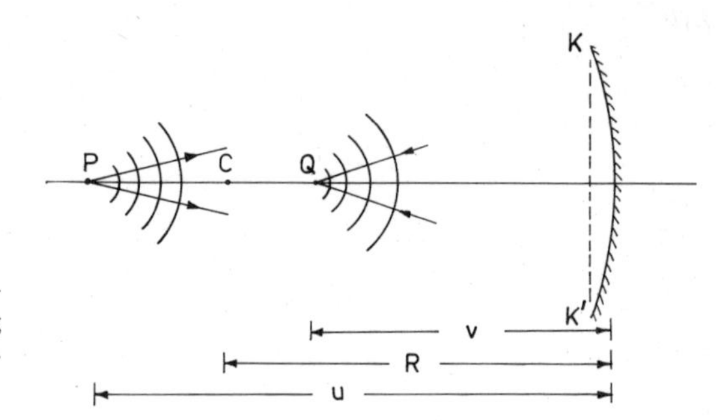

Fig. 4.23. Spherical waves emerging from the point P after suffering a reflection from the mirror converge to the point Q.

v from the mirror such that

$$\frac{1}{u}+\frac{1}{v}=\frac{2}{R} \tag{4.15-1}$$

The equation of a sphere of radius R with origin at a point on the sphere is approximately given by [see Eq. (1.5-9)]

$$z \simeq \frac{x^2+y^2}{2R} \tag{4.15-2}$$

where we have assumed $x, y \ll R$, which corresponds to the paraxial approximation. Thus, under the above approximation, the phase variation of the spherical wavefront emanating from the point P on the plane KK' is

$$\exp\left[-ik\frac{(x^2+y^2)}{2u}\right]$$

Similarly, the phase variation (on the plane KK') of the spherical wavefront converging to the point Q is

$$\exp\left[ik\frac{(x^2+y^2)}{2v}\right]$$

Thus, if p_{m} represents the effect of the mirror on the incident field, then we may write

$$\exp\left[-ik\frac{(x^2+y^2)}{2u}\right]p_{\mathrm{m}}=\exp\left[ik\frac{(x^2+y^2)}{2v}\right]$$

or

$$p_{\mathrm{m}}=\exp\left[\frac{ik}{2f}(x^2+y^2)\right] \tag{4.15-3}$$

where $f = R/2$ represents the focal length of the mirror [cf. Eq. (6.3-5)]. We have neglected the finite size of the mirror.

In order to calculate the transverse modes, we proceed as follows: Let $f(x, y)$ represent the field distribution along the plane AB shown in Fig. 4.22. This field undergoes diffraction over a distance R before getting reflected from the mirror M_2. Hence using Eqs. (4.11-3) and (4.15-3), one obtains the following field distribution on the plane CD just after the wave has undergone reflection from the mirror M_2:

$$g(x, y) = \left(\frac{i}{\lambda R} e^{-ikR} \iint_{\mathscr{A}} dx'\, dy'\, f(x', y') \times \exp\left\{ -\frac{ik}{2R}\left[(x-x')^2 + (y-y')^2\right]\right\}\right) p_{\mathrm{m}}$$

$$= \frac{i}{\lambda R} e^{-ikR} \iint_{\mathscr{A}} dx'\, dy'\, f(x', y') \exp\left\{ -\frac{ik}{2R}\left[(x-x')^2 + (y-y')^2\right]\right\} \times \exp\left[\frac{ik}{R}(x^2 + y^2)\right] \tag{4.15-4}$$

where $\mathscr{A}$ represents the area of the mirror M_1. In order that $f(x, y)$ be a mode of the resonator, we require*

$$g(x, y) = \sigma f(x, y) \tag{4.15-5}$$

where σ is some complex constant. The magnitude of σ will determine the loss suffered, while the phase of σ will determine the phase shift suffered by the field in propagating through a half-cycle. Thus, we require that the fields over the planes AB and CD differ by a constant factor. Using Eq. (4.15-5), we may write Eq. (4.15-4) as

$$\sigma f(x, y) = \frac{i}{\lambda R} e^{-ikR} \iint_{\mathscr{A}} dx'\, dy'\, f(x', y') \exp\left[-\frac{ik}{2R}(x'^2 + y'^2 - 2xx' - 2yy')\right] \times \exp\left[\frac{ik}{2R}(x^2 + y^2)\right] \tag{4.15-6}$$

If we now define a function $h(x, y)$ through the equation

$$h(x, y) = f(x, y) \exp\left[-\frac{ik}{2R}(x^2 + y^2)\right] \tag{4.15-7}$$

* A complete round trip consists of two reflections and two traversals over a distance R. We are here considering only a half-cycle. Since the mirrors M_1 and M_2 are identical, we require the field variations over the planes AB and CD to be identical.

then Eq. (4.15-6) becomes

$$\sigma h(x, y) = \frac{i}{\lambda R} e^{-ikR} \iint_{\mathscr{A}} dx'\, dy'\, h(x', y') \exp\left[\frac{ik}{R}(xx' + yy')\right] \tag{4.15-8}$$

If we introduce the dimensionless variables ξ and η through the relations

$$\xi = (2\pi/\lambda R)^{1/2}\, x', \qquad \eta = (2\pi/\lambda R)^{1/2}\, y' \tag{4.15-9}$$

we may rewrite Eq. (4.15-8) as

$$\sigma h(\xi, \eta) = \frac{i}{2\pi} e^{-ikR} \iint_{\mathscr{B}} d\xi'\, d\eta'\, h(\xi', \eta') \exp[i(\xi\xi' + \eta\eta')] \tag{4.15-10}$$

where $\mathscr{B}$ represents the modified limits of integration. For simplicity we will now consider the mirrors to be squares of sides a. Then Eq. (4.15-10) becomes

$$\sigma h(\xi, \eta) = \frac{ie^{-ikR}}{2\pi} \int_{-\xi_0}^{\xi_0} d\xi' \int_{-\eta_0}^{\eta_0} d\eta'\, h(\xi', \eta') \exp[i(\xi\xi' + \eta\eta')] \tag{4.15-11}$$

where $\xi_0 = a(2\pi/\lambda R)^{1/2}$ and $\eta_0 = a(2\pi/\lambda R)^{1/2}$. We now write $\sigma = \kappa\tau$ and try a separable solution of the form

$$h(\xi, \eta) = p(\xi)\, q(\eta) \tag{4.15-12}$$

On substitution, we find that Eq. (4.15-11) can be written as two separate equations:

$$\kappa p(\xi) = \frac{i^{1/2} e^{-ikR/2}}{(2\pi)^{1/2}} \int_{-\xi_0}^{\xi_0} d\xi' p(\xi')\, e^{i\xi\xi'} \tag{4.15-13}$$

$$\tau q(\eta) = \frac{i^{1/2} e^{-ikR/2}}{(2\pi)^{1/2}} \int_{-\eta_0}^{\eta_0} d\eta' q(\eta')\, e^{i\eta\eta'} \tag{4.15-14}$$

If we compare the integrals appearing on the right-hand side of Eqs. (4.15-13) and (4.15-14) with the Fourier integral [Eq. (B-2)], we observe that in Eqs. (4.15-13) and (4.15-14) the integration limits are finite. Such a transform is known as a finite Fourier transform. The above equations are integral equations and their solutions are prolate spheroidal functions (see, e.g., Slepian and Pollack, 1961; Boyd and Gordon, 1961). Here we assume that the fields are concentrated near the axis and hence the contributions from large values of x and y may be neglected. Thus, we may extend the limits of integration in Eqs. (4.15-13) and (4.15-14) to the region $-\infty$ to $+\infty$.

A complete set of functions that satisfy the condition that they are their own Fourier transforms (up to a constant) are the Hermite–Gauss functions [see, e.g., Gradshtein and Ryzhik (1965), and Problem 4.20]*:

$$i^m H_m(\xi)\exp(-\xi^2/2) = \frac{1}{(2\pi)^{1/2}}\int_{-\infty}^{\infty} d\xi' H_m(\xi')\exp(-\xi'^2/2)\exp(i\xi\xi') \tag{4.15-15}$$

where $H_m(\xi)$ are the Hermite polynomials (see Section 10.3). Hence in order that $p(\xi)$ and $q(\eta)$ be of the Hermite–Gauss form, we require

$$\kappa = i^m i^{1/2} e^{-ikR/2} = \exp\{-i[\tfrac{1}{2}kR - \tfrac{1}{2}(m + \tfrac{1}{2})\pi]\} \tag{4.15-16}$$

$$\tau = i^n i^{1/2} e^{-ikR/2} = \exp\{-i[\tfrac{1}{2}kR - \tfrac{1}{2}(n + \tfrac{1}{2})\pi]\} \tag{4.15-17}$$

Since $|\kappa| = 1$ and $|\tau| = 1$, one notices the absence of any diffraction losses. This is due to our assumption about large mirrors. The total phase shift in one-half round trip is

$$\kappa\tau = \exp\{-i[kR - \tfrac{1}{2}(m + n + 1)\pi]\} \tag{4.15-18}$$

Since the total phase shift in one complete traversal must be an even multiple of π, we must have

$$kR - \tfrac{1}{2}(m + n + 1)\pi = q\pi, \qquad q = 1, 2, 3, \ldots \tag{4.15-19}$$

where q refers to the longitudinal mode number. Since $k = \omega/c$, where ω is the angular frequency of the wave, we have

$$\omega_{mnq} = \pi(2q + m + n + 1)c/2R \tag{4.15-20}$$

Thus only those frequencies that satisfy the above equation are allowed to oscillate in the resonator.

The integers m and n determine the transverse field distribution of the modes and q determines the number of wavelengths between the resonator mirrors. Equation (4.15-20) for the oscillation frequencies shows that there exists more than one mode having the same frequency. Thus, all modes having the same value of $(m + n + 2q)$ have the same frequency of oscillation. These modes are hence degenerate.

We have assumed in our analysis that the radii of curvatures of the mirrors are large compared to λ. Thus the fields are nearly transverse, i.e., the z components of the electric and magnetic fields are small. Hence such modes are termed transverse electric and magnetic modes or, in short, TEM_{mn} modes. The subscripts refer to the mode numbers. (The mode num-

* Hermite–Gauss functions also appear as modes of a particular kind of inhomogeneous optical waveguides that find applications in light wave communication systems (see Chapter 10). The differential equation satisfied by the Hermite–Gauss functions is also discussed in Chapter 10 and Appendix C. The form of the few low-order Hermite–Gauss modes is shown in Fig. 10.3.

ber q is usually omitted since for optical resonators q is extremely large—about 10^4.) The transverse field variations of the modes may be written in the form*

$$E_{mn}(x, y) = AH_m(\xi)\, H_n(\eta) \exp[-\tfrac{1}{2}(\xi^2 + \eta^2)] \exp[+\tfrac{1}{2}i(\xi^2 + \eta^2)] \quad (4.15\text{-}21)$$

where we have made use of Eq. (4.15-12). The lowest-order mode corresponds to $m = n = 0$. For such a mode (TEM_{00}), the intensity variation along the transverse direction is seen to be Gaussian. [This is because $H_0(\xi) = H_0(\eta) = 1$.] This is the mode in which the laser is usually made to oscillate.

The field distribution represented by Eq. (4.15-21) is the field distribution in a transverse plane passing through the pole of the mirror. The field distribution at any other plane can be calculated using the diffraction formula derived earlier. Thus, the field in the plane midway between the two mirrors is [see Eq. (4.11-3)]

$$\begin{aligned} E_{mn} &= \frac{Ai}{\pi} e^{-ikR/2} \iint H_m(\xi')\, H_n(\eta') \exp\left[-\frac{1}{2}\xi'^2(1-i) - \frac{1}{2}\eta'^2(1-i)\right] \\ &\quad \times \exp[-i(\xi^2 + \xi'^2 - 2\xi\xi') - i(\eta^2 + \eta'^2 - 2\eta\eta')]\, d\xi'\, d\eta' \\ &= \frac{iA}{\pi} e^{-ikR/2} \exp(-i\xi^2 - i\eta^2) \int H_m(\xi') \exp\left[-\frac{1}{2}\xi'^2(1+i) + 2i\xi\xi'\right] d\xi' \\ &\quad \times \int H_n(\eta') \exp\left[-\frac{1}{2}\eta'^2(1+i) + 2i\eta\eta'\right] d\eta' \end{aligned} \quad (4.15\text{-}22)$$

Using the integral

$$\int_{-\infty}^{\infty} \exp(-x^2)\, e^{2xy} H_n(\alpha x)\, dx = \pi^{1/2} \exp(y^2)(1-\alpha^2)^{n/2} H_n\left[\frac{\alpha y}{(1-\alpha^2)^{1/2}}\right] \quad (4.15\text{-}23)$$

we obtain

$$E_{mn} = Ae^{-i(kR/2 - \pi/2)} \exp[-(\xi^2 + \eta^2)]\, H_m(2^{1/2}\xi)\, H_n(2^{1/2}\eta) \quad (4.15\text{-}24)$$

which shows that the phase along the transverse plane midway between the two mirrors is constant. Thus the phase fronts are plane midway between the two mirrors.

* Observe that as long as the mirrors are large in the transverse dimension, the modal patterns are independent of the size of the mirrors and are determined essentially by the curvatures of the mirrors.

Problem 4.17. Assuming the field in the plane $z = 0$ (chosen to be midway between the two mirrors) to be

$$E_{mn}(x, y, z = 0) = E_0 \exp[-(x^2 + y^2)/w_0^2]\, H_m\left(\frac{2^{1/2}x}{w_0}\right) H_n\left(\frac{2^{1/2}y}{w_0}\right) \quad (4.15\text{-}25)$$

where $w_0 = (R/k)^{1/2}$, show that the field at any plane z is

$$E_{mn}(x, y, z) = E_0\left[\frac{R}{(R^2 + 4z^2)^{1/2}}\right] \exp\left\{-ik\left[z + \frac{2z}{4z^2 + R^2}(x^2 + y^2)\right]\right\}$$
$$\times \exp\left\{i(m + n + 1)\left[\frac{\pi}{2} - \tan^{-1}\left(\frac{R}{2z}\right)\right]\right\} \exp\left[-\frac{x^2 + y^2}{w^2(z)}\right]$$
$$\times H_m\left(\frac{2^{1/2}x}{w(z)}\right) H_n\left(\frac{2^{1/2}y}{w(z)}\right) \quad (4.15\text{-}26)$$

where

$$w(z) = w_0\left(1 + \frac{4z^2}{R^2}\right)^{1/2} \quad (4.15\text{-}27)$$

is the beam-width parameter.

The above result shows that the beam width is a minimum midway between the mirrors and increases monotonically on either side (see Fig. 4.24). Notice that the beam width is independent of the mirror dimensions.

Problem 4.18. Using the result of Problem 4.17 show that the wavefronts at $z = \pm R/2$ are spherical with radius of curvature R. Thus, the mirrors are constant phase surfaces.

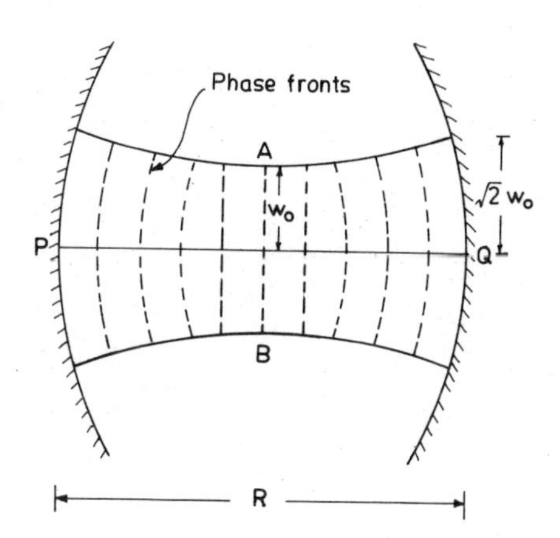

Fig. 4.24. A Gaussian beam resonating in a confocal resonator system. The beam width of the Gaussian beam in the plane widway between the two mirrors is $w_0 = (R/k)^{1/2}$. The phase front is plane midway between the mirrors and matches the mirror surfaces at the points P and Q so that the mirrors form equiphase surfaces. The solid curve shows the function $w(z)$ given by Eq. (4.15-27).

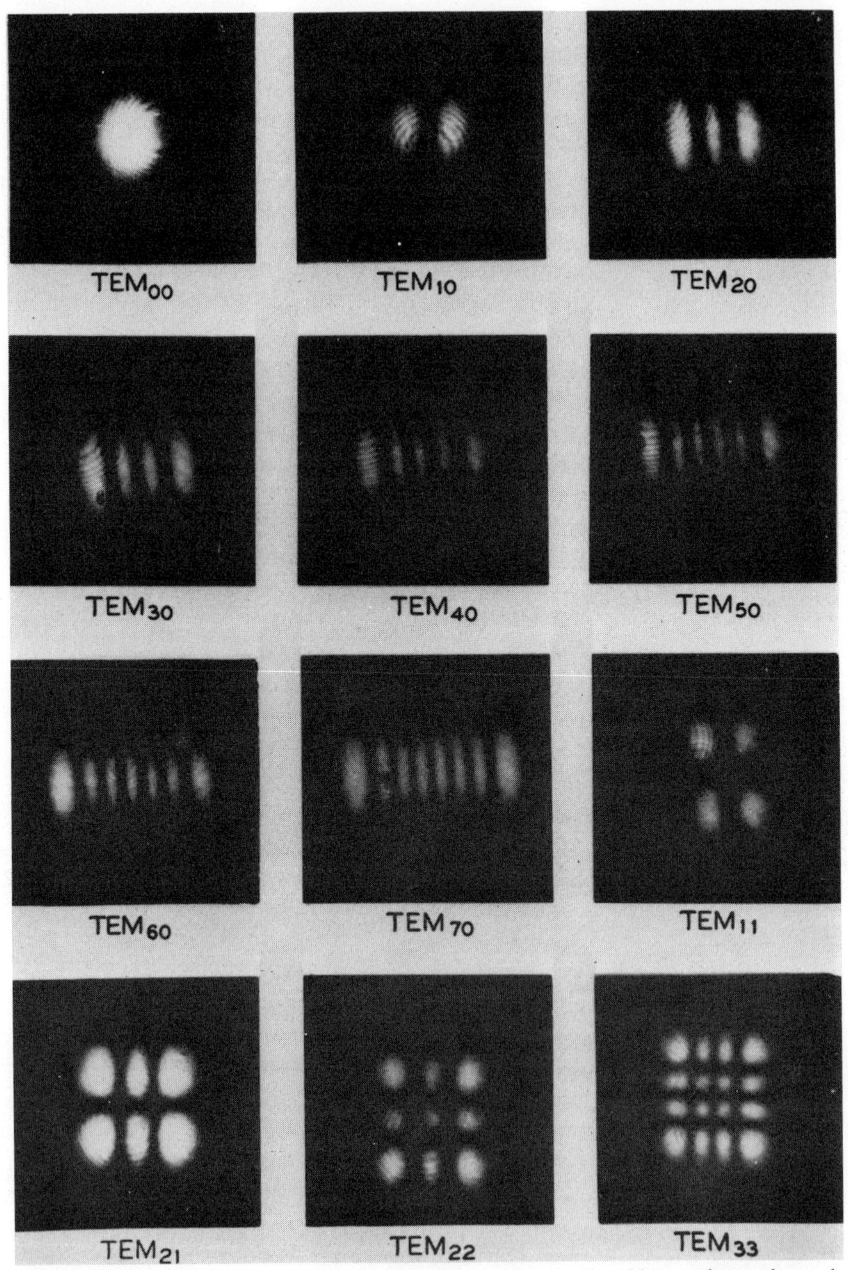

Fig. 4.25. A photograph of some of the lower-order resonator modes. Observe that as the mode order increases, the mode pattern becomes more and more spread. (After Kogelnik and Li, 1966. Photograph courtesy of Dr. H. Kogelnik.)

Figure 4.25 shows photographs of mode patterns obtained from a laser. It can be seen from the figure that as the mode numbers m and n increase, the fields extend more and more away from the axis of the resonator. This leads to higher diffraction losses for higher-order modes. One usually prefers to work with the lowest-order mode (which as we have seen is Gaussian) because it has a monotonically decreasing intensity away from the axis and there are no abrupt phase changes across the wavefront. A selection of the lowest order may be made by inserting an aperture in the resonator and decreasing its radius to a value where the losses for all higher-order modes exceed the gain from the medium.

In this section we have considered the resonator to be bound by concave mirrors of equal radii of curvature and separated by a distance equal to the radius of curvature. In general, the mirrors may be of different radii of curvature. Even in such a case the lowest-order resonant mode would be Gaussian. This follows from the fact that if one considers a Gaussian beam with waist size w_0 situated at $z = 0$ (say) and having waist sizes w_1 and w_2 and radii of curvature R_1 and R_2 at z_1 and z_2, respectively, then by placing mirrors of radii of curvature R_1 and R_2 with transverse dimensions much larger than w_1 and w_2 at z_1 and z_2, respectively, the Gaussian beam would keep bouncing back and forth on itself and would thus resonate (see Fig. 4.26).

Problem 4.19. Consider a resonator made of two mirrors of radii of curvature R_1 and R_2 and separated by a distance d (see Fig. 4.26). Determine the parameters of the Gaussian beam that would keep bouncing back and forth in such a resonator without changing in time.

Solution. Let the poles of the mirrors with radii of curvature R_1 and R_2 be at $z = z_1$ and $z = z_2$, respectively (see Fig. 4.26). We assume that a Gaussian beam

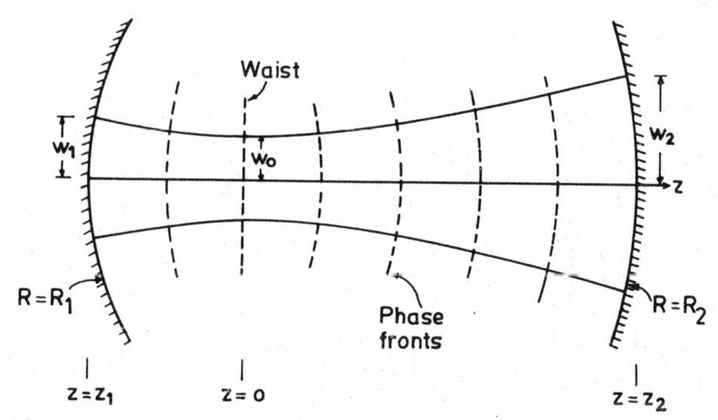

Fig. 4.26. A Gaussian beam resonating in a resonator made of mirrors of radii of curvature R_1 and R_2. For the Gaussian beam to have finite beam widths at the mirrors, the resonator parameters must satisfy the stability condition [see Eq. (4.15-43)].

of waist size w_0 resonates between the mirrors. From Eq. (4.11-19), it follows that the radius of curvature of the phase front of the Gaussian beam at $z = z_1$ and $z = z_2$ is

$$R(z_1) = z_1 + \alpha^2/z_1, \qquad R(z_2) = z_2 + \alpha^2/z_2 \tag{4.15-28}$$

where

$$\alpha^2 = k^2 w_0^4/4 = \pi^2 w_0^4/\lambda^2 \tag{4.15-29}$$

In order that the Gaussian beam may keep bouncing back and forth between the two mirrors, the radius of the phase front of the Gaussian beam at $z = z_1$ and $z = z_2$ must be equal to the radius of the mirrors R_1 and R_2, respectively. In such a case the beam would retrace its path and thus reproduce itself at any plane.

We choose a sign convention such that the radius of curvature of the mirror is positive if it is concave toward the resonator. Under such a convention, $R(z_2) = R_2$ and $R(z_1) = -R_1$, and Eq. (4.15-28) reduces to

$$z_1 + \alpha^2/z_1 = -R_1 \tag{4.15-30}$$

$$z_2 + \alpha^2/z_2 = R_2 \tag{4.15-31}$$

Since the distance between the two mirrors is given to be d, we have another equation connecting z_1 and z_2, namely,

$$z_2 - z_1 = d \tag{4.15-32}$$

We must now solve Eqs. (4.15-30), (4.15-31), and (4.15-32) for z_1, z_2, and α. From Eqs. (4.15-30) and (4.15-31), we obtain

$$\alpha^2 = -(R_1 + z_1) z_1 = (R_2 - z_2) z_2 \tag{4.15-33}$$

Using Eq. (4.15-32), we obtain

$$z_2 = \frac{d(d - R_1)}{2d - R_1 - R_2} \tag{4.15-34}$$

If we introduce two parameters g_1 and g_2 defined by

$$g_1 = 1 - d/R_1, \qquad g_2 = 1 - d/R_2 \tag{4.15-35}$$

then Eq. (4.15-34) reduces to

$$z_2 = \frac{dg_1(1 - g_2)}{g_1 + g_2 - 2g_1 g_2} \tag{4.15-36}$$

Similarly, we may obtain

$$z_1 = -\frac{dg_2(1 - g_1)}{g_1 + g_2 - 2g_1 g_2} \tag{4.15-37}$$

$$\alpha^2 = \frac{d^2 g_1 g_2 (1 - g_1 g_2)}{(g_1 + g_2 - 2g_1 g_2)^2} \tag{4.15-38}$$

Thus, given the parameters of the resonator system, i.e., given g_1, g_2 (which are simply related to R_1, R_2, and d), and d, one can determine the parameters of the

Gaussian beam that will form the lowest-order resonant mode of the resonator.

Let us now calculate the beam width of the Gaussian mode at the two mirrors. From Eqs. (4.15-29), and (4.15-38), we obtain

$$w_0^4 = \frac{\lambda^2}{\pi^2} \frac{d^2 g_1 g_2 (1 - g_1 g_2)}{(g_1 + g_2 - 2g_1 g_2)^2} \tag{4.15-39}$$

Thus, using Eq. (4.11-18), we obtain

$$w^2(z_1) = w_0^2 \left(1 + \frac{\lambda^2 z_1^2}{\pi^2 w_0^4} \right) \tag{4.15-40}$$

which, using Eqs. (4.15-37) and (4.15-39), becomes

$$w^2(z_1) = \frac{\lambda d}{\pi} \left[\frac{g_2}{g_1(1 - g_1 g_2)} \right]^{1/2} \tag{4.15-41}$$

Similarly, we obtain

$$w^2(z_2) = \frac{\lambda d}{\pi} \left[\frac{g_1}{g_2(1 - g_1 g_2)} \right]^{1/2} \tag{4.15-42}$$

It follows from Eqs. (4.15-41) and (4.15-42) that if $g_1 g_2 \to 1$ or $g_1 g_2 \to 0$, then the beam width at one or both the mirrors becomes infinite and the analysis we have carried out loses validity. It can also be seen that $g_1 g_2$ must lie between zero and unity for the existence of a stable Gaussian beam mode:

$$0 \leq g_1 g_2 = \left(1 - \frac{d}{R_1} \right) \left(1 - \frac{d}{R_2} \right) \leq 1 \tag{4.15-43}$$

This is exactly the same as the stability condition obtained by ray analysis by requiring that rays of light must keep bouncing back and forth between the two resonator mirrors without escaping from the resonator (see, e.g., Seigman, 1971). Figure 4.27 shows the stability diagram. Resonator configurations lying in the shaded region are stable configurations. For a symmetric confocal resonator, $R_1 = R_2 = d$, giving $g_1 = g_2 = 0$. Thus the origin of the stability diagram corresponds to the confocal system. A resonator configuration specified by $R_1 = R_2 = d/2$ corresponds to $g_1 = g_2 = -1$ and is represented by point B in Fig. 4.27. Such a system is termed concentric because the centers of both the mirrors coincide at the center of the resonator. Similarly, a resonator consisting of two plane mirrors has $g_1 = g_2 = 1$ and is represented by point A in Fig. 4.27.

Problem 4.20. Show that the solutions of Eq. (4.15-13), when $\xi_0 \to \infty$, are Hermite–Gauss functions.

Solution. In order to show that the solutions are Hermite–Gauss functions, we convert the integral equation to a differential equation. We first write Eq. (4.15-13) in the form

$$Ap(\xi) = \int_{-\infty}^{\infty} p(\xi') e^{i\xi\xi'} d\xi' \tag{4.15-44}$$

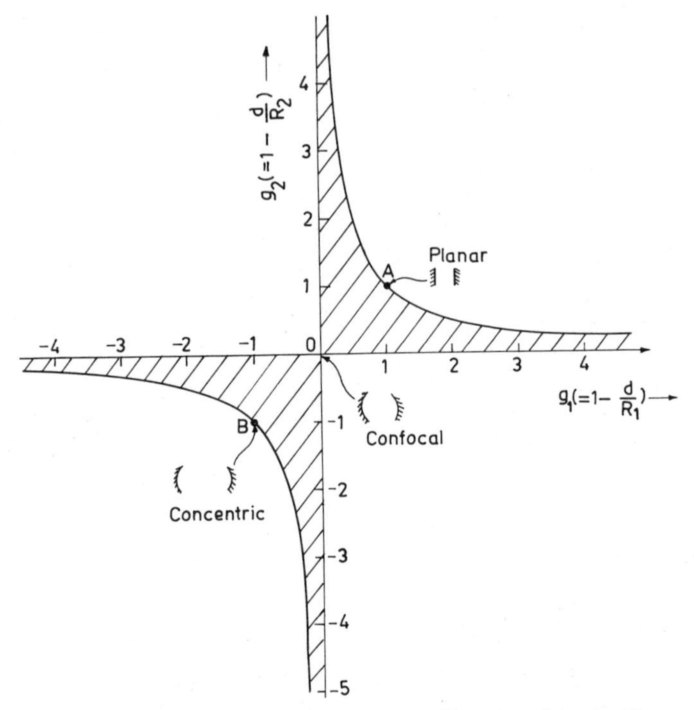

Fig. 4.27. The stability diagram. Resonator configurations lying inside the shaded region are stable configurations. Resonator configurations represented by points marked O, A, B correspond to the confocal ($R_1 = R_2 = d$), the planar ($R_1 = R_2 = \infty$), and the concentric ($R_1 = R_2 = 2d$) configurations, respectively.

where

$$A = \frac{(2\pi)^{1/2}}{i^{1/2}} \kappa e^{ikR/2} \tag{4.15-45}$$

and we have extended the integration limits to $\pm\infty$ assuming that the fields die out at very large distances. We now differentiate Eq. (4.15-44) twice with respect to ξ, and obtain

$$A \frac{d^2 p}{d\xi^2} = -\int_{-\infty}^{\infty} p(\xi')\, \xi'^2 \exp(i\xi\xi')\, d\xi' \tag{4.15-46}$$

Let us next consider the integral

$$I = \int_{-\infty}^{\infty} \frac{d^2 p(\xi')}{d\xi'^2} \exp(i\xi\xi')\, d\xi' \tag{4.15-47}$$

We assume that both $p(\xi)$ and its derivative vanish at infinity and integrate Eq.

(4.15-47) by parts to obtain

$$\int_{-\infty}^{\infty} \frac{d^2 p(\xi')}{d\xi'^2} \exp(i\xi\xi')\, d\xi' = -\xi^2 \int_{-\infty}^{\infty} p(\xi') \exp(i\xi\xi')\, d\xi'$$
$$= -A\xi^2 p(\xi) \tag{4.15-48}$$

Combining Eqs. (4.15-46) and (4.15-48), we obtain

$$A\left[\frac{d^2 p}{d\xi^2} - \xi^2 p(\xi)\right] = \int_{-\infty}^{\infty} \left[\frac{d^2 p}{d\xi'^2} - \xi'^2\right] \exp(i\xi\xi')\, d\xi' \tag{4.15-49}$$

Comparing Eqs. (4.15-44) and (4.15-49), we note that both $p(\xi)$ and $[d^2p/d\xi^2 - \xi^2 p(\xi)]$ satisfy the same equation; as such one must have

$$\frac{d^2 p}{d\xi^2} - \xi^2 p(\xi) = -Kp(\xi) \tag{4.15-50}$$

where K is some arbitrary constant. Equation (4.15-50) may be rewritten

$$\frac{d^2 p}{d\xi^2} + (K - \xi^2)\, p(\xi) = 0 \tag{4.15-51}$$

The solutions of Eq. (4.15-51) under the condition that $p(\xi)$ vanish at large distances are the Hermite–Gauss functions (see Appendix C):

$$p(\xi) = N_m H_m(\xi) \exp(-\xi^2/2) \tag{4.15-52}$$

where N_m is the normalization constant and $K = 2m + 1$; $m = 0, 1, 2, \ldots$.

If we substitute Eq. (4.15-52) into Eq. (4.15-44), we obtain

$$AH_m(\xi)\, e^{-\xi^2/2} = \int_{-\infty}^{\infty} H_m(\xi') \exp(-\xi'^2/2) \exp(i\xi\xi')\, d\xi'$$
$$= (2\pi)^{1/2}\, i^m H_m(\xi) \exp(-\xi^2/2) \tag{4.15-53}$$

where we have used Eq. (4.15-15). If we use Eq. (4.15-45), we obtain Eq. (4.15-16), which leads to the oscillation frequencies.

For a more general analysis, the reader should see Bergstein and Schachter (1967).

5

Partially Coherent Light

5.1. Introduction

When discussing the theory of interference and diffraction of waves, one usually assumes that the fields remain perfectly sinusoidal for all values of time. This is obviously an idealized situation and we know that the radiation from an ordinary light source consists of finite* size wave trains, a typical time variation of which is shown in Fig. 5.1. As can be seen from the figure, the electric fields at times t and $t + \Delta t$ will have a definite phase relationship if $\Delta t \ll \tau_c$ and will (almost) never have any phase relationship for $\Delta t \gg \tau_c$, where τ_c represents the average duration of the wave trains (see Fig. 5.1). The time τ_c is known as the coherence time of the radiation and the field is said to remain coherent for times $\sim\tau_c$. For the neon 6328 Å

* The finite value of τ_c is, in general, due to the following reasons: (1) any atom has a finite lifetime in the energy level from which it drops to the lower energy level while radiating; (2) there are collisions occurring between the atoms; and (3) since the atoms are moving randomly, a Doppler shift is also introduced.

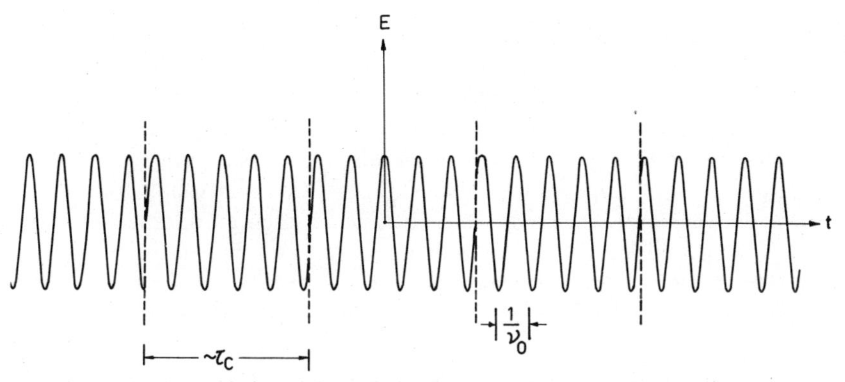

Fig. 5.1. Typical variation of the radiation field with time. The coherence time $\sim\tau_c$.

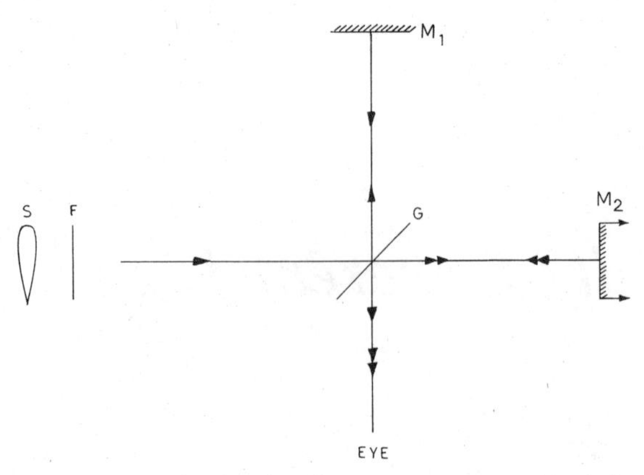

Fig. 5.2. The Michelson interferometer arrangement to study the temporal coherence of the radiation. S and F represent the source and filter, respectively.

line, $\tau_c \sim 10^{-10}$ sec. On the other hand, for a laser beam τ_c can be as large as a few milliseconds. A finite value of τ_c is related to the spectral width of the line, and if one makes a Fourier analysis of the electric field shown in Fig. 5.1 then one can show that it is essentially a superposition of waves whose frequencies lie in a region $\Delta\nu$ around the frequency $\nu = \nu_0$, where

$$\Delta\nu \sim 1/\tau_c \tag{5.1-1}$$

In order to study the time coherence of the radiation let us consider the Michelson interferometer experiment using a nearly monochromatic light source (see Fig. 5.2). For the source, we may use a neon lamp in front of which we place a filter so that the radiation corresponding to 6328 Å is allowed to fall on the beam splitter G. If the eye is in the position shown in Fig. 5.2, then circular fringes will be visible due to the interference of the beams reflected from mirrors M_1 and M_2; we are assuming here that mirrors M_1 and M_2 are at right angles to each other and are equidistant from the beam splitter G. If mirror M_2 is moved away in the direction of the arrow, then the visibility* and hence the contrast of the interference fringes will become poorer, and eventually the fringe pattern will slowly disappear. The disappearance of the fringes is due to the following phenomenon: when mirror M_2 is moved a distance d, then an additional path of $2d$ is introduced for the beam, which gets reflected by M_2; consequently, the beam reflected from M_2 interferes with that reflected from M_1, which had originated $2d/c$ sec earlier from the light source. (Here c is the velocity

* A precise definition of visibility of interference fringes will be given in Section 5.4.

of light in free space.) Clearly if $2d/c \gg \tau_c$, then the wave reflected from M_2 will have no phase relationship with the wave reflected from M_1, and no interference fringes will be seen. On the other hand, if $2d/c \ll \tau_c$, then a definite phase relationship between the two beams will exist and interference fringes with good contrast will be observed.

Associated with any radiation field, there is another type of coherence, which is known as spatial coherence. In order to understand this, we consider a light source like a neon lamp illuminating a pinhole C (see Fig. 5.3a). The points Q and P will have a definite phase relationship provided

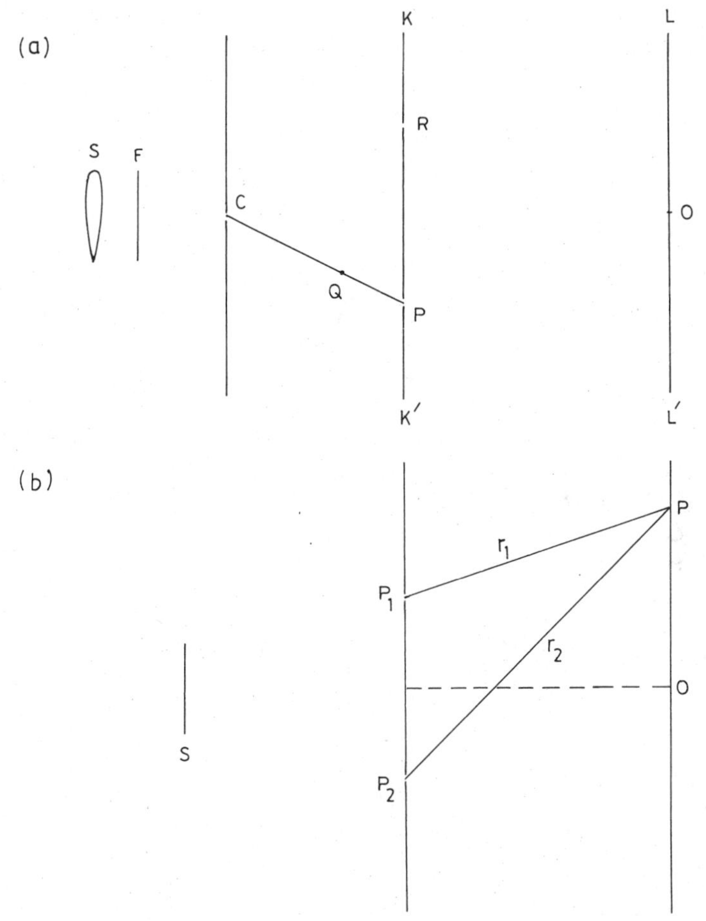

Fig. 5.3. (a) P and R represent pinholes on the screen KK'. As we increase the diameter of the hole C, the contrast of the fringes observed on the screen LL' will become poorer. (b) Field at a point P due to superposition of fields arriving from P_1 and P_2. S represents an extended source.

$QP/c \ll \tau_c$. On the other hand, the points P and R (which are equidistant from the point C) will always have a definite phase relationship and therefore if we have pinholes at the points P and R then clear interference fringes will be observed around the point O on the screen LL'. If we now increase the diameter of the hole C, then due to the finite extent of the source, the points P and R will not, in general, have a definite phase relationship, which may result in the disappearance of the fringes. The distance over which the radiation remains coherent determines the spatial coherence of the field.

In this chapter, we will study the coherence properties associated with the radiation field. We will use a scalar theory in which the polarization effects of the radiation field will be neglected. The theory to be developed is essentially a study of correlations between the fields at two space–time points. A perfect correlation would correspond to complete coherence and a zero correlation would correspond to incoherence. Intermediate states correspond to states of partial coherence.

5.2. Complex Representation

In order to study the coherence properties associated with a field, it is convenient to use the complex representation, which we will develop in this section. The scalar field will be represented by $V^{\mathrm{r}}(\mathbf{r}, t)$, which can simply be the transverse electric field associated with the electromagnetic wave. The superscript r represents the fact that the function $V^{\mathrm{r}}(\mathbf{r}, t)$ is a real function of time. If we make a Fourier decomposition, we may write

$$V^{\mathrm{r}}(t) = \int_{-\infty}^{\infty} \tilde{V}^{\mathrm{r}}(\nu)\, e^{2\pi i \nu t}\, d\nu \tag{5.2-1}$$

Thus

$$V^{\mathrm{r}*}(t) = \int_{-\infty}^{\infty} \tilde{V}^{\mathrm{r}*}(\nu)\, e^{-2\pi i \nu t}\, d\nu \tag{5.2-2}$$

where $*$ represents complex conjugation. Also

$$V^{\mathrm{r}}(t) = \int_{-\infty}^{\infty} \tilde{V}^{\mathrm{r}}(\nu)\, e^{2\pi i \nu t}\, d\nu = \int_{-\infty}^{\infty} \tilde{V}^{\mathrm{r}}(-\nu)\, e^{-2\pi i \nu t}\, d\nu \tag{5.2-3}$$

Since $V^{\mathrm{r}}(t)$ is a real function of time, $V^{\mathrm{r}*}(t) = V^{\mathrm{r}}(t)$. Consequently, from Eqs. (5.2-2) and (5.2-3), we get

$$\tilde{V}^{\mathrm{r}}(-\nu) = \tilde{V}^{\mathrm{r}*}(\nu) \tag{5.2-4}$$

which is the reality condition of $V^{\mathrm{r}}(t)$. Furthermore,

$$V^{\mathrm{r}}(t) = \int_{-\infty}^{\infty} \tilde{V}^{\mathrm{r}}(\nu)\, e^{2\pi i\nu t}\, d\nu = \int_{0}^{\infty} \tilde{V}^{\mathrm{r}}(\nu)\, e^{2\pi i\nu t}\, d\nu + \int_{-\infty}^{0} \tilde{V}^{\mathrm{r}}(\nu)\, e^{2\pi i\nu t}\, d\nu$$

$$= \int_{0}^{\infty} \tilde{V}^{\mathrm{r}}(\nu)\, e^{2\pi i\nu t}\, d\nu + \int_{0}^{\infty} \tilde{V}^{\mathrm{r}*}(\nu)\, e^{-2\pi i\nu t}\, d\nu \qquad (5.2\text{-}5)$$

where we have made use of Eq. (5.2-4). If we write

$$\tilde{V}^{\mathrm{r}}(\nu) = \alpha(\nu) \exp[-i\beta(\nu)] \qquad (5.2\text{-}6)$$

where $\alpha(\nu)$ and $\beta(\nu)$ are real functions of ν, then Eq. (5.2-5) may be written in the form

$$V^{\mathrm{r}}(t) = 2 \int_{0}^{\infty} \alpha(\nu) \cos[2\pi\nu t - \beta(\nu)]\, d\nu \qquad (5.2\text{-}7)$$

Notice that Eq. (5.2-7) involves an integration over positive frequencies only. We now define a function $V^{\mathrm{i}}(t)$, which is obtained from $V^{\mathrm{r}}(t)$ by changing the phase of all Fourier components of $V^{\mathrm{r}}(t)$ by $\pi/2$. Thus*

$$V^{\mathrm{i}}(t) = 2 \int_{0}^{\infty} \alpha(\nu) \sin[2\pi\nu t - \beta(\nu)]\, d\nu \qquad (5.2\text{-}8)$$

Using Eqs. (5.2-7) and (5.2-8), we define the analytic signal $V(t)$ associated with $V^{\mathrm{r}}(t)$ by

$$V(t) = V^{\mathrm{r}}(t) + iV^{\mathrm{i}}(t) = 2 \int_{0}^{\infty} \alpha(\nu)\, e^{2\pi i\nu t - i\beta(\nu)}\, d\nu$$

$$= 2 \int_{0}^{\infty} \tilde{V}^{\mathrm{r}}(\nu)\, e^{2\pi i\nu t}\, d\nu \qquad (5.2\text{-}9)$$

Thus, given $V^{\mathrm{r}}(t)$, we can obtain $V(t)$ by suppressing all the negative frequency components and multiplying the positive frequency components by a factor of 2. We will represent the field $V^{\mathrm{r}}(t)$ by its analytic signal $V(t)$ remembering that

$$V^{\mathrm{r}}(t) = \mathrm{Re}[V(t)] \qquad (5.2\text{-}10)$$

* The functions $V^{\mathrm{r}}(t)$ and $V^{\mathrm{i}}(t)$ are related functions and it can be shown that they form a Hilbert transform pair (see, e.g., Titchmarsh, 1948):

$$V^{\mathrm{i}}(t) = \frac{1}{\pi} P \int_{-\infty}^{+\infty} \frac{V^{\mathrm{r}}(t')}{t' - t}\, dt', \qquad V^{\mathrm{r}}(t) = -\frac{1}{\pi} P \int_{-\infty}^{+\infty} \frac{V^{\mathrm{i}}(t')}{t' - t}\, dt'$$

where P denotes the principal value integral at $t' = t$. An example of such a pair is ($\cos t$, $-\sin t$).

where Re[⋯] denotes the real part of [⋯]. Obviously, if $V_1(t)$ and $V_2(t)$ represent the analytic signals associated with $V_1^r(t)$ and $V_2^r(t)$, then

$$V_1^r(t) + V_2^r(t) = \mathrm{Re}[V_1(t) + V_2(t)] \tag{5.2-11}$$

However,

$$V_1^r(t)\, V_2^r(t) \neq \mathrm{Re}[V_1(t)\, V_2(t)] \tag{5.2-12}$$

5.3. Mutual Coherence Function and Degree of Coherence

Let us consider the field at a point P due to the superposition of the fields arriving from two points P_1 and P_2 (see Fig. 5.3b). If $V_1(t)$ represents the field at P_1, then the field at point P (at time t) due to the radiation from point P_1 can be written $K_1 V_1(t - r_1/c)$, where K_1 is a proportionality constant which has been shown in Section 4.4 to be an imaginary quantity. In an exactly similar manner, the field at P due to the radiation from the point P_2 is given by $K_2 V_2(t - r_2/c)$. Here r_1 and r_2 denote the distances PP_1 and PP_2, respectively. Thus, if $V(P, t)$ represents the field at P at time t, then

$$V(P, t) = K_1 V_1\left(t - \frac{r_1}{c}\right) + K_2 V_2\left(t - \frac{r_2}{c}\right) \tag{5.3-1}$$

Since the optical periods are extremely small in comparison to the resolution time of an optical detector,* the detector will record only the time-averaged intensity:

$$I(P) = \langle [\mathrm{Re}\, V(t)]^2 \rangle \tag{5.3-2}$$

The angular brackets denote time averaging, which is defined by†

$$\langle f(t) \rangle = \lim_{T\to\infty} \frac{1}{2T} \int_{-T}^{T} f(t)\, dt = \lim_{T\to\infty} \frac{1}{2T} \int_{-\infty}^{\infty} f_T(t)\, dt. \tag{5.3-3}$$

where the truncated function $f_T(t)$ is given by

$$f_T(t) \equiv \begin{cases} f(t) & |t| \le T \\ 0 & |t| > T \end{cases} \tag{5.3-4}$$

* The optical periods are of the order of 10^{-14} sec and the detectors rarely have a resolution time less than 10^{-10} sec.

† It may be mentioned here that when we discuss a phenomenon like optical beats we have to carry out a time averaging over times that are much greater than optical periods but much smaller than the period of the beats (see Problem 5.2).

Now, for any two complex functions f and g,

$$\operatorname{Re} f \operatorname{Re} g = \tfrac{1}{2}\operatorname{Re}[fg + f^*g] \tag{5.3-9}$$

If f and g have time variations of the form $\exp(2\pi i\nu_1 t)$ and $\exp(2\pi i\nu_2 t)$, then fg will have a time variation of the form $\exp[2\pi i(\nu_1 + \nu_2)t]$, and for optical frequencies it is extremely difficult to record such rapid variations. Since the time average of fg is zero, we may write

$$I = \lim_{T\to\infty} \frac{1}{2T}\int_{-\infty}^{\infty} \frac{1}{2}\operatorname{Re}[V_T^*(t)\, V_T(t)]\, dt = \tfrac{1}{2}\langle V^*(t)\, V(t)\rangle \tag{5.3-6}$$

Thus the average intensity will be given by

$$\begin{aligned} I(P) &= \langle I(P,t)\rangle = \tfrac{1}{2}\langle V^*(P,t)\, V(P,t)\rangle \\ &= \tfrac{1}{2}|K_1|^2 \langle |V_1(t - r_1/c)|^2\rangle + \tfrac{1}{2}|K_2|^2 \langle |V_2(t - r_2/c)|^2\rangle \\ &\quad + \tfrac{1}{2}[K_1^* K_2 \langle V_1^*(t - r_1/c)\, V_2(t - r_2/c)\rangle \\ &\quad + K_1 K_2^* \langle V_1(t - r_1/c)\, V_2^*(t - r_2/c)\rangle] \end{aligned} \tag{5.3-7}$$

We assume here that the field is stationary, i.e., the time average of a quantity is independent of the origin of time. Under this assumption, we can write

$$I_1 = \tfrac{1}{2}|K_1|^2 \langle |V_1(t - r_1/c)|^2\rangle = \tfrac{1}{2}|K_1|^2 \langle |V_1(t)|^2\rangle \tag{5.3-8}$$

$$I_2 = \tfrac{1}{2}|K_2|^2 \langle |V_2(t - r_2/c)|^2\rangle = \tfrac{1}{2}|K_2|^2 \langle |V_2(t)|^2\rangle \tag{5.3-9}$$

where I_1 and I_2 are the intensities that would be produced at P by P_1 and P_2 independently. From the assumption of stationarity it also follows that $\langle V_1(t - r_1/c)\, V_2^*(t - r_2/c)\rangle$ depends only on $(r_2 - r_1)/c$. Thus, we may write

$$I(P) = I_1 + I_2 + \operatorname{Re}[K_1 K_2^* \Gamma_{12}(\tau)] \tag{5.3-10}$$

where $\tau = (1/c)(r_2 - r_1)$ and

$$\Gamma_{12}(\tau) = \langle V_1(t - r_1/c)\, V_2^*(t - r_2/c)\rangle = \langle V_1(t + \tau)\, V_2^*(t)\rangle \tag{5.3-11}$$

is known as the mutual coherence function. Since K_1 and K_2 are imaginary quantities, we can write

$$I(P) = I_1 + I_2 + |K_1|\,|K_2| \operatorname{Re}[\Gamma_{12}(\tau)] \tag{5.3-12}$$

We can also define a normalized form of $\Gamma_{12}(\tau)$, namely,

$$\gamma_{12}(\tau) = \frac{\Gamma_{12}(\tau)}{[\Gamma_{11}(0)]^{1/2}\,[\Gamma_{22}(0)]^{1/2}} = \frac{\langle V_1(t + \tau)\, V_2^*(t)\rangle}{[\langle |V_1(t)|^2\rangle \langle |V_2(t)|^2\rangle]^{1/2}} \tag{5.3-13}$$

$\gamma_{12}(\tau)$ is called the complex degree of coherence. In terms of $\gamma_{12}(\tau)$, Eq. (5.3-12) can be written

$$I(P) = I_1 + I_2 + 2(I_1 I_2)^{1/2} \operatorname{Re}[\gamma_{12}(\tau)] \tag{5.3-14}$$

If we write

$$\gamma_{12}(\tau) = |\gamma_{12}(\tau)| \exp[i\alpha_{12}(\tau) + 2\pi i \nu_0 \tau] \tag{5.3-15}$$

where

$$\alpha_{12}(\tau) = \arg[\gamma_{12}(\tau)] - 2\pi\nu_0\tau \tag{5.3-16}$$

then Eq. (5.3-14) can also be written

$$I(P) = I_1 + I_2 + 2(I_1 I_2)^{1/2} |\gamma_{12}(\tau)| \cos[\alpha_{12}(\tau) + 2\pi\nu_0\tau] \tag{5.3-17}$$

Problem 5.1. Show that $|\gamma_{12}(\tau)| \leq 1$. [*Hint*: Make use of the Schwarz inequality, $|\int f(\xi)\, g^*(\xi)\, d\xi|^2 \leq \int |f(\xi)|^2\, d\xi \int |g(\xi)|^2\, d\xi$.]

5.4. *Quasi-Monochromatic Sources*

In the analysis given in Section 5.3 we have not made any assumption regarding the nature of the radiation. In most practical cases one deals with sources that emit radiation over a band of frequencies (say $\Delta\nu$) about a mean frequency ν_0 with the condition that $\Delta\nu \ll \nu_0$. Such sources are called quasi-monochromatic sources (as opposed to monochromatic sources, for which $\Delta\nu = 0$).

For quasi-monochromatic radiation we may assume that $|\gamma_{12}(\tau)|$ and $\alpha_{12}(\tau)$ are slowly varying functions of τ, so that the variation of intensity around the point P is primarily due to the variation of $2\pi\nu_0 t$ in the argument of the cosine term [see Eq. (5.3-17)] and hence the intensity variation with τ is approximately sinusoidal. In such a case, we find that the intensity maxima and minima are given by

$$I_{\max} = I_1 + I_2 + 2(I_1 I_2)^{1/2} |\gamma_{12}| \tag{5.4-1}$$

$$I_{\min} = I_1 + I_2 - 2(I_1 I_2)^{1/2} |\gamma_{12}| \tag{5.4-2}$$

Thus the visibility, which is defined by the expression

$$\mathcal{V} = \frac{I_{\max} - I_{\min}}{I_{\max} + I_{\min}} \tag{5.4-3}$$

is given by

$$\mathcal{V} = \frac{2(I_1 I_2)^{1/2}}{I_1 + I_2} |\gamma_{12}| \tag{5.4-4}$$

It is obvious from Eq. (5.4-4) that the visibility (and hence the contrast) is maximum when $|\gamma_{12}| = 1$, which implies complete coherence of the fields arriving at the point P from the points P_1 and P_2. On the other hand, if $|\gamma_{12}| = 0$, $I_{\max} = I_{\min}$ and no fringe pattern is observed. This corresponds to complete incoherence, i.e., there is no definite phase relationship between the waves arriving at the point P from the points P_1 and P_2 (see Fig. 5.3b). The field is said to be partially coherent when $0 < |\gamma_{12}| < 1$.

When the intensities I_1 and I_2 are adjusted to be equal, then we find

$$\mathcal{V} = |\gamma_{12}| \tag{5.4-5}$$

Thus, visibility is equal to the modulus of the complex degree of coherence. We will now determine the variation of $|\gamma_{12}(\tau)|$ with τ for a quasi-monochromatic radiation. Let us define the truncated function $V_{1T}(t)$ by the following equation:

$$V_{1T}(t) = \begin{cases} V_1(t) & |t| < T \\ 0 & |t| > T \end{cases} \tag{5.4-6}$$

$V_{1T}(t)$ is the analytic signal associated with $V^{r}_{1T}(t)$, and if $\tilde{V}^{r}_{1T}(\nu)$ is the Fourier transform of $V^{r}_{1T}(t)$, then as shown in Section 5.2 we can write

$$V_{1T}(t) = \int_0^\infty \tilde{V}_{1T}(\nu)\, e^{2\pi i\nu t}\, d\nu \tag{5.4-7}$$

where $\tilde{V}_{1T}(\nu) = 2\tilde{V}^{r}_{1T}(\nu)$. Similarly, we have

$$V_{2T}(t) = \int_0^\infty \tilde{V}_{2T}(\nu)\, e^{2\pi i\nu t}\, d\nu \tag{5.4-8}$$

Thus

$$\Gamma_{12}(\tau) = \langle V_1(t+\tau)\, V_2^*(t)\rangle = \lim_{T\to\infty} \frac{1}{2T}\int_{-\infty}^{\infty} V_{1T}(t+\tau)\, V_{2T}^*(t)\, dt$$

$$= \lim_{T\to\infty} \frac{1}{2T}\int_{-\infty}^{\infty} dt \int_0^\infty\int_0^\infty \tilde{V}_{1T}(\nu)\, \tilde{V}_{2T}^*(\nu')\, e^{2\pi i(\nu-\nu')t} e^{2\pi i\nu\tau}\, d\nu\, d\nu' \tag{5.4-9}$$

where we have used Eqs. (5.4-7) and (5.4-8). Since

$$\int_{-\infty}^{+\infty} e^{2\pi i(\nu-\nu')t}\, dt = \delta(\nu-\nu') \tag{5.4-10}$$

where $\delta(x)$ represents the Dirac delta function (see Appendix A), we obtain

$$\Gamma_{12}(\tau) = \int_0^\infty \tilde{\Gamma}_{12}(\nu)\, e^{2\pi i\nu\tau}\, d\nu \tag{5.4-11}$$

where

$$\tilde{\Gamma}_{12}(\nu) = \lim_{T\to\infty} \left[\frac{\tilde{V}_{1T}(\nu)\, \tilde{V}^*_{2T}(\nu)}{2T} \right] \tag{5.4-12}$$

Since the radiation is quasi-monochromatic, the quantity $\tilde{\Gamma}_{12}(\nu)$ is appreciable only over a region $\Delta\nu\,(\approx 1/\tau_c)$ around the frequency $\nu = \nu_0$, where τ_c represents the coherence time of the radiation. Thus $\tilde{\Gamma}_{12}(\nu)$ can be assumed to be finite only in the frequency interval $|\nu - \nu_0| \lesssim \Delta\nu$. Consequently

$$\begin{aligned} \Gamma_{12}(\tau) &= \exp(2\pi i\nu_0\tau) \int_0^\infty \tilde{\Gamma}_{12}(\nu) \exp[2\pi i(\nu - \nu_0)\tau]\, d\nu \\ &\simeq \exp(2\pi i\nu_0\tau) \int_{\nu_0 - \Delta\nu/2}^{\nu_0 + \Delta\nu/2} \tilde{\Gamma}_{12}(\nu) \exp[2\pi i(\nu - \nu_0)\tau]\, d\nu \end{aligned} \tag{5.4-13}$$

The exponential function in the integrand of Eq. (5.4-13) can be assumed to be a constant in the region of integration if $\tau\,\Delta\nu \ll 1$. This implies that the path differences involved are small in comparison to $c/\Delta\nu$. In this approximation we obtain

$$\Gamma_{12}(\tau) \simeq \exp(2\pi i\nu_0\tau) \int_0^\infty \tilde{\Gamma}_{12}(\nu)\, d\nu = \Gamma_{12}(0) \exp(2\pi i\nu_0\tau) \tag{5.4-14}$$

where we have used the fact that

$$\Gamma_{12}(0) = \int_0^\infty \tilde{\Gamma}_{12}(\nu)\, d\nu \tag{5.4-15}$$

Thus, the dependence of $\Gamma_{12}(\tau)$ on τ is through the factor $\exp[2\pi i\nu_0\tau]$, ν_0 being the central frequency of the quasi-monochromatic radiation. In a similar manner we can obtain

$$\gamma_{12}(\tau) = \gamma_{12}(0) \exp(2\pi i\nu_0\tau) \tag{5.4-16}$$

It may be mentioned explicitly that in deriving Eqs. (5.4-14) and (5.4-16), we have made two important assumptions: (a) that $\tilde{\Gamma}_{12}(\nu)$ is significantly different from zero over values of ν satisfying $|\nu - \nu_0| \ll \nu_0$, and (b) that the path differences involved are such that τ satisfies the relation $\tau\,\Delta\nu \ll 1$.

Thus under the above approximation, the intensity law for quasi-monochromatic radiation becomes

$$I(P) = I_1 + I_2 + 2(I_1 I_2)^{1/2} |\mu_{12}| \cos(\beta_{12} + 2\pi\nu_0\tau) \tag{5.4-17}$$

where $\mu_{12} = \gamma_{12}(0)$ and $\beta_{12} = \alpha_{12}(0)$. If $I_1 = I_2$, the visibility is just $|\mu_{12}|$.

Thus by measuring I_1, I_2, and the visibility, one can determine $|\mu_{12}|$.

We also observe that the positions of the maxima are given by

$$\beta_{12} + 2\pi\nu_0\tau = 2m\pi \tag{5.4-18}$$

Hence by making measurements of the positions of the maxima, one can indeed measure the phase of the complex degree of coherence, β_{12}; for various methods of measurement of Γ_{12}, see Françon and Mallick (1967).

We will now study the relation between $\Gamma_{12}(\tau)$ and the temporal and spatial coherence properties of the radiation. Let us consider a Young-type experiment in which a source S is placed in front of a screen with two holes. We observe the interference fringes in another screen placed behind it (see Fig. 5.3b). First let us assume that the source is essentially a point source with a finite spectral width. Then the disturbances at P_1 and P_2 are the same [i.e., $V_1(t) = V_2(t)$] and the mutual-coherence function becomes

$$\Gamma_{12}(\tau) = \Gamma_{11}(\tau) = \langle V_1(t+\tau)\, V_1^*(t)\rangle \tag{5.4-19}$$

$\Gamma_{11}(\tau)$ is known as the self-coherence function. Since the decrease in visibility with increase in τ is attributed to the temporal coherence of the beam, $\Gamma_{11}(\tau)$ is a direct measure of the temporal coherence. The Fourier transform of $\Gamma_{11}(\tau)$ gives the spectral density of the radiation (see Problem 5.4). Thus by measuring $\Gamma_{11}(\tau)$ for various values of τ, one can obtain the spectral distribution of the source.

If we now consider a source of finite spatial extent as shown in Fig. 5.3b, then as the source dimension is increased, the visibility of the interference fringes becomes poorer. If we restrict ourselves to the point O (which corresponds to $\tau = 0$), then by measuring visibility we get a measure of $\Gamma_{12}(0)$, which is given by

$$\Gamma_{12}(0) = J_{12} = \langle V_1(t)\, V_2^*(t)\rangle \tag{5.4-20}$$

Thus $\Gamma_{12}(0)$, which is called the mutual intensity, corresponds to the spatial coherence of the radiation.

It may be mentioned that, in general, $\Gamma_{12}(\tau)$ measures both spatial and temporal coherence properties; only in some limiting cases is it possible to separate the effects of the two.

Problem 5.2. Consider the interference between two quasi-monochromatic beams whose central frequencies are ν_1 and ν_2. Calculate the maximum allowable values of the half-widths of each source and also the frequency difference for the detection of optical beats.

Solution. Let $V_1(t)$ and $V_2(t)$ be the fields at a point at a time t for the two beams:

$$V_1(t) = V_{01} \exp[i(2\pi\nu_1 t + \varphi_1)] \tag{5.4-21}$$

$$V_2(t) = V_{02} \exp[i(2\pi\nu_2 t + \varphi_2)] \tag{5.4-22}$$

Since the fields are assumed to be quasi-monochromatic, V_{01}, V_{02}, φ_1, and φ_2 are slowly varying functions of time. The total amplitude at any time t is given by

$$V(t) = V_{01}\exp[i(2\pi\nu_1 t + \varphi_1)] + V_{02}\exp[i(2\pi\nu_2 t + \varphi_2)] \tag{5.4-23}$$

If T_0 represents the integration time of the detector, the mean intensity observed is

$$\begin{aligned} I &= \frac{1}{T_0}\int_{t-T_0/2}^{t+T_0/2}\left[\frac{1}{2}V^*(t)\,V(t)\right]dt \\ &= \frac{1}{T_0}\int_{t-T_0/2}^{t+T_0/2}\left\{\frac{1}{2}V_{01}^2 + \frac{1}{2}V_{02}^2 + V_{01}V_{02}\cos[2\pi(\nu_1-\nu_2)t + \varphi_1 - \varphi_2]\right\}dt \\ &= I_1 + I_2 + 2(I_1 I_2)^{1/2}\frac{1}{T_0}\int_{t-T_0/2}^{t+T_0/2}\cos[2\pi(\nu_1-\nu_2)t + \varphi_1 - \varphi_2]\,dt \end{aligned} \tag{5.4-24}$$

where $I_1 = \frac{1}{2}V_{01}^2$ and $I_2 = \frac{1}{2}V_{02}^2$ are the intensities of either of the two beams in the absence of the other.

Let us first consider the case when T_0 is much greater than the coherence time of either of the beams. Then $\varphi_1 - \varphi_2$ will change randomly within the time limits and hence the integral appearing in Eq. (5.4-24) will be zero. Thus, in such a case $I = I_1 + I_2$, i.e., the two beams add incoherently.

Let us now consider the case when T_0 is much less than the coherence time of the two beams (i.e., $T_0 \ll 1/\Delta\nu_1, 1/\Delta\nu_2$) and further $(\nu_1 - \nu_2)\,T_0 \ll 1$, i.e., the frequency difference between the two beams is much less than the inverse of the integration time of the detector. For such a case, during the detection time, the cosine term appearing in the integrand can be assumed to be constant, and we obtain

$$I(t) = I_1 + I_2 + 2(I_1 I_2)^{1/2}\cos[2\pi(\nu_1-\nu_2)t + \varphi_1 - \varphi_2] \tag{5.4-25}$$

Thus, the detector will record the optical beats, which has indeed been observed using two independent laser beams (Lipsett and Mandel, 1963). On the other hand, if $(\nu_1 - \nu_2)\,T_0 \gg 1$, then the detector will again record the sum of the two intensities.

Problem 5.3. Equation (5.4-11) can be written for the self-coherence function as

$$\Gamma_{11}(\tau) = \int_0^\infty \tilde{\Gamma}_{11}(\nu)\,e^{2\pi i\nu\tau}\,d\nu \tag{5.4-26}$$

where $\tilde{\Gamma}_{11}(\nu)$ is the spectral density of the radiation.

(a) Show that the complex degree of coherence between the fields at one point at two different times is

$$\gamma_{11}(\tau) = \int_0^\infty \tilde{\Gamma}_{11}(\nu)\,e^{2\pi i\nu\tau}\,d\nu \bigg/ \int_0^\infty \tilde{\Gamma}_{11}(\nu)\,d\nu \tag{5.4-27}$$

(A finite time difference can, for example, be introduced by amplitude division of the beam and by allowing the two beams to travel different pathlengths before they are made to interfere as in the Michelson interferometer arrangement.)

(b) Consider a laser source oscillating at two nearby frequencies differing by about 10^8 Hz. Calculate $|\gamma_{11}(\tau)|$ assuming that each of the two frequencies is infinitely narrow* (i.e., each of them corresponds to a Dirac delta function distribution) and that the intensity of the radiation at the frequency ν_1 is I_1 and that of the radiation at the frequency ν_2 is I_2.

Solution. (a) By definition [see Eq. (5.3-13)]

$$\gamma_{11}(\tau) = \Gamma_{11}(\tau)/\Gamma_{11}(0) \tag{5.4-28}$$

which, using Eq. (5.4-26), gives us Eq. (5.4-27).

(b) Since each oscillating frequency has been assumed to be infinitely narrow, the spectral density of the radiation $\tilde{\Gamma}_{11}(\nu)$ can be written (see Problem 5.4)

$$\tilde{\Gamma}_{11}(\nu) = I_1\,\delta(\nu - \nu_1) + I_2\,\delta(\nu - \nu_2) \tag{5.4-29}$$

Substituting this into Eq. (5.4-28) and carrying out the integration, we obtain

$$\gamma_{11}(\tau) = \exp(2\pi i\nu_1\tau)\,[1 + r\exp(2\pi i\,\Delta\nu\tau)]/(1 + r) \tag{5.4-30}$$

where $r = I_1/I_2$ and $\Delta\nu = \nu_1 - \nu_2$. Hence

$$|\gamma_{11}(\tau)| = \frac{1}{1 + r}[1 + r^2 + 2r\cos(2\pi\,\Delta\nu\tau)]^{1/2} \tag{5.4-31}$$

Thus $|\gamma_{11}(\tau)|$ depends not only on $\Delta\nu$ but also on r. For example, for $r = 1$, i.e., when the two frequencies of oscillation are equally intense, then

$$|\gamma_{11}(\tau)| = |\cos(\pi\,\Delta\nu\tau)| \tag{5.4-32}$$

Thus $|\gamma_{11}(\tau)|$ varies periodically with τ and for $\tau = 1/2\,\Delta\nu$, the complex degree of coherence is zero. If one could tolerate a degree of coherence of $1/2^{1/2}$, τ should be restricted to within $1/4\,\Delta\nu$. Hence for $\Delta\nu \sim 10^8$ Hz, the coherence length is reduced to about 300 cm!

Problem 5.4. For a perfectly monochromatic source emitting radiation at a frequency ν_0, we may write

$$V(t) = V_0\exp(2\pi i\nu_0 t) \tag{5.4-33}$$

Show that $\tilde{\Gamma}_{11}(\nu) = V_0^2\,\delta(\nu - \nu_0)$ and thus that $\tilde{\Gamma}_{11}(\nu)$ is a measure of the spectral distribution of the source.

Solution. Since $\tilde{\Gamma}_{11}(\nu)$ is the Fourier transform of $\Gamma_{11}(\tau)$, we will first determine $\Gamma_{11}(\tau)$. By definition,

$$\Gamma_{11}(\tau) = \lim_{T\to\infty}\frac{1}{2T}\int_{-T}^{+T} V(t + \tau)\,V^*(t)\,dt = V_0^2\exp(2\pi i\nu_0\tau) \tag{5.4-34}$$

Thus

$$\tilde{\Gamma}_{11}(\nu) = \int_{-\infty}^{+\infty}\Gamma_{11}(\tau)\,e^{-2\pi i\nu\tau}\,d\tau = V_0^2\,\delta(\nu - \nu_0) \tag{5.4-35}$$

* For a laser oscillating at a single frequency with a spectral profile that is approximately Gaussian and of half-width $\sim 10^4$ Hz, the coherence length is about 30 km; consequently, in this problem, we neglect the spectral width of each line.

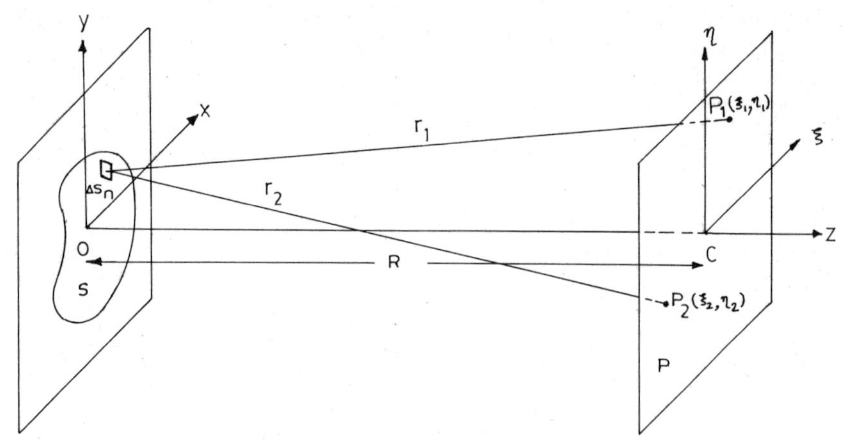

Fig. 5.4. S represents an incoherent quasi-monochromatic planar source and P represents the plane in which we wish to determine the complex degree of coherence.

5.5. Van Cittert–Zernike Theorem

In this section, we will determine the complex degree of coherence in a plane illuminated by an incoherent quasi-monochromatic planar source. For simplicity we will assume that the plane of the source is parallel to the plane in which we are going to determine the complex degree of coherence. We also assume that the medium between the source and the screen is homogeneous.

Let S represent the source and let P be the plane in which we wish to determine the complex degree of coherence (see Fig. 5.4). Let us divide the plane of the source into a large number of elemental areas $\Delta S_1, \Delta S_2, \ldots$. The intensity produced at an arbitrary point by the elemental area ΔS_n will be proportional to ΔS_n itself; thus the field (the square of which is proportional to intensity) produced by the elemental area will be proportional to $(\Delta S_n)^{1/2}$, and if $V_1(t)$ and $V_2(t)$ represent the fields at the points P_1 and P_2, then we may write

$$V_i(t) = \sum_n u_{in}(t)\,(\Delta S_n)^{1/2}, \qquad i = 1, 2 \tag{5.5-1}$$

Thus, the mutual intensity [see Eq. (5.4.20)] is given by

$$\begin{aligned} J_{12} = \langle V_1(t)\, V_2^*(t)\rangle &= \left\langle \sum_n \sum_m u_{1n}(t)\, u_{2m}^*(t)\,(\Delta S_n)^{1/2}\,(\Delta S_m)^{1/2} \right\rangle \\ &= \sum_n \langle u_{1n}(t)\, u_{2n}^*(t)\rangle\, \Delta S_n + \sum_n \sum_{\substack{m \\ n\neq m}} \langle u_{1n}(t)\, u_{2m}^*(t)\rangle \\ &\quad \times (\Delta S_n)^{1/2}\,(\Delta S_m)^{1/2} \end{aligned} \tag{5.5-2}$$

For an incoherent source, fields due to two different elemental areas have no definite phase relationship. Thus $\langle u_{1n} u_{2m}^*(t) \rangle = 0$ for $m \neq n$. Furthermore, we may write

$$u_{1n} = A_n\left(t - \frac{r_{1n}}{c}\right) \frac{\exp[2\pi i \nu_0 (t - r_{1n}/c)]}{r_{1n}} \tag{5.5-3}$$

$$u_{2n} = A_n\left(t - \frac{r_{2n}}{c}\right) \frac{\exp[2\pi i \nu_0 (t - r_{2n}/c)]}{r_{2n}} \tag{5.5-4}$$

Since the source is essentially continuous, we substitute from Eqs. (5.5-3) and (5.5-4) into Eq. (5.5-2) and replace the sum by an integral. Thus, we have

$$J_{12} = \iint \left\langle A(x, y, t)\, A^*\left(x, y, t - \frac{r_2 - r_1}{c}\right)\right\rangle \frac{1}{r_1 r_2} \times \exp\left[2\pi i \nu_0 \frac{r_2 - r_1}{c}\right] dx\, dy \tag{5.5-5}$$

where we have used the fact that the field is stationary. Furthermore, if $(r_2 - r_1)$ is much less than the coherence length of the radiation, then we may write

$$\left\langle A(x, y, t)\, A^*\left(x, y, t - \frac{r_2 - r_1}{c}\right)\right\rangle \simeq \langle A(x, y, t)\, A^*(x, y, t)\rangle \tag{5.5-6}$$

Thus

$$J_{12} = \iint I(x, y) \frac{1}{r_1 r_2} \exp\left[2\pi i \nu_0 \frac{r_2 - r_1}{c}\right] dx\, dy \tag{5.5-7}$$

where $I(x, y) \equiv \langle A(x, y, t)\, A^*(x, y, t)\rangle$. Hence

$$\mu_{12} = \frac{J_{12}}{J_{11}^{1/2} J_{22}^{1/2}} = \frac{1}{(I_1 I_2)^{1/2}} \iint I(x, y) \frac{1}{r_1 r_2} \exp\left[2\pi i \nu_0 \frac{r_2 - r_1}{c}\right] dx\, dy \tag{5.5-8}$$

where $I_1 = \iint I(x, y)\,(1/r_1^2)\, dx\, dy$, etc. Equation (5.5-8), relating the complex degree of coherence and the intensity distribution across the source, is called the Van Cittert–Zernike theorem. This relation is similar to the diffraction pattern produced at the point P_1 due to a spherical wave (with an amplitude distribution proportional to $I(x, y)$ and converging toward the point P_2 when it undergoes diffraction at an aperture of the same size and shape as the source.

It is interesting to observe that the radiation that was incoherent when it started from the source becomes partially coherent as it propagates.

Physically this is due to the fact that each point of the source contributes to each set of points in any subsequent plane.

We choose the z axis along the normal from any conveniently chosen point on the source to the plane of observation. Let (x, y) represent the coordinates of a point on the source and let (ξ_1, η_1) and (ξ_2, η_2) be the coordinates of P_1 and P_2 (see Fig. 5.4). Let R denote the distances between the source and the plane of observation. We assume that $\xi_1, \eta_1, \xi_2, \eta_2, x, y \ll R$. Then under the above assumption, we can write

$$r_1 = [(\xi_1 - x)^2 + (\eta_1 - y)^2 + R^2]^{1/2} \simeq R\left[1 + \frac{(\xi_1 - x)^2}{2R^2} + \frac{(\eta_1 - y)^2}{2R^2}\right] \tag{5.5-9}$$

with a similar equation for r_2. Substituting the above values of r_1 and r_2 into Eq. (5.5-8) and replacing $r_1 r_2$ in the denominator by R^2, we obtain

$$\mu_{12} = e^{i\chi} \frac{\iint I(x, y) \exp[-(2\pi i \nu_0/c)(px + qy)]\, dx\, dy}{\iint I(x, y)\, dx\, dy} \tag{5.5-10}$$

where

$$\chi = \frac{\pi}{R} \frac{\nu_0}{c} [(\xi_2^2 - \xi_1^2) + (\eta_2^2 - \eta_1^2)]$$
$$p = \frac{\xi_2 - \xi_1}{R}, \qquad q = \frac{\eta_2 - \eta_1}{R} \tag{5.5-11}$$

Thus we find that, under the above approximation, the complex degree of coherence is proportional to the normalized Fourier transform of the source intensity distribution.

We will now consider an important application of Eq. (5.5-10) in the determination of the angular diameter of stars. Let us assume that the star is a circular source (of radius a) of incoherent quasi-monochromatic radiation with a uniform intensity distribution across its disk (see Fig. 5.5a). If the points P_1 and P_2 are situated symmetrically about the axis, then $\xi_1 = \xi_2$ and $\eta_1 = \eta_2$, implying $\chi = 0$. Thus Eq. (5.5-10) reduces to

$$\mu_{12} = \frac{\iint_{\mathscr{A}} \exp[-(2\pi i \nu_0/c)(px + qy)]\, dx\, dy}{\iint_{\mathscr{A}} dx\, dy} \tag{5.5-12}$$

where $\mathscr{A}$ represents the area of the source. Carrying out the integration in Eq. (5.5-12), we obtain (see Section 4.7)

$$\mu_{12} = 2J_1(v)/v \tag{5.5-13}$$

where

$$v = \frac{2\pi\nu_0}{c}(p^2 + q^2)^{1/2} a = \frac{2\pi}{\lambda} \frac{a}{R} \rho = \frac{\pi}{\lambda} \rho\theta \tag{5.5-14}$$

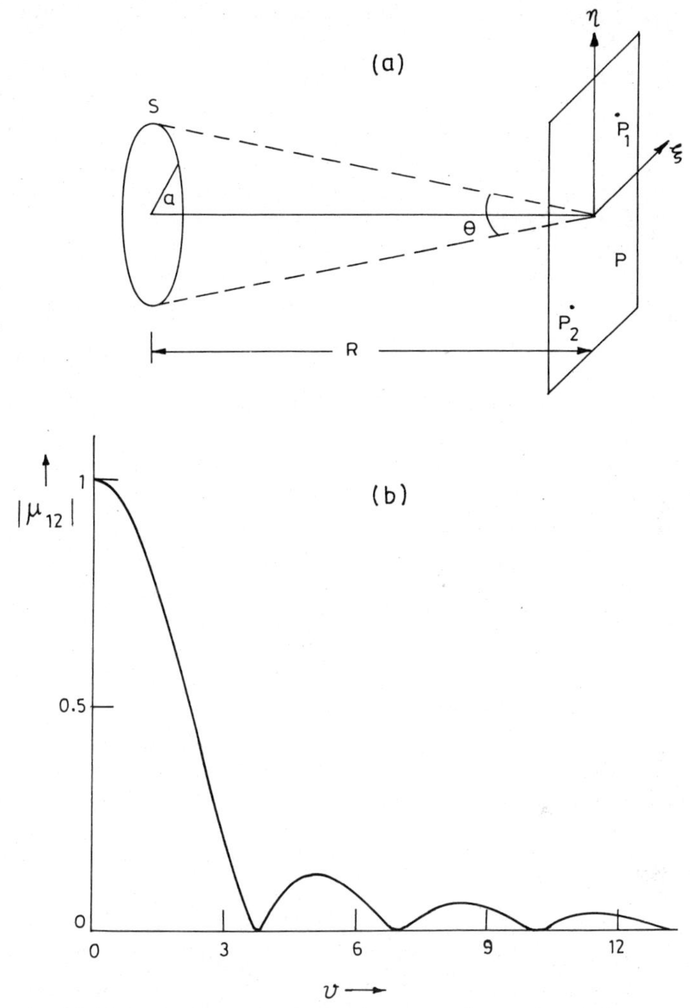

Fig. 5.5. (a) S represents a circular incoherent quasi-monochromatic source of radius a subtending an angle θ at the plane of observation. (b) Variation of $|\mu_{12}|$ with v.

where ρ is the distance between P_1 and P_2 and θ $(= 2a/R)$ is the angle subtended by the source at the plane of observation (see Fig. 5.5a). The variation of $|\mu_{12}|$ with v is shown in Fig. 5.5b. The first zero of $J_1(v)$ (and hence of μ_{12}) occurs at $v = 3.833$. Thus, if we have two pinholes at P_1 and P_2 that are separated by a distance such that $v = 3.833$, then there is no phase correlation of the fields between the points P_1 and P_2. Hence, if we have two pinholes at P_1 and P_2 and a screen placed at its back, then no interference pattern will be observed when

$$P_1P_2 = 3.833\left(\frac{c}{\pi \nu_0 \theta}\right), \quad 7.016\left(\frac{c}{\pi \nu_0 \theta}\right), \quad \text{etc.} \tag{5.5-15}$$

where 3.833, 7.016, etc., correspond to the zeroes of $J_1(v)$. Clearly, points lying in a circle of area less than πr_0^2 (where $r_0 \lesssim \lambda/\pi\theta$) will have fairly definite phase correlation. This area is known as the coherence area.

Thus, we see that by knowing the value of ρ at which the visibility goes to zero, we can determine θ, the angular diameter of the source. This is the principle behind the measurement of the angular diameter of stars using the Michelson stellar interferometer.

The main setup of the interferometer is shown in Fig. 5.6. A screen having two pinholes P_1 and P_2 that are equidistant from the axis is placed symmetrically in front of a lens. Mirrors M_1 and M_2 reflect the starlight into mirrors M_3 and M_4, which in turn reflect the light into pinholes P_1 and P_2, respectively. An interference pattern is formed in the focal plane of the lens. The distance ρ in Eq. (5.5-14) now corresponds to the distance between mirrors M_1 and M_2. Thus when the separation between M_1 and M_2 is varied, the visibility of the interference fringes obtained at the point P changes. By measuring the variation of the visibility as a function of the

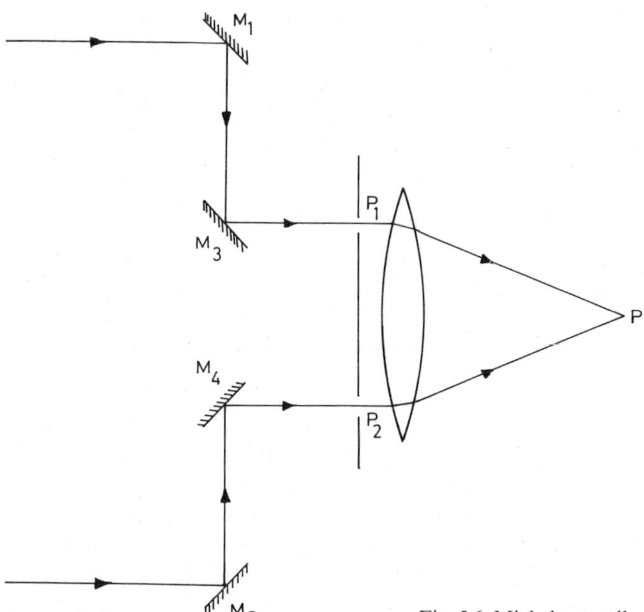

Fig. 5.6. Michelson stellar interferometer setup.

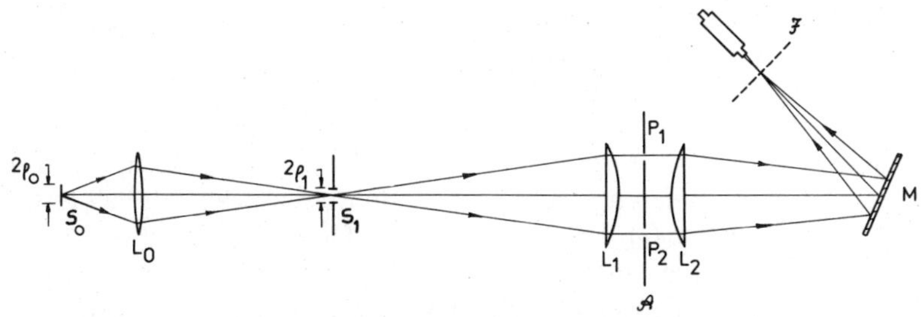

Fig. 5.7. The optical diffractometer arrangement for studying the variation of μ_{12} with v (adapted from Thompson and Wolf, 1957).

separation between M_1 and M_2, one can obtain μ_{12}, from which the angular diameter of the star can be determined. The need for the mirror combination M_1M_3, M_2M_4 can be seen from the following consideration. Michelson measured the angular diameter of the star Betelgeuse to be 0.047 sec of arc. Assuming $\lambda_0 = 5500$ Å, the separation between the two pinholes (when the mirrors are absent) for minimum visibility is about 2.9 m. It is extremely difficult and costly to make telescopes of such diameters. To overcome this difficulty, Michelson set up the mirror arrangement shown in Fig. 5.6. In this arrangement ρ is the separation between the two mirrors, and since ρ can be varied over large values (to about 6 m), the resulting arrangement can measure even smaller angular diameters.

Figure 5.7 shows an experimental arrangement called the diffractometer employed by Thompson and Wolf (1957) and Thompson (1958) for studying the variation of μ_{12} with the diameter of the source and the separation between points P_1 and P_2. S_0 is an incoherent quasi-monochromatic source of light* and the lens L_0 focuses the rays emanating from S_0 onto the hole S_1. The rays emanating from S_1 are made parallel by a lens L_1. These rays are made to pass through a set of two holes placed in the plane $\mathcal{A}$, and the lens L_2 forms the Fraunhofer diffraction pattern in its focal plane. The diameter of the hole S_1 was large compared to λ/α, which made the hole S_1 act as an incoherent source†; here α is the angle subtended by the lens at S_1 (see Fig. 5.7).

* For the source S_0, Thompson and Wolf (1957) used a mercury lamp in combination with a filter that selected the yellow doublet with mean wavelength $\bar{\lambda} = 5790$ Å. The passband of the filter was about 400 Å centered on $\bar{\lambda}$.

† This can be understood as follows: If S_0 were a point source then in the image plane, one would obtain the Airy pattern (see Section 4.7) and the radius of the first Airy ring would be $1.22\lambda/\alpha$. Clearly, for an extended source, the hole S_1 will act as an incoherence source if its radius is much greater than $1.22\lambda/\alpha$. In the experiment by Thompson (1958), the diameter of the Airy disk was about 4 μm, which was (1/25)th the diameter of the smallest pinhole and (1/120)th of the largest.

The intensity distribution on the plane F is given by [see Eq. (5.4-17)]

$$I = I_1 + I_2 + 2(I_1 I_2)^{1/2} |\mu_{12}| \cos(\beta_{12} + \varphi) \qquad (5.5\text{-}16)$$

where I_1 and I_2 represent the intensities produced, respectively, by P_1 and P_2 independently; μ_{12} is the complex degree of coherence [see Eq. (5.5-13)]; and φ is the phase difference between the interfering beams. It can also be seen that $\beta_{12} = 0$ if $J_1(v)/v > 0$ and $\beta_{12} = \pi$ if $J_1(v)/v < 0$.

It has already been shown (see Section 4.7) that if ψ is the angle made with the normal to the pinholes P_1 and P_2 at which the rays proceed to reach the point P, then*

$$I_1(P) = I_2(P) = I_0[2J_1(u)/u]^2 \qquad (5.5\text{-}17)$$

where

$$u = \frac{2\pi\nu_0}{c} b \sin\psi \qquad (5.5\text{-}18)$$

b being the radius of the holes at P_1 and P_2. Thus Eq. (5.5-16) reduces to

$$I(P) = 2I_0\left[\frac{2J_1(u)}{u}\right]^2\left[1 + \left|\frac{2J_1(v)}{v}\right| \cos\left(\beta_{12} - \frac{2\pi\nu_0}{c}\rho\sin\psi\right)\right] \qquad (5.5\text{-}19)$$

where we have used the relation $\varphi = (2\pi\nu_0/c)\,\rho\sin\psi$.

It can be seen from Eq. (5.5-14) that by varying either a, the radius of the sources, or ρ, the separation between P_1 and P_2, one can vary the value of v and thus the complex degree of coherence. It can also be seen from Eq. (5.5-19) that when $\beta_{12} = 0$, then the central fringe corresponding to $\psi = 0$ is a bright fringe and when $\beta_{12} = \pi$ the central fringe is a dark fringe. Thus, as the value of a (or ρ) is increased, the visibility goes on decreasing, the central fringe being a bright fringe. At $v = 3.833$ the visibility becomes zero, i.e., the interference fringes disappear. As v is increased further, the visibility again starts increasing; however, for $7.016 < v < 3.833$, $\beta_{12} = \pi$ and the central fringe now becomes a dark fringe. The visibility exhibits an oscillatory character.

In Fig. 5.8 we give the results of the experiment performed by Thompson (1958). It can be clearly seen that in Fig. 5.8b, which corresponds to a value of v between the first and second zeroes of $2J_1(v)/v$, the central fringe is dark. Theoretical calculations are also shown in the figure. Figure 5.9 shows the variation of μ_{12} as a function of ρ for various values of the source radius a.

* It may be mentioned that Eq. (5.5-17) is valid only when the spatial dimension of the source S_1 is small (see Problem 5.5).

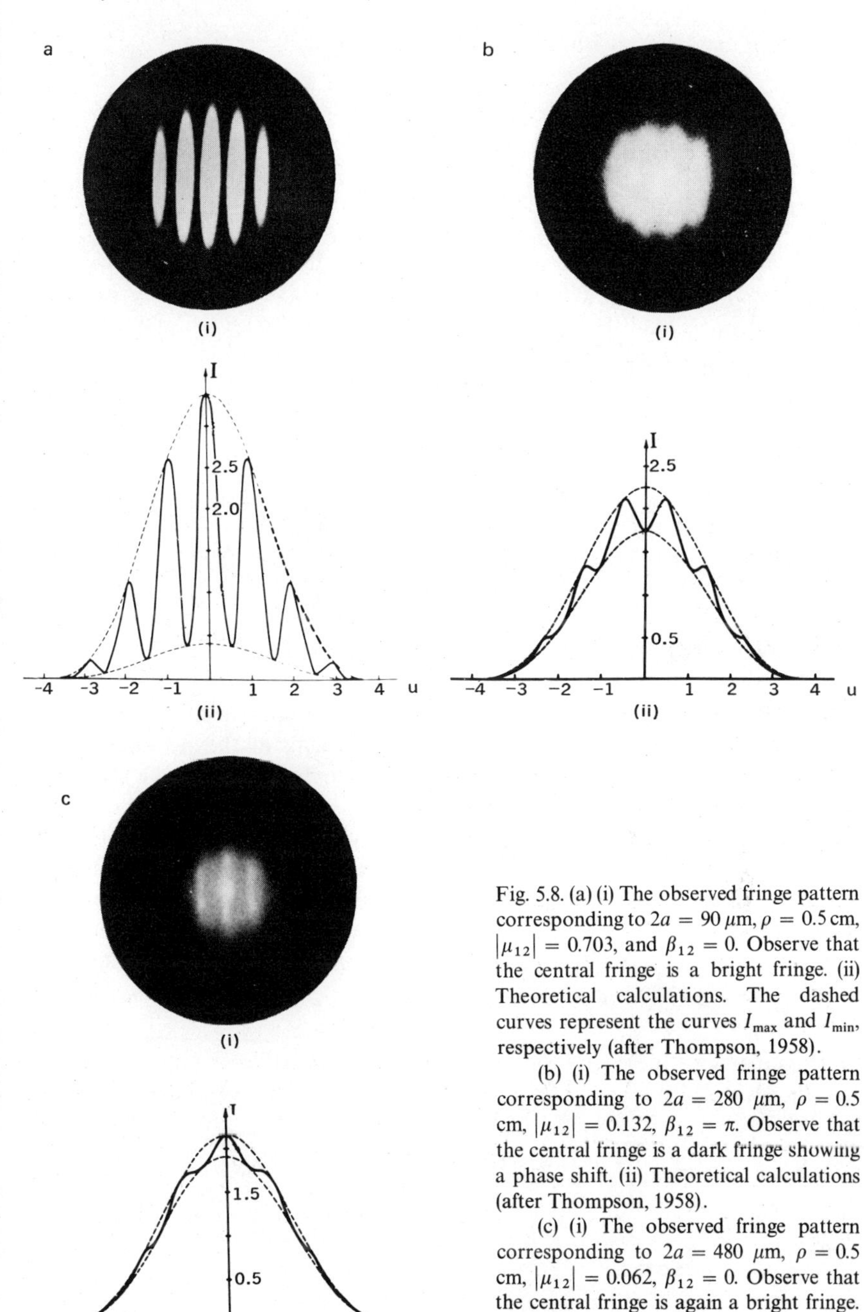

Fig. 5.8. (a) (i) The observed fringe pattern corresponding to $2a = 90\ \mu\text{m}$, $\rho = 0.5$ cm, $|\mu_{12}| = 0.703$, and $\beta_{12} = 0$. Observe that the central fringe is a bright fringe. (ii) Theoretical calculations. The dashed curves represent the curves $I_{\max}$ and $I_{\min}$, respectively (after Thompson, 1958).

(b) (i) The observed fringe pattern corresponding to $2a = 280\ \mu\text{m}$, $\rho = 0.5$ cm, $|\mu_{12}| = 0.132$, $\beta_{12} = \pi$. Observe that the central fringe is a dark fringe showing a phase shift. (ii) Theoretical calculations (after Thompson, 1958).

(c) (i) The observed fringe pattern corresponding to $2a = 480\ \mu\text{m}$, $\rho = 0.5$ cm, $|\mu_{12}| = 0.062$, $\beta_{12} = 0$. Observe that the central fringe is again a bright fringe. (ii) Theoretical calculations (after Thompson, 1958).

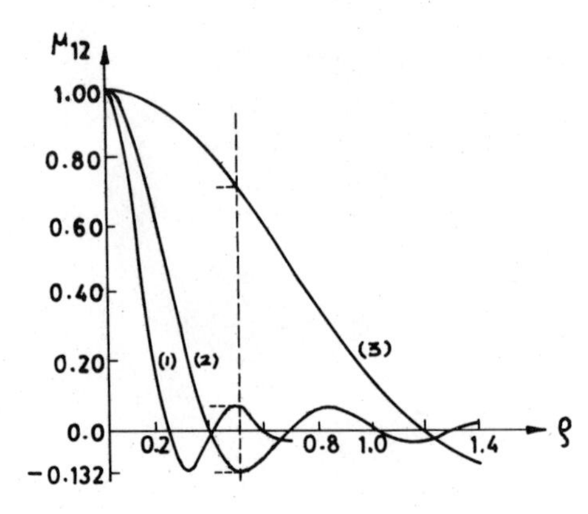

Fig. 5.9. Variation of μ_{12} with ρ for (1) $a = 480\ \mu$m, (2) $a = 280\ \mu$m, and (3) $a = 90\ \mu$m. The vertical dashed line corresponds to $\rho = 0.5$ cm (after Thompson, 1958).

Problem 5.5. Show that $I_1(P)$ as given by Eq. (5.5-17) is valid if $\theta \ll \lambda/2b$, where θ is the angular diameter of the hole S_1 as seen from P_1 (or P_2). For the experimental arrangement of Thompson (1958) show that this condition is indeed satisfied.

5.6. Differential Equations Satisfied by $\Gamma_{12}(\tau)$

In this section we will show that the mutual coherence function $\Gamma_{12}(\tau)$ satisfies two wave equations. Thus, one can speak of the propagation of mutual coherence from one plane to another, analogous to the propagation of the electromagnetic field itself.

Since $\Gamma_{12}(\tau)$ is defined in terms of the analytic signals $V_1(t)$ and $V_2(t)$ [see Eq. (5.3-11)], we must first determine the differential equations satisfied by the analytic signal. Since $V^{\mathrm{r}}(t)$ is the real amplitude, it satisfies the scalar wave equation

$$\nabla^2 V^{\mathrm{r}} - \frac{1}{c^2}\frac{\partial^2 V^{\mathrm{r}}}{\partial t^2} = 0 \tag{5.6-1}$$

If we substitute the expression for $V^{\mathrm{r}}(t)$ from Eq. (5.2-1) into Eq. (5.6-1), we obtain

$$\nabla^2 V^{\mathrm{r}} - \frac{1}{c^2}\frac{\partial^2 V^{\mathrm{r}}}{\partial t^2} = \int_{-\infty}^{+\infty}\left[\nabla^2 \tilde{V}^{\mathrm{r}}(\nu) + \frac{4\pi^2\nu^2}{c^2}\tilde{V}^{\mathrm{r}}(\nu)\right]e^{2\pi i\nu t}\,d\nu = 0 \tag{5.6-2}$$

If Eq. (5.6-2) has to be true for all values of t, then $\tilde{V}^{\mathrm{r}}(\nu)$ must satisfy the Helmholtz equation

$$\nabla^2 \tilde{V}^{\mathrm{r}}(\nu) + \frac{4\pi^2\nu^2}{c^2}\tilde{V}^{\mathrm{r}}(\nu) = 0 \tag{5.6-3}$$

Now using Eq. (5.2-9), we obtain

$$\nabla^2 V(t) - \frac{1}{c^2}\frac{\partial^2 V}{\partial t^2} = 2\int_0^\infty \left[\nabla^2 \tilde{V}^{\mathrm{r}}(\nu) + \frac{4\pi^2\nu^2}{c^2}\tilde{V}^{\mathrm{r}}(\nu)\right] e^{2\pi i\nu t}\, d\nu = 0 \quad (5.6\text{-}4)$$

Thus, the analytic signal $V(t)$ also satisfies the same equation as $V^{\mathrm{r}}(t)$. Hence,

$$\nabla_1^2 V_1(t) = \frac{1}{c^2}\frac{\partial^2 V_1}{\partial t^2}, \qquad \nabla_2^2 V_2(t) = \frac{1}{c^2}\frac{\partial^2 V_2}{\partial t^2} \quad (5.6\text{-}5)$$

where ∇_1^2 and ∇_2^2 are the Laplacian operators referred to the coordinates of V_1 and V_2, respectively. Let us write the mutual coherence function as

$$\Gamma_{12}(t_1, t_2) = \langle V_1(t - t_1)\, V_2^*(t - t_2)\rangle \quad (5.6\text{-}6)$$

Thus we have

$$\nabla_1^2\Gamma_{12}(t_1,t_2) = \nabla_1^2\langle V_1(t - t_1)\, V_2^*(t - t_2)\rangle = \langle[\nabla_1^2 V_1(t - t_1)]\, V_2^*(t - t_2)\rangle \quad (5.6\text{-}7)$$

since V_2 is independent of the coordinates of V_1. Thus, using Eq. (5.6-5), we get

$$\begin{aligned}\nabla_1^2\Gamma_{12}(t_1, t_2) &= \frac{1}{c^2}\left\langle \frac{\partial^2 V_1(t - t_1)}{\partial(t - t_1)^2}\, V_2^*(t - t_2)\right\rangle \\ &= \frac{1}{c^2}\left\langle \frac{\partial^2 V_1(t - t_1)}{\partial t_1^2}\, V_2^*(t - t_2)\right\rangle \\ &= \frac{1}{c^2}\frac{\partial^2}{\partial t_1^2}\Gamma_{12}(t_1, t_2) \end{aligned} \quad (5.6\text{-}8)$$

In an exactly similar manner, we obtain

$$\nabla_2^2\Gamma_{12}(t_1, t_2) = \frac{1}{c^2}\frac{\partial^2}{\partial t_2^2}\Gamma_{12}(t_1, t_2) \quad (5.6\text{-}9)$$

For stationary fields $\Gamma_{12}(t_1, t_2)$ depends only on the time difference $t_2 - t_1 = \tau$, and since

$$\frac{\partial^2}{\partial t_1^2} = \frac{\partial^2}{\partial \tau^2} = \frac{\partial^2}{\partial t_2^2}$$

we obtain

$$\nabla_i^2\Gamma_{12}(t_1, t_2) = \nabla_i^2\Gamma_{12}(\tau) = \frac{1}{c^2}\frac{\partial^2\Gamma_{12}(\tau)}{\partial\tau^2}, \qquad i = 1, 2 \quad (5.6\text{-}10)$$

and thus, $\Gamma_{12}(\tau)$ satisfies two scalar wave equations.

For quasi-monochromatic fields and for small time delays, we have

$$\Gamma_{12}(\tau) = J_{12}\exp(2\pi i\nu_0\tau) \quad (5.6\text{-}11)$$

J_{12} being independent of τ. Thus substituting Eq. (5.6-11) into Eq. (5.6-10), we find that J_{12} satisfies two Helmholtz equations:

$$\nabla_i^2 J_{12} + \frac{4\pi^2 \nu_0^2}{c^2} J_{12} = 0, \qquad i = 1, 2 \tag{5.6-12}$$

These equations form the basis of a more rigorous theory of partial coherence. We will not go into the details of the rigorous theory; interested readers are referred to Beran and Parrent (1964) and Born and Wolf (1975).

5.7. Partial Polarization

5.7.1. The Coherency Matrix

In previous sections, we made use of the scalar approximation and specified the field by a scalar quantity. We studied the correlation between the values of the field at two different space–time points. In the present section we consider a plane wave propagating along the z direction and study its polarization property. The electric vector of such a field will lie in the x-y plane and therefore we need consider only the x and y components of the electric field.

By definition, a monochromatic wave is completely polarized, i.e., the tip of the electric vector rotates along the circumference of an ellipse. For a linearly polarized wave the ellipse degenerates into a straight line, while for a circularly polarized wave it degenerates into a circle. For unpolarized light, the state of polarization changes in a random manner with time. The polarized and unpolarized waves form two extreme cases and, in general, light is partially polarized so that the motion of the tip of the electric vector is neither completely regular nor completely irregular.

We will restrict ourselves to quasi-monochromatic radiation, i.e., the spectral width of the wave $\Delta\nu$ will be assumed to be much less than ν_0, the central frequency of the wave:

$$\Delta\nu \ll \nu_0$$

As in previous sections, we will use the analytic signal representation of the field. Let $E_x(\mathbf{r}, t)$ and $E_y(\mathbf{r}, t)$ represent the analytic signals associated with the x and y components of the real fields. We form the following column vector $\mathscr{E}(\mathbf{r}, t)$ with elements $E_x(\mathbf{r}, t)$ and $E_y(\mathbf{r}, t)$:

$$\mathscr{E}(\mathbf{r}, t) = \begin{pmatrix} E_x(\mathbf{r}, t) \\ E_y(\mathbf{r}, t) \end{pmatrix} \tag{5.7-1}$$

The Hermitian conjugate of $\mathscr{E}(\mathbf{r}, t)$ is the row matrix

$$\mathscr{E}^{\dagger}(\mathbf{r}, t) = \begin{pmatrix} E_x^*(\mathbf{r}, t) & E_y^*(\mathbf{r}, t) \end{pmatrix} \tag{5.7-2}$$

where the superscript † is used for the Hermitian conjugate and $*$ for the complex conjugate. From $\mathscr{E}(\mathbf{r}, t)$ and $\mathscr{E}^{\dagger}(\mathbf{r}, t)$ we may define a 2×2 square matrix $\mathbf{J}$ as

$$\begin{aligned}\mathbf{J} &= \langle \mathscr{E}(\mathbf{r}, t)\, \mathscr{E}^{\dagger}(\mathbf{r}, t) \rangle \\ &= \left\langle \begin{pmatrix} E_x(\mathbf{r}, t) \\ E_y(\mathbf{r}, t) \end{pmatrix} \begin{pmatrix} E_x^*(\mathbf{r}, t) & E_y^*(\mathbf{r}, t) \end{pmatrix} \right\rangle \\ &= \begin{pmatrix} J_{xx} & J_{xy} \\ J_{yx} & J_{yy} \end{pmatrix} \end{aligned} \tag{5.7-3}$$

where, as before, angular brackets denote time averaging (see Section 5.3) and

$$\begin{aligned} J_{xx} &= \langle E_x(\mathbf{r}, t)\, E_x^*(\mathbf{r}, t) \rangle \\ J_{xy} &= \langle E_x(\mathbf{r}, t)\, E_y^*(\mathbf{r}, t) \rangle \\ J_{yx} &= \langle E_y(\mathbf{r}, t)\, E_x^*(\mathbf{r}, t) \rangle \\ J_{yy} &= \langle E_y(\mathbf{r}, t)\, E_y^*(\mathbf{r}, t) \rangle \end{aligned} \tag{5.7-4}$$

The matrix $\mathbf{J}$ is called the *coherency matrix* and the components J_{xy} and J_{yx} measure the correlations existing between the x and y components of the electric vector at a point. Since $J_{xy} = J_{yx}^*$, we see that the matrix $\mathbf{J}$ is Hermitian, i.e.,

$$\mathbf{J} = \mathbf{J}^{\dagger} \tag{5.7-5}$$

Analogous to the complex degree of coherence $\gamma_{12}(\tau)$, which measures the degree of coherence between the field at two points, we introduce the quantity

$$\mu_{xy} = \frac{J_{xy}}{(J_{xx} J_{yy})^{1/2}} \tag{5.7-6}$$

which measures the degree of coherence between the two components of the field at a point. From the Schwarz inequality, one may easily prove that

$$0 \le |\mu_{xy}| \le 1 \tag{5.7-7}$$

As we will see shortly, the coherency matrix $\mathbf{J}$ is used because its elements can be measured.

It may be noted here that the total electric field at a point is

$$\mathbf{E} = E_x \hat{\mathbf{x}} + E_y \hat{\mathbf{y}} \tag{5.7-8}$$

where $\hat{\mathbf{x}}$ and $\hat{\mathbf{y}}$ represent unit vectors along the x and y directions, respectively. Thus the intensity is (see Section 5.3)

$$\begin{aligned} I &= \tfrac{1}{2}\langle \mathbf{E}\cdot\mathbf{E}^*\rangle \\ &= \tfrac{1}{2}\langle E_x E_x^*\rangle + \tfrac{1}{2}\langle E_y E_y^*\rangle \\ &= \tfrac{1}{2}\,\mathrm{Tr}\,\mathbf{J} \end{aligned} \tag{5.7-9}$$

where Tr represents the trace of the matrix, i.e., the sum of the diagonal elements of the matrix. The quantities $\frac{1}{2}J_{xx}$ and $\frac{1}{2}J_{yy}$ represent the intensity corresponding to the components of the field along the x and y directions.

Let us now consider the following experiment: We first introduce a phase shift of δ in the y component of the field with respect to the x component (say, with the help of a compensator). This beam is now sent through a polarizer oriented at an angle θ with the x axis. In such a case, the electric vector along the direction inclined at an angle θ to the x axis is

$$E = E_x \cos\theta + E_y e^{i\delta}\sin\theta \tag{5.7-10}$$

Therefore the intensity is

$$\begin{aligned} I(\theta,\delta) &= \tfrac{1}{2}\langle \mathbf{E}\cdot\mathbf{E}^*\rangle \\ &= \tfrac{1}{2}J_{xx}\cos^2\theta + \tfrac{1}{2}J_{yy}\sin^2\theta + \tfrac{1}{2}(J_{xy}e^{-i\delta} + J_{yx}e^{i\delta})\cos\theta\sin\theta \\ &= \tfrac{1}{2}J_{xx}\cos^2\theta + \tfrac{1}{2}J_{yy}\sin^2\theta + (J_{xx}J_{yy})^{1/2}|\mu_{xy}|\cos\theta\sin\theta\cos(\varphi_{xy}-\delta) \end{aligned} \tag{5.7-11}$$

where we have used Eq. (5.7-6) and

$$\mu_{xy} = |\mu_{xy}|\,e^{i\varphi_{xy}} \tag{5.7-12}$$

Observe that $I(\theta,\delta)$ depends on the various elements of the coherency matrix. In Section 5.7.2 we will use Eq. (5.7-11) for determining the values of the elements of the coherency matrix by determining the intensity of the beam for six different sets of values of θ and δ.

Let us now obtain the coherency matrix for unpolarized light. Unpolarized light is characterized by the fact that $I(\theta,\delta)$ is a constant, independent of θ and δ. For such a case, the x and y components of light are completely uncorrelated. For $I(\theta,\delta)$ to be a constant, we obtain from Eq. (5.7-11),

$$\mu_{xy} = 0 \qquad \text{or} \qquad J_{xy} = 0 \tag{5.7-13}$$

and

$$J_{xx} = J_{yy} \tag{5.7-14}$$

If I represents the total intensity, then

$$2I = J_{xx} + J_{yy} = 2J_{xx} = 2J_{yy} \tag{5.7-15}$$

Thus, the coherency matrix for unpolarized light can be written

$$\mathbf{J} = I\begin{pmatrix} 1 & 0 \\ 0 & 1 \end{pmatrix} \tag{5.7-16}$$

i.e., it is a multiple of the unit matrix.

Let us now consider a monochromatic field. As we observed earlier, such a field is completely polarized. Let

$$E_x = e_x \exp(2\pi i\nu_0 t + i\psi_x) \tag{5.7-17}$$

$$E_y = e_y \exp(2\pi i\nu_0 t + i\psi_y) \tag{5.7-18}$$

where e_x, e_y, ψ_x, and ψ_y are constants independent of time. For such a case, the coherency matrix becomes

$$\mathbf{J} = \begin{pmatrix} e_x^2 & e_x e_y \exp[i(\psi_x - \psi_y)] \\ e_x e_y \exp[-i(\psi_x - \psi_y)] & e_y^2 \end{pmatrix} \tag{5.7-19}$$

We observe that

$$\det \mathbf{J} = 0 \tag{5.7-20}$$

Consider a quasi-monochromatic field specified by

$$E_x = e_x(t) \exp[2\pi i\nu_0 t + i\psi_x(t)] \tag{5.7-21}$$

$$E_y = e_y(t) \exp[2\pi i\nu_0 t + i\psi_y(t)] \tag{5.7-22}$$

where $e_x(t)$, $e_y(t)$, $\psi_x(t)$, and $\psi_y(t)$ are slowly varying functions of time. If the time dependence is such that the ratio of $e_x(t)$ to $e_y(t)$ and the phase difference $[\psi_x(t) - \psi_y(t)]$ are constants (independent of time) then the light would be polarized and it can be seen that Eq. (5.7-20) is again satisfied. (For example, if quasi-monochromatic light is passed through a polarizer, such a state would result.) Thus the coherency matrix for a completely polarized radiation satisfies the condition

$$\det \mathbf{J} = J_{xx}J_{yy} - |J_{xy}|^2 = 0 \tag{5.7-23}$$

Problem 5.6. Obtain the coherency matrices representing a monochromatic wave that is (a) linearly polarized and (b) circularly polarized.

Solution. (a) If the polarization direction makes an angle θ with the x axis, then

$$E_x = a\cos\theta \exp(2\pi i\nu_0 t + i\varphi)$$

$$E_y = a\sin\theta \exp(2\pi i\nu_0 t + i\varphi)$$

where a and φ are constants. Hence

$$\mathbf{J} = \begin{pmatrix} a^2\cos^2\theta & a^2\cos\theta\sin\theta \\ a^2\cos\theta\sin\theta & a^2\sin^2\theta \end{pmatrix} \tag{5.7-24}$$

Since the intensity of the wave is equal to $\frac{1}{2}a^2$, we may also write

$$\mathbf{J} = 2I\begin{pmatrix} \cos^2\theta & \cos\theta\sin\theta \\ \cos\theta\sin\theta & \sin^2\theta \end{pmatrix} \tag{5.7-25}$$

For a light wave polarized along the x direction, $\theta = 0$ and

$$\mathbf{J} = 2I\begin{pmatrix} 1 & 0 \\ 0 & 0 \end{pmatrix} \tag{5.7-26}$$

(b) For a right circularly polarized wave, we may write

$$E_x = a'\exp(2\pi i\nu_0 t + i\varphi), \qquad E_y = ia'\exp(2\pi i\nu_0 t + i\varphi)$$

giving us

$$\mathbf{J} = \begin{pmatrix} a'^2 & -ia'^2 \\ ia'^2 & a'^2 \end{pmatrix}$$

$$= I\begin{pmatrix} 1 & -i \\ i & 1 \end{pmatrix} \tag{5.7-27}$$

where $I = a'^2$ is the intensity of the wave. A left circularly polarized wave is specified by

$$\mathbf{J} = I\begin{pmatrix} 1 & i \\ -i & 1 \end{pmatrix} \tag{5.7-28}$$

5.7.2. Degree of Polarization

We obtained in Section 5.7.1 the coherency matrices for completely polarized and completely unpolarized waves. As noted earlier, a wave is, in general, partially polarized. Let us now determine whether it is possible to write uniquely, a partially polarized wave as a superposition of a completely polarized and a completely unpolarized wave. Let us write

$$\mathbf{J} = \begin{pmatrix} J_{xx} & J_{xy} \\ J_{yx} & J_{yy} \end{pmatrix} = \mathbf{J}_1 + \mathbf{J}_2 \tag{5.7-29}$$

where

$$\mathbf{J}_1 = \begin{pmatrix} D & 0 \\ 0 & D \end{pmatrix} \tag{5.7-30}$$

$$\mathbf{J}_2 = \begin{pmatrix} A & C \\ C^* & B \end{pmatrix} \tag{5.7-31}$$

The matrix $\mathbf{J}_1$ is the coherency matrix representing the unpolarized part of the radiation and the matrix $\mathbf{J}_2$, with the condition

$$AB - |C|^2 = 0 \tag{5.7-32}$$

is the coherency matrix representing the polarized part of the radiation. From Eqs. (5.7-29)–(5.7-31) we can write

$$A + D = J_{xx}, \qquad C = J_{xy}, \qquad C^* = J_{yx}, \qquad B + D = J_{yy} \tag{5.7-33}$$

and

$$AB = |J_{xy}|^2 \tag{5.7-34}$$

We have to evaluate five quantities, A, B, D and the real and imaginary parts of C. Since we have five equations, the five unknowns can be determined uniquely.

The intensity associated with the polarized part is [see Eq. (5.7-9)]

$$I_p = \tfrac{1}{2}(A + B) \tag{5.7-35}$$

and that associated with the unpolarized part is

$$I_u = D \tag{5.7-36}$$

Hence the total intensity is

$$I_t = \tfrac{1}{2}(A + B + 2D) \tag{5.7-37}$$

We now define the degree of polarization P as the ratio of the intensity of the polarized part to that of the total intensity:

$$P = \frac{I_p}{I_t} = \frac{A + B}{A + B + 2D} \tag{5.7-38}$$

Using Eqs. (5.7-33) and (5.7-34), we obtain

$$P = \left[1 - \frac{4 \det \mathbf{J}}{(\mathrm{Tr}\,\mathbf{J})^2}\right]^{1/2} \tag{5.7-39}$$

where $\det \mathbf{J}$ represents the determinant of the coherency matrix

$$\det \mathbf{J} = J_{xx}J_{yy} - |J_{xy}|^2 \tag{5.7-40}$$

$$\mathrm{Tr}\,\mathbf{J} = J_{xx} + J_{yy} \tag{5.7-41}$$

Since J_{xx} and J_{yy} are positive quantities and $J_{xy}J_{yx}$ is also a positive quantity,

$$\det \mathbf{J} = J_{xx}J_{yy} - J_{xy}J_{yx} \leq J_{xx}J_{yy} \tag{5.7-42}$$

Also since

$$(J_{xx} - J_{yy})^2 \geq 0$$

or

$$(J_{xx} + J_{yy})^2 \geq 4J_{xx}J_{yy}$$

we obtain

$$\det \mathbf{J} \le \tfrac{1}{4}(J_{xx} + J_{yy})^2$$

or

$$4 \det \mathbf{J} \le (\operatorname{Tr} \mathbf{J})^2 \tag{5.7-43}$$

Thus

$$0 \le P \le 1 \tag{5.7-44}$$

For a polarized wave, $\det \mathbf{J} = 0$ and one obtains $P = 1$. For an unpolarized wave, $J_{xy} = J^*_{yx} = 0$ and $J_{xx} = J_{yy}$, giving $P = 0$. For P lying between 0 and 1, the wave is said to be partially polarized.

Observe that a rotation of the x and y axes leaves the quantities $\det \mathbf{J}$ and $\operatorname{Tr} \mathbf{J}$ unchanged. Thus the degree of polarization P is independent of the choice of the coordinate system.

Problem 5.7. Show that for a wave made up of many independent waves, the coherency matrix is the sum of the individual coherency matrices.

5.7.3. *Measurement of the Elements of* $\mathbf{J}$

The usefulness of the coherency matrix lies in the fact that its elements can be measured. Given a wave, we observed that if a phase change of δ is introduced in the y component relative to that of the x component, and that if we pass the radiation through a polarizer inclined at an angle θ to the x direction, then the intensity $I(\theta, \delta)$ is [see Eq. (5.7-11)],

$$I(\theta, \delta) = \tfrac{1}{2} J_{xx} \cos^2\theta + \tfrac{1}{2} J_{yy} \sin^2\theta + \tfrac{1}{2} J_{xy} e^{-i\delta} \cos\theta \sin\theta + \tfrac{1}{2} J_{yx} e^{i\delta} \cos\theta \sin\theta \tag{5.7-45}$$

By choosing specific values of θ and δ, and making six measurements, the elements of the coherency matrix can be determined. Thus, we have

$$J_{xx} = 2I(0, 0) \tag{5.7-46}$$

$$J_{yy} = 2I(\tfrac{1}{2}\pi, 0) \tag{5.7-47}$$

$$J_{xy} = [I(\tfrac{1}{4}\pi, 0) - I(\tfrac{3}{4}\pi, 0)] + i[I(\tfrac{1}{4}\pi, \tfrac{1}{2}\pi) - I(\tfrac{3}{4}\pi, \tfrac{1}{2}\pi)] \tag{5.7-48}$$

$$J_{yx} = [I(\tfrac{1}{4}\pi, 0) - I(\tfrac{3}{4}\pi, 0)] - i[I(\tfrac{1}{4}\pi, \tfrac{1}{2}\pi) - I(\tfrac{3}{4}\pi, \tfrac{1}{2}\pi)] \tag{5.7-49}$$

Hence by using a compensator and a polarizer, it is possible to determine the elements of the coherency matrix experimentally.

Problem 5.8. Show that $|\mu_{xy}| \le P$ and the equality sign will hold when $J_{xx} = J_{yy}$, i.e., if the time-averaged intensities along the x and y directions are equal. Thus, the maximum value of $|\mu_{xy}|$ is just equal to the degree of polarization.

Hint:

$$1 - P^2 = \frac{4J_{xx}J_{yy}}{(J_{xx} + J_{yy})^2}[1 - |\mu_{xy}|^2] \tag{5.7-50}$$

5.7.4. Optical Devices

Having discussed the various properties of the coherency matrix, we will now determine how the coherency matrix becomes altered as the partially polarized beam passes through various devices such as a polarizer, a rotator, a compensator, or an absorber. These are elements found in many optical systems and the matrix treatment forms an elegant approach for the evaluation of the effect of these elements on the polarization property of the incident radiation. Since we are restricted to quasi-monochromatic approximation, we will assume that all the frequency components are altered similarly, although the elements through which the radiation passes are frequency dependent.

Let $\mathscr{E}$ represent the column vector of the incident field and $\mathscr{E}'$ that of the emerging field. Then if E'_x and E'_y represent the components of the emerging wave, we may write

$$\begin{aligned} E'_x &= AE_x + BE_y \\ E'_y &= CE_x + DE_y \end{aligned} \tag{5.7-51}$$

where the elements have been assumed to act linearly on the wave components. Equation (5.7-51) can be rewritten as

$$\mathscr{E}' = \mathscr{D}\mathscr{E} \tag{5.7-52}$$

where

$$\mathscr{D} = \begin{pmatrix} A & B \\ C & D \end{pmatrix} \tag{5.7-53}$$

represents the matrix corresponding to the element. Thus, the coherency matrix for the emerging wave is

$$\begin{aligned} \mathbf{J}' &= \langle \mathscr{E}'\mathscr{E}'^{\dagger} \rangle \\ &= \langle \mathscr{D}\mathscr{E}\mathscr{E}^{\dagger}\mathscr{D}^{\dagger} \rangle \\ &= \mathscr{D}\mathbf{J}\mathscr{D}^{\dagger} \end{aligned} \tag{5.7-54}$$

where, as before

$$\mathbf{J} = \langle \mathscr{E}\mathscr{E}^{\dagger} \rangle \tag{5.7-55}$$

represents the coherency matrix of the incident wave. Hence if the matrix $\mathscr{D}$ corresponding to each element is determined, the coherency matrix corresponding to the radiation passing through the element is simply given by Eq. (5.7-54).

Let us consider some specific examples.

Polarizer. A (linear) polarizer is an element which passes only the component of the field along a particular direction. Let this direction make an angle θ with the x axis. The incident field is specified by electric field components E_x and E_y along the x and y directions, respectively. As the polarizer passes only the field component in the direction inclined at θ to the x axis, the field emerging from the polarizer has a magnitude

$$E = E_x \cos\theta + E_y \sin\theta \tag{5.7-56}$$

and is inclined at an angle θ to the x axis. The x and y components of this field form the x and y components of the emerging field. Thus

$$\begin{aligned} E'_x &= E\cos\theta = E_x\cos^2\theta + E_y \sin\theta\cos\theta \\ E'_y &= E\sin\theta = E_x\cos\theta\sin\theta + E_y\sin^2\theta \end{aligned} \tag{5.7-57}$$

Thus, the polarizer can be represented by the matrix

$$\mathscr{D}_{\mathrm{P}} = \begin{pmatrix} \cos^2\theta & \sin\theta\cos\theta \\ \sin\theta\cos\theta & \sin^2\theta \end{pmatrix} \tag{5.7-58}$$

Compensator. The compensator is a device that introduces a relative phase difference between the two electric field components. If δ represents the relative phase difference, one may write

$$E'_x = E_x e^{i\delta/2}, \qquad E'_y = E_y e^{-i\delta/2} \tag{5.7-59}$$

and thus the matrix representing the compensator is

$$\mathscr{D}_{\mathrm{C}} = \begin{pmatrix} e^{i\delta/2} & 0 \\ 0 & e^{-i\delta/2} \end{pmatrix} \tag{5.7-60}$$

Rotator. A rotator is a device that produces a rotation of the plane of polarization. Clearly, if θ represents the angle of rotation, then the electric field components of the emerging beam are

$$E'_x = E_x\cos\theta + E_y\sin\theta, \qquad E'_y = -E_x\sin\theta + E_y\cos\theta \tag{5.7-61}$$

so that the matrix representing the rotator can be written

$$\mathscr{D}_R = \begin{pmatrix} \cos\theta & \sin\theta \\ -\sin\theta & \cos\theta \end{pmatrix} \tag{5.7-62}$$

Absorber. An absorber acts to decrease the field strength and if ζ_x and ζ_y represent the absorption coefficients for the x and y components, respectively, then we may write

$$E'_x = E_x e^{-\zeta_x/2}, \qquad E'_y = E_y e^{-\zeta_y/2} \tag{5.7-63}$$

Hence the operator for the absorber is

$$\mathscr{D}_{\mathrm{A}} = \begin{pmatrix} e^{-\zeta_x/2} & 0 \\ 0 & e^{-\zeta_y/2} \end{pmatrix} \tag{5.7-64}$$

In each of the above cases, the coherency matrix for the emerging wave is given by Eq. (5.7-54), with $\mathscr{D}$ representing the corresponding matrix. The effect of a set of these elements may easily be evaluated by successive application of the above matrices.

Problem 5.9. Consider a system consisting of a compensator followed by a polarizer. Obtain the coherency matrix of the radiation emerging from the system and show that the intensity is given by Eq. (5.7-11).

Problem 5.10. Noting that the trace of the coherency matrix is twice the intensity of the radiation, show that the intensity remains unchanged if the radiation passes through a compensator or a rotator, and that the intensity changes if it passes through a polarizer or an absorber.

6

Fourier Optics I. Spatial Frequency Filtering

6.1. Introduction

We will show in this chapter that the amplitude distribution in the back focal plane of an aberrationless lens is the Fourier transform* of the amplitude distribution in the front focal plane. This remarkable property has many interesting applications and forms the fundamental principle underlying the subject of spatial frequency filtering.

Just as in the time domain, where the Fourier transform of a time-varying signal gives the temporal frequency content in the signal, the Fourier transform of a spatially varying signal (namely, the object) gives the spatial frequency content in the object. For example, an object with an amplitude variation of the form

$$f(x) = A \cos(2\pi x/a) \tag{6.1-1}$$

is said to possess a spatial frequency $1/a$ (spatial period is a). Since the Fourier transform of a cosine function is a sum of two delta functions (see Section 6.4), an object with an amplitude distribution of the form Eq. (6.1-1), when placed in the front focal plane of a lens, produces two bright spots at $u = 1/a$ and $u = -1/a$ in the back focal plane of the lens. Here u is the Fourier transform variable and, as will be shown later, $u = x/\lambda f$, where f is the focal length of the lens and λ is the wavelength of the radiation in the medium surrounding the lens. Thus the two spots are produced at $x = \lambda f/a$ and $x = -\lambda f/a$. In general, the spatial frequency content of the object is displayed in the back focal plane of the lens. If we now place a second lens such that its front focal plane is coincident with the back

* Before going through this chapter, the reader should be familiar with Appendix B on Fourier transforms.

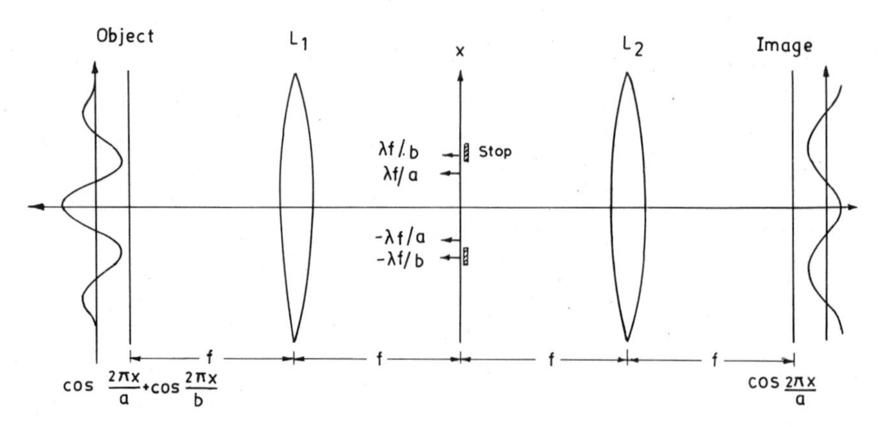

Fig. 6.1. When an object with its transmittance proportional to Eq. (6.1-2) is placed in the front focal plane of a lens, then in its back focal plane one obtains four bright spots, situated at $x = \pm\lambda f/a$ and $x = \pm\lambda f/b$. If appropriate stops are put in this plane to prevent the light coming from the two spots at $x = \pm\lambda f/b$, then in the back focal plane of the second lens one obtains a field pattern proportional to $\cos(2\pi x/a)$. Thus the frequency component $1/b$ has been filtered.

focal plane of the first lens, then on the back focal plane of the second lens we will obtain an image with amplitude variation of the form given by Eq. (6.1-1).* In order to appreciate this property of the lens, we consider an object field distribution of the form

$$f(x) = A\cos(2\pi x/a) + B\cos(2\pi x/b) \tag{6.1-2}$$

Thus, in the back focal plane of the first lens, we obtain four spots at $x = +\lambda f/a$, $-\lambda f/a$, $+\lambda f/b$, and $-\lambda f/b$. If we now put appropriate stops to prevent the light coming from the points $x = \pm\lambda f/b$ to reach the second lens, then in the back focal plane of the second lens we will obtain a field pattern proportional to $\cos(2\pi x/a)$. Thus we have filtered out the frequency component $1/b$ (see Fig. 6.1). This is the basic principle of spatial frequency filtering.

In Section 6.2 we recapitulate the formulas for Fresnel and Fraunhofer diffraction. In Section 6.3 we study the effect of a thin lens on the phase distribution of an incident field. In Section 6.4 we explicitly demonstrate the Fourier-transforming property of a lens. In Section 6.5 we consider a number of applications of the Fourier-transforming property of the lens.

* This follows from the fact that the Fourier transform of the Fourier transform of a function is the original function itself except for an inversion [see Eq. (B-5)].

6.2. Fraunhofer and Fresnel Diffraction Approximations

In Section 4.11 we showed that if $f(x, y)$ represents the field distribution in a plane, say $z = 0$, then the field distribution $g(x, y)$ in a plane z is, under the Fresnel approximation, given by

$$g(x, y) = -\frac{\exp(-ikz)}{i\lambda z} \iint_{-\infty}^{+\infty} f(x_1, y_1) \times \exp\left\{ -\frac{ik}{2z}[(x - x_1)^2 + (y - y_1)^2] \right\} dx_1\, dy_1 \quad (6.2\text{-}1)$$

where $k = 2\pi/\lambda$ and the integration is performed over the initial plane $z = 0$. Now, the convolution of two functions $f_1(x)$ and $f_2(x)$ is defined through the following relation (see Appendix B):

$$f_1(x) * f_2(x) \equiv \int_{-\infty}^{\infty} f_1(x')\, f_2(x - x')\, dx' \quad (6.2\text{-}2)$$

Thus

$$g(x, y) = f(x, y) * h(x, y) \quad (6.2\text{-}3)$$

where

$$h(x, y) = -\frac{\exp(-ikz)}{i\lambda z} \exp\left[-\frac{ik}{2z}(x^2 + y^2) \right] \quad (6.2\text{-}4)$$

Consequently the effect of propagation over space is to convolute the initial field distribution with $h(x, y)$. Returning to Eq. (6.2-1), we see that under Fraunhofer approximation (i.e., neglecting terms quadratic in x_1 and y_1) the field $g(x, y)$ is given by

$$g(x, y) = -\frac{e^{-ikz}}{i\lambda z} \exp\left[-\frac{ik}{2z}(x^2 + y^2) \right] \iint_{-\infty}^{+\infty} f(x_1, y_1) \times \exp\left[\frac{2\pi i}{\lambda z}(xx_1 + yy_1) \right] dx_1\, dy_1 \quad (6.2\text{-}5)$$

It can be seen that the integral represents the Fourier transform of $f(x, y)$. Thus

$$g(x, y) = -\frac{e^{-ikz}}{i\lambda z} \exp\left[-\frac{ik}{2z}(x^2 + y^2) \right] F\left(\frac{x}{\lambda z}, \frac{y}{\lambda z} \right) \quad (6.2\text{-}6)$$

where

$$F(u) = \mathscr{F}[f(x)] = \int_{-\infty}^{+\infty} f(x)\, e^{2\pi i u x}\, dx \tag{6.2-7}$$

represents the Fourier transform of $f(x)$.* The symbol $\mathscr{F}$ represents the Fourier transform of its argument. The function $F(x/\lambda z, y/\lambda z)$ represents the Fourier transform of $f(x, y)$ evaluated at the spatial frequencies $u = x/\lambda z$ and $v = y/\lambda z$.

6.3. Effect of a Thin Lens on an Incident Field Distribution†

A lens is called thin if a ray incident at a point (x, y) on one surface of the lens emerges at approximately the same height from the other surface, z being along the axis of the lens. Thus, if absorption and reflection are neglected, a thin lens just introduces a phase delay, which is a function of the coordinates x and y. We will now determine this phase delay.

Consider an object point O at a distance d_1 from an aberrationless thin lens of focal length f (see Fig. 6.2). We know that the lens, under geometrical optics approximation, images the point O at a point I that is at a distance d_2 $[=d_1 f/(d_1 - f)]$ from the lens. The phase factor corresponding to the disturbance emanating from the point O is simply $\exp(-ikr)$, where r is the distance measured from the point O. The phase distribution in a transverse plane at a distance d_1 from the point O (i.e., immediately in front of the lens) is given by

$$\exp[-ik(x^2 + y^2 + d_1^2)^{1/2}] \simeq \exp\left[-ik\left(d_1 + \frac{x^2 + y^2}{2d_1}\right)\right] \tag{6.3-1}$$

where in writing the last expression, we have assumed $x, y \ll d_1$, i.e., we have confined ourselves to a region close to the axis of the lens. This is known as the paraxial approximation. Since the image is formed at I, the incident spherical wave emerges as another spherical wave of radius d_2, which under the paraxial approximation is

$$\exp\left[+ik\left(d_2 + \frac{x^2 + y^2}{2d_2}\right)\right] \tag{6.3-2}$$

* We will denote the Fourier transform of a function, say $f(x)$, by either $\mathscr{F}[f(x)]$ or by the corresponding capital letter, namely, F.

† In Section 4.15, we evaluated the effect of a concave mirror of radius of curvature R on an incident field distribution.

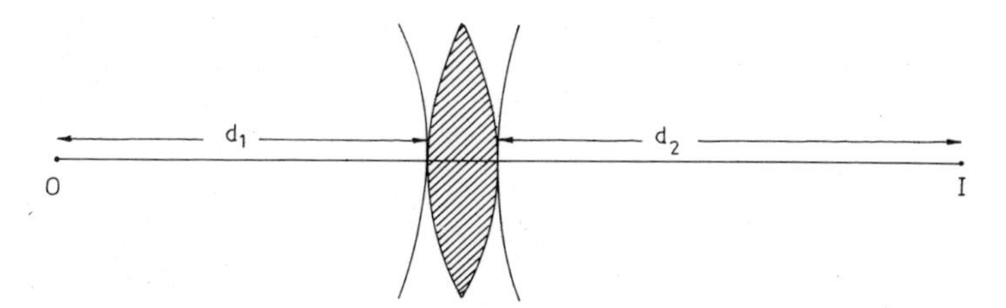

Fig. 6.2. Spherical waves emanating from an object point O, after refraction through a convex lens, emerge as spherical waves converging to the image point I.

The positive sign refers to the fact that we have a converging spherical wave. Thus if p_L represents the factor that when multiplied to the incident phase distribution gives the phase distribution of the emergent wave, then

$$\exp\left[ik\left(d_2 + \frac{x^2 + y^2}{2d_2}\right)\right] = \exp\left[-ik\left(d_1 + \frac{x^2 + y^2}{2d_1}\right)\right] p_L$$

or

$$p_L = \exp\left[ik(d_1 + d_2)\right]\exp\left[\frac{ik}{2}\left(\frac{1}{d_1} + \frac{1}{d_2}\right)(x^2 + y^2)\right] \qquad (6.3\text{-}3)$$

(The subscript L on p corresponds to the fact that we are referring to a lens.) Since the point I is the image of the point O, d_1 and d_2 satisfy the lens law

$$\frac{1}{d_1} + \frac{1}{d_2} = \frac{1}{f} \qquad (6.3\text{-}4)$$

Hence, neglecting the first factor in Eq. (6.3-3), because it is independent of x and y, we obtain

$$p_L = \exp\left[ik\frac{(x^2 + y^2)}{2f}\right] \qquad (6.3\text{-}5)$$

Thus the effect of a thin lens on an incident field is to multiply the incident phase distribution by a factor that is given by Eq. (6.3-5). For a plane wave incident along the axis, the emerging disturbance will simply be p_L, which can be seen to be the paraxial approximation of a spherical wavefront of radius f. Thus, as expected, an incident plane wave is focused to a point that is at a distance f from the lens. Although the formula for p_L has been derived assuming a convex lens, the effects of other types of lenses are also given by Eq. (6.3-5). For example, for a concave lens, f is negative; thus an incident plane wave emerges from the lens as a diverging spherical wave and hence forms a virtual image at a distance f from the lens.

In order to take into account the finite transverse dimension of the lens, we must multiply the emergent phase distribution [Eq. (6.3-2)] by a function $p(x, y)$, which is defined by the equation

$$p(x, y) = \begin{cases} 1 & \text{for } x^2 + y^2 < a^2 \\ 0 & \text{for } x^2 + y^2 > a^2 \end{cases} \tag{6.3-6}$$

where a is the radius of the lens. The function $p(x, y)$ is called the pupil function of the lens. Thus the effect of a lens of finite radius a is to multiply the initial phase distribution by the factor

$$p(x, y) \exp\left[\frac{ik}{2f}(x^2 + y^2)\right]$$

Problem 6.1. The varying thickness of a lens is responsible for the introduction of the phase shift, which depends on x and y. Show, by an explicit calculation of the thickness variation, that the lens introduces a phase shift given by Eq. (6.3-5) (apart from a constant phase factor). Assume the refractive index of the surrounding medium to be unity.

Solution. Let R_1 and R_2 be the radii of curvatures of the two surfaces of the lens and let n be the refractive index of the medium of the lens. A ray incident along the axis at the point A traverses a distance t_0 (see Fig. 6.3a). The phase shift introduced on such a ray is knt_0. On the other hand, a paraxial ray passing through

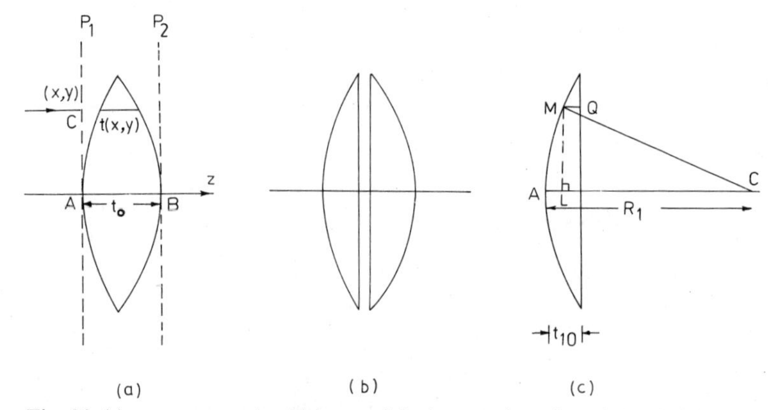

Fig. 6.3. (a) t_0 represents the thickness of the lens on the axis and $t(x, y)$ the thickness at an arbitrary value of x and y; the z axis is along the axis of the lens. (b) Lens broken in two parts. (c) Thickness of first portion.

the point C [whose transverse coordinates are (x, y)] suffers a phase shift of

$$k[t_0 - t(x, y)] + knt(x, y) = kt_0 + k(n-1)t(x, y)$$

in traveling from the plane P_1 to P_2 (see Fig. 6.3a); here $t(x, y)$ represents the thickness of the lens and we have assumed that the rays are traveling approximately parallel to the axis.

In order to calculate the function $t(x, y)$, we break the lens into two parts as shown in Fig. 6.3b. Thus the thickness of the first portion as a function of x and y is, from Fig. 6.3c,

$$\begin{aligned} t_1(x, y) = MQ = t_{10} - AL &= t_{10} - (R_1 - LC) = t_{10} - R_1 + (MC^2 - ML^2)^{1/2} \\ &= t_{10} - R_1 + [R_1^2 - (x^2 + y^2)]^{1/2} \end{aligned} \tag{6.3-7}$$

where t_{10} is the thickness defined in Fig. 6.3c. Under paraxial approximation, we can expand the square root in Eq. (6.3-7) to obtain

$$t_1(x, y) \simeq t_{10} - \frac{x^2 + y^2}{2R_1} \tag{6.3-8}$$

Similarly, the thickness of the second portion is

$$t_2(x, y) \simeq t_{20} + \frac{x^2 + y^2}{2R_2} \tag{6.3-9}$$

where R_2 is negative for a concave surface; thus the total thickness function becomes

$$t(x, y) = t_1(x, y) + t_2(x, y) \simeq t_0 - \frac{x^2 + y^2}{2}\left(\frac{1}{R_1} - \frac{1}{R_2}\right) \tag{6.3-10}$$

where $t_0 = t_{10} + t_{20}$. Thus the phase shift introduced by the lens is

$$\begin{aligned} -\left\{kt_0 + k(n-1)\left[t_0 - \frac{x^2 + y^2}{2}\left(\frac{1}{R_1} - \frac{1}{R_2}\right)\right]\right\} \\ = -knt_0 + k(n-1)\frac{x^2 + y^2}{2}\left(\frac{1}{R_1} - \frac{1}{R_2}\right) \\ = -knt_0 + k\frac{x^2 + y^2}{2f} \end{aligned} \tag{6.3-11}$$

where the focal length f is related to n, R_1, and R_2 by

$$\frac{1}{f} = (n-1)\left(\frac{1}{R_1} - \frac{1}{R_2}\right) \tag{6.3-12}$$

Thus apart from a constant phase factor, the lens multiplies any incident field by the same phase factor as given by Eq. (6.3-5).

6.4. Lens as a Fourier-Transforming Element

Consider an object having a field distribution $f(x, y)$ placed at a distance d_1 from a lens L (see Fig. 6.4). The function $f(x, y)$ could represent the transmission function of the object. If a plane wave illuminates this object, then the field distribution immediately behind the object is just $f(x, y)$. Let us find the field distribution* in a second plane, which is at a distance d_2 from the lens. Now, the field distribution in plane P_3 (see Fig. 6.4) that is produced due to the Fresnel diffraction of the field $f(x, y)$ from plane P_1 to P_3 is given by [see Eq. (6.2-3)]

$$\frac{1}{\lambda d_1} f(x, y) * \exp[-i\alpha(x^2 + y^2)]$$

where $\alpha = k/2d_1$ and the constant phase factor $[i \exp(-ikd_1)]$ has been omitted. Using the analysis of Section 6.3, the field distribution in plane P_4 is given by

$$\frac{1}{\lambda d_1} \{f(x, y) * \exp[-i\alpha(x^2 + y^2)]\} \exp[i\beta(x^2 + y^2)]$$

where $\beta = k/2f$. Finally, using Eq. (6.2-3), the field distribution at plane P_2 is given by

$$g(x, y) = \frac{1}{\lambda^2 d_1 d_2}(\{f(x, y) * \exp[-i\alpha(x^2 + y^2)]\} \exp[i\beta(x^2 + y^2)]) * \exp[-i\gamma(x^2 + y^2)] \quad (6.4\text{-}1)$$

or explicitly

$$g(x, y) = \frac{1}{\lambda^2 d_1 d_2} \iiiint_{-\infty}^{\infty} f(\xi, \eta) \exp\{-i\alpha[(\xi' - \xi)^2 + (\eta' - \eta)^2]\} \times \exp[i\beta(\xi'^2 + \eta'^2)] \exp\{-i\gamma[(x - \xi')^2 + (y - \eta')^2]\}\, d\xi\, d\eta\, d\xi'\, d\eta' \quad (6.4\text{-}2)$$

where $\gamma = k/2d_2$. Let us consider some important specific cases.

(1) If plane P_2 is the image plane of plane P_1 then d_1, d_2, and f are related through the lens law [see Eq. (6.3-4)]. Thus

$$\alpha + \gamma = \beta \quad (6.4\text{-}3)$$

* For the present we neglect the effect of the pupil function $p(x, y)$; thus $p(x, y) = 1$ for all values of x and y. This essentially implies that the effect of the field beyond the lens is negligible.

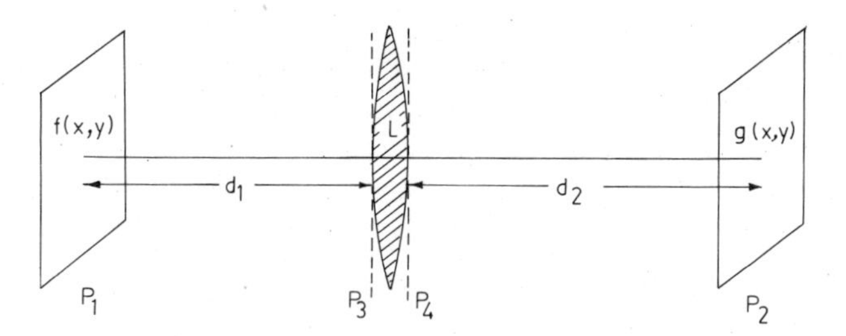

Fig. 6.4. A field distribution $f(x, y)$ placed at a distance d_1 in front of a lens produces a field distribution $g(x, y)$ in a plane P_2 at a distance d_2 from the lens. The field $f(x, y)$ first undergoes Fresnel diffraction from plane P_1 to P_3, then it gets multiplied by a phase factor due to the presence of the lens, and the resultant field again undergoes Fresnel diffraction from plane P_4 to P_2 to produce the field distribution $g(x, y)$.

Using Eq. (6.4-3), Eq. (6.4-2) simplifies to

$$g(x, y) = \frac{1}{\lambda^2 d_1 d_2} \exp[-i\gamma(x^2 + y^2)] \iint_{-\infty}^{\infty} d\xi\, d\eta\, f(\xi, \eta) \exp[-i\alpha(\xi^2 + \eta^2)]$$
$$\times \iint_{-\infty}^{\infty} d\xi'\, d\eta' \exp\left\{2\pi i\left[\xi'\left(\frac{\xi}{\lambda d_1} + \frac{x}{\lambda d_2}\right) + \eta'\left(\frac{\eta}{\lambda d_1} + \frac{y}{\lambda d_2}\right)\right]\right\} \tag{6.4-4}$$

The second double integral is just (see Appendix A)

$$\lambda^2 d_1^2\, \delta\left(\xi + \frac{x}{M}\right) \delta\left(\eta + \frac{y}{M}\right) \tag{6.4-5}$$

where $\delta(x)$ is the Dirac delta function and $M = d_2/d_1$ is the magnification of the system. Substituting this in Eq. (6.4-4) and after simplication, we obtain

$$g(x, y) = \frac{1}{M} f\left(-\frac{x}{M}, -\frac{y}{M}\right) \exp\left[-i\frac{\pi d_1}{\lambda d_2 f}(x^2 + y^2)\right] \tag{6.4-6}$$

The intensity distribution is therefore given by

$$|g(x, y)|^2 = \frac{1}{M^2}\left|f\left(-\frac{x}{M}, -\frac{y}{M}\right)\right|^2 \tag{6.4-7}$$

Thus the intensity distribution in plane P_2 is exactly the same as that in the object plane except for a magnification M and inversion (negative sign). Plane P_2 is the geometrical image plane of plane P_1. The perfect resemblance between the distributions in planes P_1 and P_2 is obtained only when the pupil function is unity in the entire region of integration. This is normally not so for two reasons: (a) any lens is of finite size, and (b) the lens, in general, will suffer from aberrations and this will make $p(x, y)$ a complex function of x and y, with unit magnitude, and the phase depending on x and y (see Chapter 7). Thus the image and the object are not, in general, identical.

(2) As a second case, consider d_1 and d_2 such that

$$d_1 = d_2 = f, \qquad \text{i.e.,} \qquad \alpha = \beta = \gamma \tag{6.4-8}$$

Then simplifying Eq. (6.4-2), we obtain

$$g(x, y) = \frac{1}{\lambda^2 f^2} \iint d\xi \, d\eta \, f(\xi, \eta) \, I(\xi, \eta) \exp[-i\alpha(\xi^2 + \eta^2 + x^2 + y^2)]$$

where

$$\begin{aligned} I(\xi, \eta) &= \int d\xi' \exp\{-i\alpha[\xi'^2 - 2(x + \xi)\xi']\} \\ &\quad \times \int d\eta' \exp\{-i\alpha[\eta'^2 - 2(y + \eta)\eta']\} \\ &= \exp\{i\alpha[(x + \xi)^2 + (y + \eta)^2]\} \int_{-\infty}^{\infty} d\zeta \exp(-i\alpha\zeta^2) \\ &\quad \times \int_{-\infty}^{\infty} d\zeta' \exp(-i\alpha\zeta'^2) \\ &= \frac{\pi}{i\alpha} \exp\{i\alpha[(x + \xi)^2 + (y + \eta)^2]\} \end{aligned} \tag{6.4-9}$$

and we have used Eq. (4.10-9). Thus

$$\begin{aligned} g(x, y) &= \frac{1}{i\lambda f} \iint d\xi \, d\eta \, f(\xi, \eta) \exp\left[\frac{2\pi i}{\lambda f}(x\xi + y\eta)\right] \\ &= \frac{1}{i\lambda f} F\left(\frac{x}{\lambda f}, \frac{y}{\lambda f}\right) \end{aligned} \tag{6.4-10}$$

Thus, the field distribution at the back focal plane of a lens is the Fourier transform of the field distribution in its front focal plane evaluated at the spatial frequencies $u = x/\lambda f$ and $v = y/\lambda f$. For this reason the back focal plane of a lens is also called the Fourier transform plane, or spatial frequency plane because the spatial frequencies are displayed in that plane. This can be understood from the following example: Consider an object with the transmission function

$$f(x) = A\cos\frac{2\pi x}{a} = \frac{A}{2}\left[\exp\left(\frac{2\pi ix}{a}\right) + \exp\left(-\frac{2\pi ix}{a}\right)\right] \quad (6.4\text{-}11)$$

placed in the front focal plane of a lens. This is just an amplitude object with spatial period a or spatial frequency $1/a$. The field pattern in the Fourier transform plane is proportional to the Fourier transform of $f(x)$, namely,

$$F(u) = \frac{A}{2}\left[\delta\left(u + \frac{1}{a}\right) + \delta\left(u - \frac{1}{a}\right)\right] \quad (6.4\text{-}12)$$

where $u = x/\lambda f$ and we have used the fact that $\mathscr{F}[\exp(2\pi ix/a)] = \delta(u + 1/a)$. Consequently, in the Fourier transform plane, two bright spots are obtained at $u = 1/a$ (i.e., $x = \lambda f/a$) and $u = -1/a$ corresponding to the spatial frequency of the object (see Fig. 6.1). Thus the pattern in the Fourier transform plane corresponds to the spatial frequency spectrum of the object.

As another example, we consider a long slit of width a (placed in the front focal plane) and illuminated by a plane wavefront. Thus, the field distribution in the front focal plane is $\mathrm{rect}(x/a)$, where

$$\mathrm{rect}(x/a) \equiv \begin{cases} 1 & |x| < a/2 \\ 0 & |x| > a/2 \end{cases} \quad (6.4\text{-}13)$$

is known as the rectangle function. Its Fourier transform is

$$\mathscr{F}\left[\mathrm{rect}\left(\frac{x}{a}\right)\right] = \int_{-a/2}^{a/2} e^{2\pi iux}\,dx = \frac{\sin(\pi au)}{\pi u} = a\,\mathrm{sinc}(au) \quad (6.4\text{-}14)$$

Hence the intensity distribution in the back focal plane is

$$I = I_0\,\mathrm{sinc}^2(au) = I_0 \sin^2(\beta)/\beta^2 \quad (6.4\text{-}15)$$

where $\beta = \pi au = \pi ax/\lambda f$. Equation (6.4-15) represents the Fraunhofer diffraction pattern of a slit.

Problem 6.2. Show that when $d_2 = f$ but $d_1 \neq f$, the field distribution in plane P_2 (the back focal plane of the lens) is the Fourier transform of the object field distribution multiplied by a phase curvature term. However, the intensity distribution in the back focal plane is the modulus squared of the Fourier transform of the object distribution.

Problem 6.3. A plane wave falls on a lens; the emerging spherical wave illuminates an object [with transmittance $f(x, y)$] placed at a distance d from the back focal plane. Calculate the field distribution in the back focal plane of the lens.

Solution. After passing through the lens, the plane wave becomes a converging wave focusing to the focal point at a distance f from the lens. At a distance d from the focal plane, the phase distribution of the spherical wave will be of the form $\exp[(ik/2d)(x^2 + y^2)]$ and hence the field distribution to the right of the object would be $f(x, y)\exp[(ik/2d)(x^2 + y^2)]$. This field undergoes Fresnel diffraction from this plane to the back focal plane. Hence the field distribution $g(x, y)$ in the back focal plane is given by

$$\begin{aligned} g(x, y) &= \frac{1}{\lambda d}\left\{\exp\left[\frac{ik}{2d}(x^2 + y^2)\right]f(x, y)\right\} * \exp\left[-\frac{ik}{2d}(x^2 + y^2)\right] \\ &= \frac{1}{\lambda d}\exp\left[-\frac{ik}{2d}(x^2 + y^2)\right]\iint f(x', y')\exp\left[\frac{2\pi i}{\lambda d}(xx' + yy')\right]dx'\,dy' \\ &= \frac{1}{\lambda d}\exp\left[-\frac{ik}{2d}(x^2 + y^2)\right]F\left(\frac{x}{\lambda d}, \frac{y}{\lambda d}\right) \end{aligned} \tag{6.4-16}$$

Thus, in the back focal plane we again obtain the Fourier transform of the object field distribution (there is a phase curvature term also.)

We next consider a grating-like object and show how the intensity distribution in the image plane can be altered by placing filters in the Fourier transform plane. We first introduce the sampling function

$$\text{comb}\left(\frac{x}{\tau}\right) \equiv \sum_{m=-\infty}^{+\infty} \delta\left(\frac{x}{\tau} - m\right) = \tau \sum \delta(x - m\tau) \tag{6.4-17}$$

Its Fourier transform is [see Eq. (B-13)]

$$\mathscr{F}\left[\text{comb}\left(\frac{x}{\tau}\right)\right] = \tau\,\text{comb}\,(u\tau) = \tau \sum_{m=-\infty}^{\infty} \delta(u\tau - m) \tag{6.4-18}$$

The transmission function of a diffraction grating consisting of an infinite number of slits of width a separated by a distance τ (as shown in Fig. 6.5a) is represented by [see Eq. (6.4-13)]

$$\sum_{m=-\infty}^{+\infty} \text{rect}\left(\frac{x - m\tau}{a}\right) \tag{6.4-19}$$

which is the convolution of $\text{rect}\,(x/a)$ with $\text{comb}\,(x/\tau)$:

$$\begin{aligned} \text{rect}\left(\frac{x}{a}\right) * \text{comb}\left(\frac{x}{\tau}\right) &= \tau \int \text{rect}\left(\frac{x'}{a}\right)\sum_m \delta(x - m\tau - x')\,dx' \\ &= \tau \sum_m \text{rect}\left(\frac{x - m\tau}{a}\right) \end{aligned} \tag{6.4-20}$$

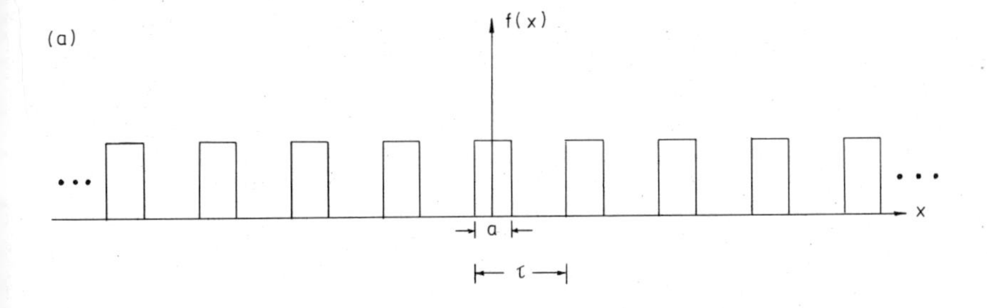

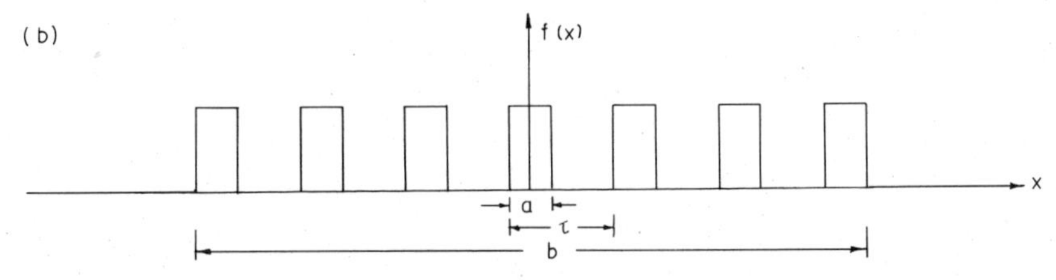

Fig. 6.5. (a) Diffraction grating represented by Eq. (6.4-21) consisting of slits of width a separated by a distance τ. (b) Diffraction grating of finite extension b represented by Eq. (6.4-23).

Thus the diffraction grating can be described by the function

$$\frac{1}{\tau}\operatorname{rect}\left(\frac{x}{a}\right) * \operatorname{comb}\left(\frac{x}{\tau}\right) \tag{6.4-21}$$

If b represents the size of the grating (see Fig. 6.5b), then the transmission function of the grating is given by

$$\sum_{m=-\infty}^{+\infty} \operatorname{rect}\left(\frac{x - m\tau}{a}\right)\operatorname{rect}\left(\frac{x}{b}\right) \tag{6.4-22}$$

which is equal to

$$\frac{1}{\tau}\left[\operatorname{rect}\left(\frac{x}{a}\right) * \operatorname{comb}\left(\frac{x}{\tau}\right)\right]\operatorname{rect}\left(\frac{x}{b}\right) \tag{6.4-23}$$

If the diffraction grating is placed in the front focal plane of a lens, then the field pattern in the Fourier transform plane is proportional to

$$\begin{aligned} F(u) &= \frac{1}{\tau}\mathscr{F}\left\{\left[\operatorname{rect}\left(\frac{x}{a}\right) * \operatorname{comb}\left(\frac{x}{\tau}\right)\right]\operatorname{rect}\left(\frac{x}{b}\right)\right\} \\ &= \frac{1}{\tau}\mathscr{F}\left[\operatorname{rect}\left(\frac{x}{a}\right) * \operatorname{comb}\left(\frac{x}{\tau}\right)\right] * \mathscr{F}\left[\operatorname{rect}\left(\frac{x}{b}\right)\right] \end{aligned} \tag{6.4-24}$$

or

$$F(u) = \frac{1}{\tau}\left\{\mathscr{F}\left[\operatorname{rect}\left(\frac{x}{a}\right)\right]\mathscr{F}\left[\operatorname{comb}\left(\frac{x}{\tau}\right)\right]\right\} * \mathscr{F}\left[\operatorname{rect}\left(\frac{x}{b}\right)\right] \tag{6.4-25}$$

where we have used Eqs. (B-9) and (B-11). Thus

$$\begin{aligned}
F(u) &= \frac{1}{\tau}\left[a \operatorname{sinc}(au)\, \tau \operatorname{comb}(u\tau)\right] * b \operatorname{sinc}(bu) \\
&= \frac{ab}{\tau}\left[\operatorname{sinc}(au) \sum_m \delta\left(u - \frac{m}{\tau}\right)\right] * \operatorname{sinc}(bu) \\
&= \frac{ab}{\tau}\sum_m \operatorname{sinc}\left(\frac{am}{\tau}\right) \delta\left(u - \frac{m}{\tau}\right) * \operatorname{sinc}(bu) \\
&= \frac{ab}{\tau}\sum_m \operatorname{sinc}\left(\frac{am}{\tau}\right) \operatorname{sinc}\left[b\left(u - \frac{m}{\tau}\right)\right] \\
&= \frac{ab}{\tau}\left\{\operatorname{sinc}(bu) + \operatorname{sinc}\left(\frac{a}{\tau}\right)\operatorname{sinc}\left[b\left(u - \frac{1}{\tau}\right)\right]\right. \\
&\qquad \left. + \operatorname{sinc}\left(\frac{a}{\tau}\right)\operatorname{sinc}\left[b\left(u + \frac{1}{\tau}\right)\right] + \cdots\right\}
\end{aligned} \tag{6.4-26}$$

where $u = x/\lambda f$. This is the diffraction pattern of the grating (see Fig. 6.6). If a second lens is placed such that the above pattern lies in the front focal plane of that lens (see Fig. 6.7), then the field distribution in the back focal

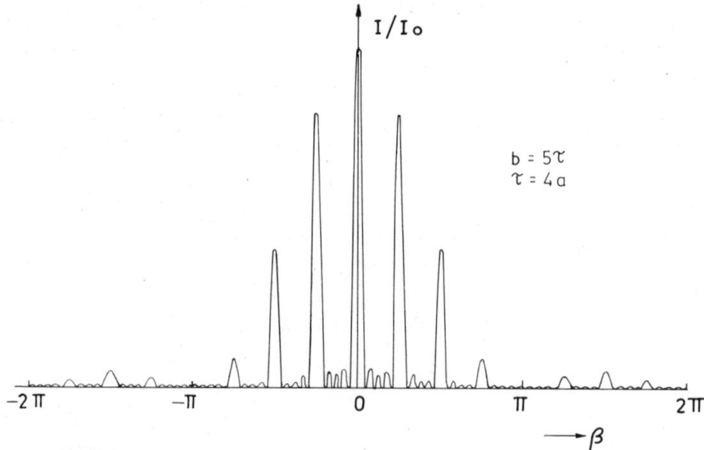

Fig. 6.6. Diffraction pattern of a grating. If a grating shown in Fig. 6.5b is kept in the front focal plane of a lens, then at the back focal plane one obtains the pattern shown here. Here, $\beta = \pi a u$.

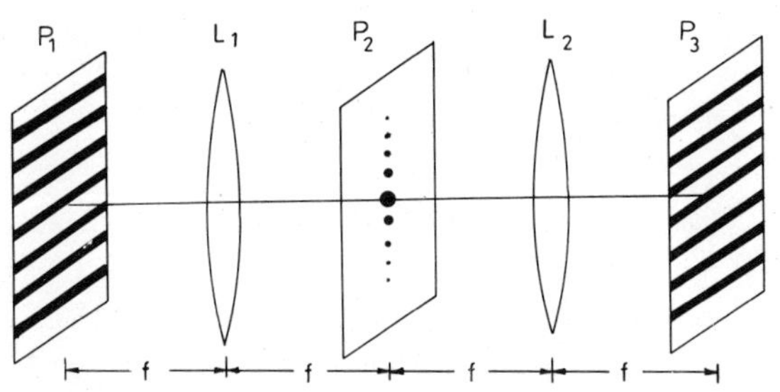

Fig. 6.7. If a diffraction grating is placed in the front focal plane P_1 of a lens L_1, then its Fourier transform is obtained in its back focal plane P_2. (The amplitude distribution in the plane P_2 is the same as shown in Fig. 6.6.) If plane P_2 lies on the front focal plane of another lens L_2, then on its back focal plane P_3, the original grating pattern is obtained.

plane of that lens will be the Fourier transform of Eq. (6.4-26). Since the Fourier transform of the Fourier transform gives us the original function (we use a reflected coordinate system in the image plane to take care of the inversion that is produced due to double Fourier transformation [see Eq. (B-5)]), the original field distribution is reproduced in the back focal plane of the second lens. (We have implicitly assumed that the pupil function of the lens is identically equal to unity.) Let us now consider the effect of putting different apertures in the Fourier transform plane P_2 of the first lens, where the spatial frequencies of the initial field distribution are displayed (see Fig. 6.8a). We will see that the image distribution (field distribution in the back focal plane P_3 of the second lens) can be altered in a significant way. We assume that $b \gg \tau$, so that the various orders do not overlap and therefore different orders can be effectively filtered.

(1) Let a slit that passes only the zero order be placed in the Fourier transform plane P_2 (see Fig. 6.8). This implies that only the pattern that is near the axis in the Fourier transform plane [namely, $(ab/\tau)\,\mathrm{sinc}\,(bu)$] and that is centered at $u = 0$ is allowed to be imaged by the second lens. The transmission function of the slit may be written

$$T(u) = \begin{cases} 1 & |u| < 1/b \text{ or } |x| < \lambda f/b \\ 0 & \text{otherwise} \end{cases} \tag{6.4-27}$$

Then the field distribution immediately to the right of the slit is given by

$$F(u)\,T(u) \simeq \frac{ab}{\tau}\,\mathrm{sinc}\,(bu) \tag{6.4-28}$$

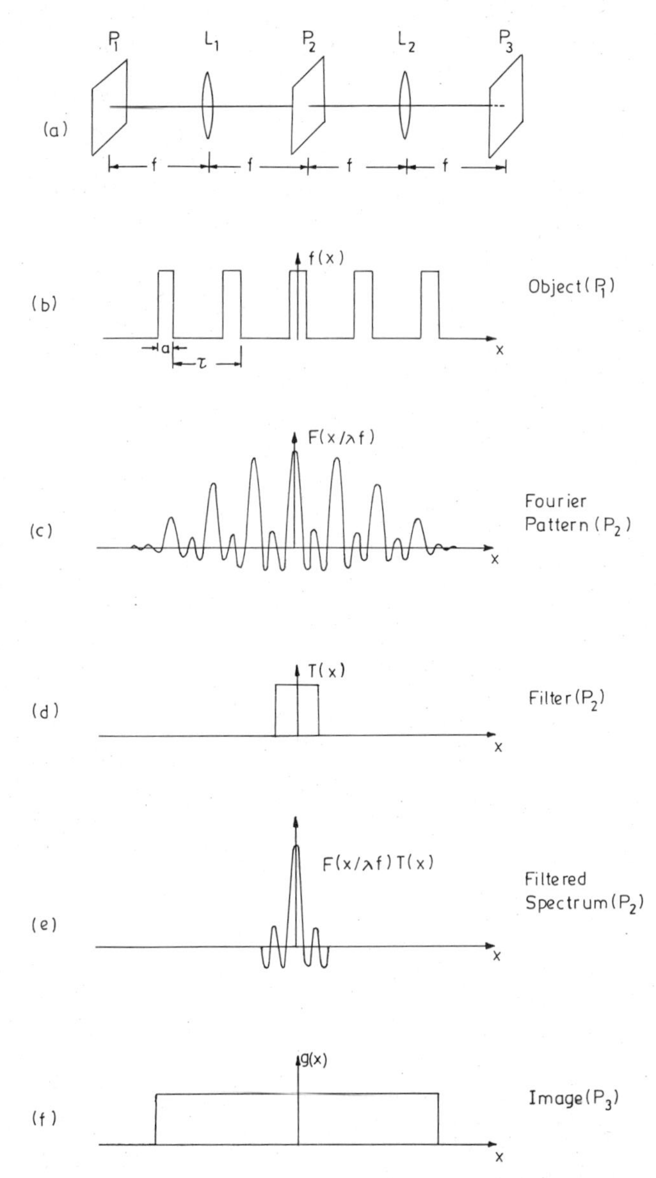

Fig. 6.8. (a) Arrangement for spatial frequency filtering. (b) Transmittance of the object placed in plane P_1 [see (a)]. (c) Fourier transform obtained in the back focal plane P_2. Slits of various widths are used as filters and placed in plane P_2. (d) Transmittance of the filter, which is a slit of width $\lambda f/b$. (e) Filtered distribution, which acts as the spectrum for the image to be formed by the lens L_2. (f) Finally, in the back focal plane of the second lens, the filtered image is obtained.

Thus the field distribution in the image plane is

$$\mathscr{F}\left[\frac{ab}{\tau}\operatorname{sinc}(bu)\right] = \frac{a}{\tau}\operatorname{rect}\left(\frac{x}{b}\right) \tag{6.4-29}$$

The complete process of filtering is shown in Fig. 6.8. Notice that the image contains no information about the periodicity present in the object.

(2) We now consider a slit that passes only the zero and the two first orders (see Fig. 6.9). The filtered spatial frequency spectrum is approximately

$$\begin{aligned}\frac{ab}{\tau}\bigg\{&\operatorname{sinc}(bu) + \operatorname{sinc}\left(\frac{a}{\tau}\right)\operatorname{sinc}\left[b\left(u-\frac{1}{\tau}\right)\right]\\ &+ \operatorname{sinc}\left(\frac{a}{\tau}\right)\operatorname{sinc}\left[b\left(u+\frac{1}{\tau}\right)\right]\bigg\}\end{aligned} \tag{6.4-30}$$

Thus the pattern in the image plane is the Fourier transform of the above

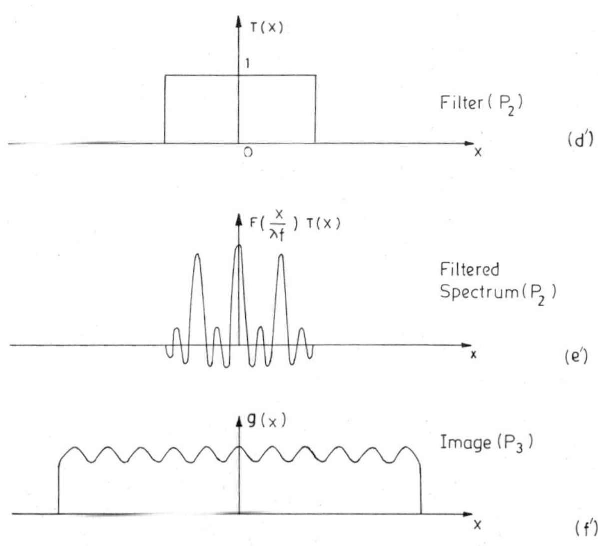

Fig. 6.9. If we are using the same object pattern as shown in Fig. 6.8b and if the filter is such that only the zero and the two first orders pass through the optical system [the corresponding filter is shown in (d')], then the filtered spectrum is of the form shown in (e'). The corresponding image pattern [shown in (f')] is seen to have the same periodicity as the object but is very much different from the object.

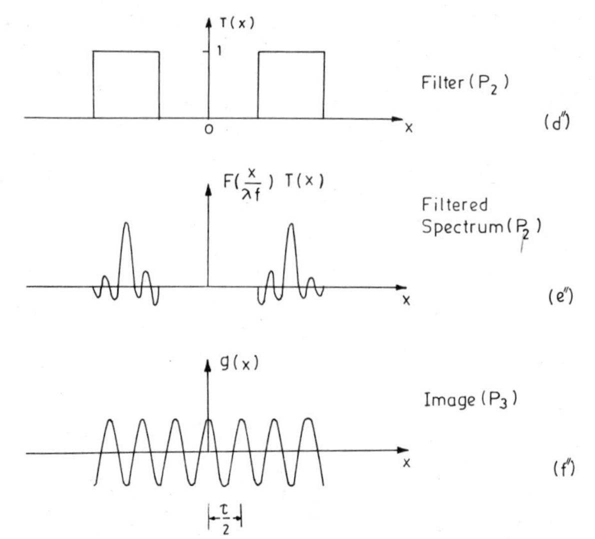

Fig. 6.10. If we again use the same object pattern as shown in Fig. 6.8b and if the filter is such that only the two second orders pass through the optical system [the corresponding filter is shown in (d″)], then the filtered spectrum is of the form shown in (e″). The corresponding image pattern [shown in (f″)] has a periodicity $\tau/2$, half of that present in the object!

pattern, which is given by

$$\frac{a}{\tau}\left[\operatorname{rect}\left(\frac{x}{b}\right)+\operatorname{sinc}\left(\frac{a}{\tau}\right)\operatorname{rect}\left(\frac{x}{b}\right)e^{2\pi i x/\tau}+\operatorname{sinc}\left(\frac{a}{\tau}\right)\operatorname{rect}\left(\frac{x}{b}\right)e^{-2\pi i x/\tau}\right]$$

$$=\frac{a}{\tau}\operatorname{rect}\left(\frac{x}{b}\right)\left[1+2\operatorname{sinc}\left(\frac{a}{\tau}\right)\cos\frac{2\pi x}{\tau}\right] \tag{6.4-31}$$

where we have used the fact that

$$\mathscr{F}\left\{b\operatorname{sinc}\left[b\left(u-\frac{m}{\tau}\right)\right]\right\}=\operatorname{rect}\left(\frac{x}{b}\right)\exp\left(2\pi i\frac{mx}{\tau}\right)$$

[see Eq. (B-4)]. The image has the same periodicity ($=\tau$) as the object but the amplitude distribution is not the same as that in the object.

(3) Similarly, if only the second orders on both sides of the zero order are passed then the spatial frequency spectrum forming the image is given by (see Fig. 6.10)

$$\frac{ab}{\tau}\operatorname{sinc}\left(\frac{2a}{\tau}\right)\left\{\operatorname{sinc}\left[b\left(u-\frac{2}{\tau}\right)\right]+\operatorname{sinc}\left[b\left(u+\frac{2}{\tau}\right)\right]\right\} \tag{6.4-32}$$

The image pattern formed by such a distribution is given by its Fourier transform, namely,

$$\frac{2a}{\tau}\operatorname{sinc}\left(\frac{2a}{\tau}\right)\operatorname{rect}\left(\frac{x}{b}\right)\cos\left(\frac{4\pi x}{\tau}\right) \tag{6.4-33}$$

The periodicity of such a distribution is $\tau/2$, half that present in the object!

(4) Let us now consider the effect of removing the zero order and at the same time passing all the other orders through the optical system. The filtered spectrum is [see Eq. (6.4-26)]

$$F^M(u) = \frac{ab}{\tau}\sum_{m=0,\pm1,\ldots}\operatorname{sinc}\left(\frac{am}{\tau}\right)\operatorname{sinc}\left[b\left(u-\frac{m}{\tau}\right)\right] - \frac{ab}{\tau}\operatorname{sinc}(bu) \tag{6.4-34}$$

Thus the image is

$$\begin{aligned} g(x) = \mathscr{F}[F^M(u)] &= \mathscr{F}\left\{\frac{ab}{\tau}\sum_m \operatorname{sinc}\left(\frac{am}{\tau}\right)\operatorname{sinc}\left[b\left(u-\frac{m}{\tau}\right)\right]\right\} \\ &\quad - \mathscr{F}\left[\frac{ab}{\tau}\operatorname{sinc}(bu)\right] \\ &= \frac{1}{\tau}\left[\operatorname{rect}\left(\frac{x}{a}\right) * \operatorname{comb}\left(\frac{x}{\tau}\right)\right]\operatorname{rect}\left(\frac{x}{b}\right) \\ &\quad - \frac{a}{\tau}\operatorname{rect}\left(\frac{x}{b}\right) \end{aligned} \tag{6.4-35}$$

where we have used the fact that

$$\begin{aligned} &\mathscr{F}\left\{\frac{ab}{\tau}\sum_m \operatorname{sinc}\left(\frac{am}{\tau}\right)\operatorname{sinc}\left[b\left(u-\frac{m}{\tau}\right)\right]\right\} \\ &= \frac{1}{\tau}\mathscr{F}\left(\mathscr{F}\left\{\left[\operatorname{rect}\left(\frac{x}{a}\right) * \operatorname{comb}\left(\frac{x}{\tau}\right)\right]\operatorname{rect}\left(\frac{x}{b}\right)\right\}\right) \\ &= \frac{1}{\tau}\left[\operatorname{rect}\left(\frac{x}{a}\right) * \operatorname{comb}\left(\frac{x}{\tau}\right)\right]\operatorname{rect}\left(\frac{x}{b}\right) \end{aligned} \tag{6.4-36}$$

Observe that in $g(x)$ the first term is normalized to unity. Let us consider some interesting specific cases.

(a) When $a/\tau = \frac{1}{2}$, i.e., the width of the slit is equal to the width of the space between the slits, then the amplitude distribution in the image plane is shown in Fig. 6.11c (we are considering a coherent optical system). Since it is the intensity that can be observed, we find that for this particular case the image is characterized by uniform intensity and has no periodicity (see Fig. 6.11d).

(b) When $a > \tau/2$, i.e., the width of the slit is greater than the spacing between the slits, then the effect of filtering the zero order is shown in Fig. 6.12. Again, since it is the intensity distribution that can be observed, we obtain the interesting result that the bright portions of the original object look darker in the image and the dark portions of the object look brighter in the image; the image consists of bright portions of width $(\tau - a)$ and dark portions of width a, as compared to the object, which contains bright portions of width a and dark portions of width $(\tau - a)$. Thus there is a reversal of contrast in this case.

The above examples enumerate the effect of filtering certain spatial frequencies from the spatial frequency plane before the image is formed.

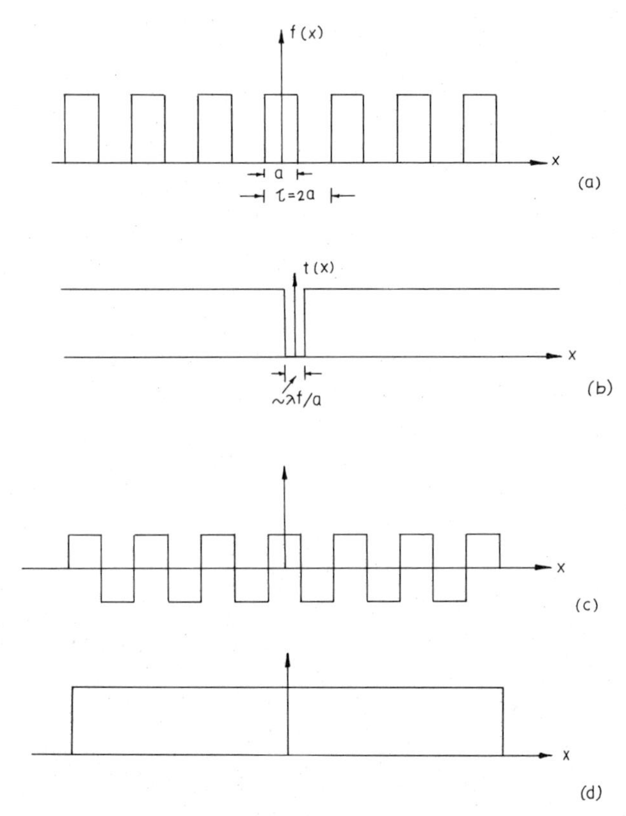

Fig. 6.11. (a) Amplitude distribution of the object, (b) transmittance of the filter, (c, d) resultant amplitude and the intensity distributions in the image plane. Observe the interesting fact that the image shows no periodicity.

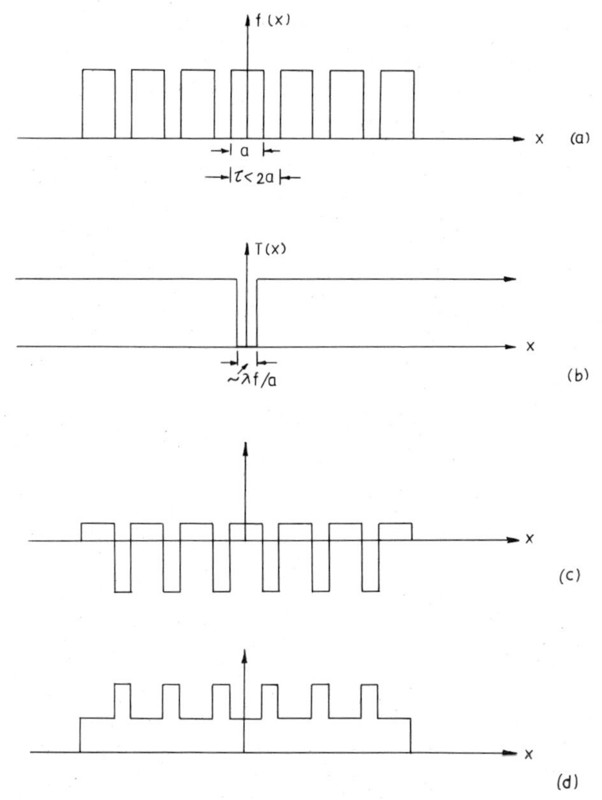

Fig. 6.12. Effect of removing the zero order from the spectrum of a diffraction grating with $a > \tau/2$ (i.e., the width of the slit of the grating is greater than the distance between the slits). Here one obtains the phenomenon of reversal of contrast; the bright and dark portions have interchanged their positions in the image as compared to that in the object.

This control over the image formation by filtering is made use of in improving the quality of images, reduction of "noise," etc. These will be discussed in Section 6.5.

Problem 6.3. Show that if $a < \tau/2$, then there is no contrast reversal. What happens when $a = \tau$?

Problem 6.4. Starting from Eq. (6.4-25) show that the Fraunhofer diffraction pattern of a grating is given by Eq. (4.13-13).

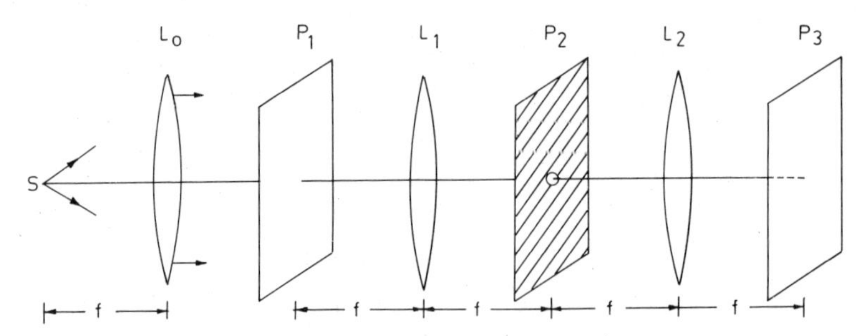

Fig. 6.13. Configuration of the optical system for spatial frequency filtering operations. A transparency whose transmittance is proportional to the object function to be filtered is placed in plane P_1 (the front focal plane of lens L_1). Lens L_1 forms the Fourier transform of the object in plane P_2, which contains a transparency with transmittance proportional to the filter function. The filtered function is again Fourier transformed by lens L_2 to produce the filtered image in plane P_3.

6.5. *Spatial Frequency Filtering and Its Applications**

In Section 6.4 we saw that by placing some slits and stops in the spatial frequency plane, we could alter the form of the image. In this section we will study the various forms of filters and their application to image processing.

Spatial frequency filtering is an operation by which one removes (or passes preferentially) certain desired spatial frequencies by placing filters in the Fourier transform plane. We restrict ourselves to coherent optical systems.† The configuration of the optical system is shown in Fig. 6.13. A point source S placed in front of a lens L_0 produces a plane wavefront which is made to fall on the object‡ which is to be filtered. The object is kept at the front focal plane P_1 of another lens L_1, which displays the frequency spectrum of the object in the back focal plane P_2. Plane P_2 is also the front focal plane of lens L_2, which Fourier transforms the spectrum in plane P_2 to form an image in plane P_3. The filters that filter the spatial frequencies are placed at plane P_2 and the filtered image is displayed in plane P_3.

The various kinds of filters that one encounters in spatial frequency filtering techniques are the following:

* For a nice review of the different techniques and interesting experimental results, see Van der Lugt (1968). [See also Tippett *et al.* (1965).]

† The effect of coherence on imaging will be discussed in Section 7.5.

‡ Alternatively, one can illuminate the object with a parallel beam coming out of a laser.

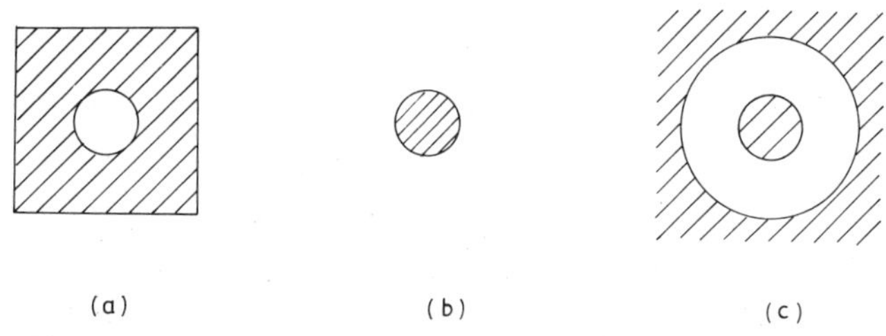

Fig. 6.14. (a) Low-pass filter, (b) high-pass filter, (c) band-pass filter.

1. Binary Filters. These are the simplest kind of filters; stops and slits are examples of this class of filters. The transmittance of such filters is either unity or zero. They can be further subdivided into the following categories:

a. Low-Pass Filters. These filters pass only the low frequencies, i.e., those frequency components lying near the axis (see Fig. 6.14a). These find application in filtering high-frequency noise. Thus for example, if one looks closely at printed photographs in a newspaper, one can see numerous dots to be printed on the image (see Fig. 6.15a). The dot structure in the photograph can be removed by placing a circular aperture centered around zero frequency in the Fourier transform plane as shown in Fig. 6.13. The circular aperture acts as a low-pass filter and the high spatial frequencies associated with the dot structure are filtered out. On the other hand, the spatial frequencies associated with the object are concentrated around

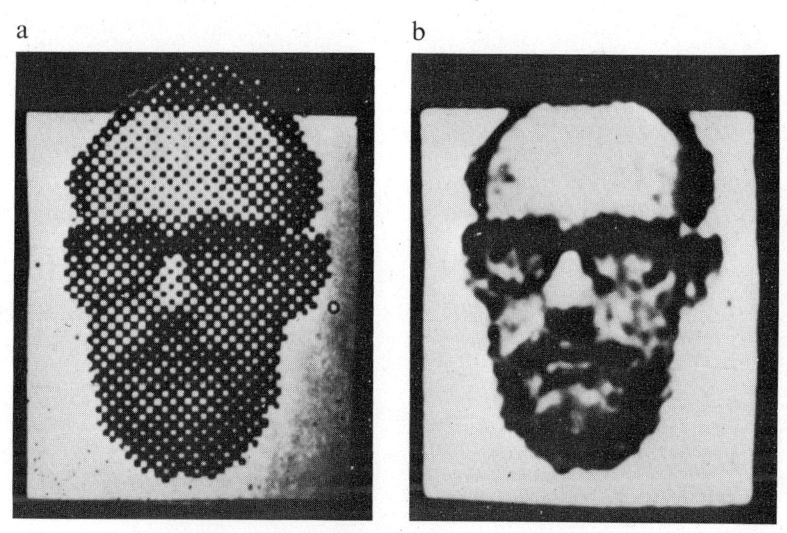

Fig. 6.15. Halftone dot structure removal using a low-pass filter (after Phillips, 1969). Photograph courtesy of Dr. R. A. Phillips.

zero frequency, which passes through the filter. The filtered image does not have the dot structure (see Fig. 6.15b). It must be noted that through this process the high-frequency components of the object are also stopped and hence the image is not exactly the same as the noise-free object.

b. High-Pass Filters. These filters allow the high frequencies to pass through while the low frequencies are stopped (see Fig. 6.14b). Such filters are used in edge enhancement.

c. Band-Pass Filters. These filters pass certain desired frequencies only (see Fig. 6.14c). Such filters are very useful in cases where the noise-free object is regular with specified spatial frequencies and is superimposed with random noise. In such cases, in order to decrease the noise it is necessary to pass only the spatial frequencies of the noise-free object. Since the noise is random, the image signal-to-noise ratio will be much lower than in the object.

2. *Amplitude Filters.* These filters possess a continuous variation of the transmittance and do not introduce any phase variations in the incident field distribution. The required variation can be produced by recording the appropriate distribution in a film. These find applications in contrast enhancement, where one attenuates the zero order partially to cut down the direct light.

3. *Phase Filters.* Using these filters only the phase of the frequency spectrum is altered. Phase contrast microscopy employs this principle; this will be discussed in Section 6.5.1.

4. *Complex Filters.* Filters that act on both the amplitude and the phase of different frequency components are included in this class. They are difficult to fabricate. Holographic methods are used for producing such filters, which find applications in character recognition problems.

We will now discuss specific applications of the spatial frequency filtering techniques.

6.5.1. Phase Contrast Microscopy

It is, in general, difficult to determine the structure of an object like a biological cell by means of an ordinary microscope. If we illuminate this cell from below, then the variation of absorption of light along the section of the cell is extremely small and when we view this through the microscope the details are not observed. However, since different portions of the cell have different optical thicknesses (either through difference in refractive index or through difference in thickness), the phase variations on the surface of the cell are not uniform. An ordinary microscope records only the intensity and as such the phase variations are not manifested in the observations.* It should be mentioned that such objects can be made partially visible through the process of staining. However, often it is not possible to stain them without causing their death or changing their structure. In such cases one uses the method of phase contrast microscopy (introduced by Zernike in 1942), which allows us to record the phase variation in the object as an intensity variation.† In order to understand the principle of phase contrast microscopy, we consider a phase object whose transmittance may be written

$$f(x, y) = \exp[-i\varphi(x, y)] \tag{6.5-1}$$

Under normal observation one observes the intensity, which is proportional to $|f(x, y)|^2$; this, as can be seen from Eq. (6.5-1), is independent of x and y. If we place such an object on the front focal plane of a lens, then the amplitude distribution in the back focal plane will be the Fourier transform of $f(x, y)$, i.e.,

$$F(u, v) = \mathscr{F}[f(x, y)] = \iint e^{-i\varphi(x,y)} e^{2\pi i(ux + vy)}\, dx\, dy \tag{6.5-2}$$

where $u = x/\lambda f$ and $v = y/\lambda f$. If the phase variation $\varphi(x, y)$ is small, then we may write‡

$$\exp[-i\varphi(x, y)] \simeq 1 - i\varphi(x, y) \tag{6.5-3}$$

On substituting Eq. (6.5-3) into Eq. (6.5-2) and carrying out the integration, we get

$$F(u, v) = \delta(u)\,\delta(v) - i\Phi(u, v) \tag{6.5-4}$$

where $\Phi(u, v) = \mathscr{F}[\varphi(x, y)]$. If one uses a phase filter that introduces a phase retardation of $\pi/2$ on the zero-frequency component with respect to the other frequency components,‡‡ then the filtered spectrum is

$$F^{M}(u, v) = -i\,\delta(u)\,\delta(v) - i\Phi(u, v) \tag{6.5-5}$$

where it is assumed that the zero-frequency components of $\Phi(u, v)$ are small.

* It must be noted that due to the finite aperture of the optical system (microscope) or some defocusing, some details of the object may be seen even with a normal microscope.

† It may be mentioned that there are other methods that can be used to study the phase variation, e.g., the Schlieren method, but these suffer from the drawback that the intensity variations do not correspond directly to the phase variations in the object (see, e.g., Goodman, 1968, p. 193).

‡ Notice that, even under this approximation, the intensity distribution is proportional to $|1 - i\varphi(x, y)|^2 \simeq 1$, where the term quadratic in $\varphi(x, y)$ is neglected.

‡‡ Such a filter can be a small $\lambda/4(n - 1)$ thick film placed in front of the point $x = 0$, $y = 0$, in the back focal plane.

The intensity pattern in the image plane is given by

$$|g(x, y)|^2 = |\mathscr{F}[F^{M}(u, v)]|^2 = -i - i\varphi(x, y)|^2 \simeq 1 + 2\varphi(x, y) \qquad (6.5\text{-}6)$$

Thus the intensity distribution in the image is linearly related to the phase variation in the object.

The contrast in the image can be improved by attenuating the zero-frequency terms in addition to having a phase shift of $\pi/2$. In such a case, the filtered spectrum is

$$F^{M}(u, v) = -i\alpha\,\delta(u)\,\delta(v) - i\Phi(u, v) \qquad (6.5\text{-}7)$$

where α is the attenuating factor. The image intensity distribution is now given by*

$$|g(x, y)|^2 = \alpha^2 + 2\alpha\varphi(x, y) \qquad (6.5\text{-}8)$$

The contrast defined by

$$C = \frac{\text{intensity at } (x, y) - \text{background intensity}}{\text{background intensity}}$$

is given by

$$C = \frac{\alpha^2 + 2\alpha\varphi(x, y) - \alpha^2}{\alpha^2} = \frac{2\varphi(x, y)}{\alpha} \qquad (6.5\text{-}9)$$

Without the attenuating factor, the contrast would be $2\varphi(x, y)$. Thus by choosing *small* values of α, the contrast can be improved. The above phase retardation by $\pi/2$ results in positive phase contrast, i.e., thicker portions (which introduce a larger phase difference) look brighter than the thinner portions. If the phase delay is $3\pi/2$, then the image intensity distribution is

$$|g(x, y)|^2 = 1 - 2\varphi(x, y) \qquad (6.5\text{-}10)$$

This is termed negative phase contrast. In this case thicker portions look darker compared to thinner portions. Thus by introducing a phase shifter in the Fourier transform plane, one can transform the phase variations into intensity variations.

We will next discuss some mathematical operations that can be performed using the Fourier-transforming property of optical systems.

6.5.2. Cross-Correlation

Cross-correlation operations arise in statistics, target detection, etc. In such problems, one is interested in evaluating the integral

$$\varphi_{12}(x_0, y_0) = \iint f_1(x, y)\, f_2(x + x_0, y + y_0)\, dx\, dy \qquad (6.5\text{-}11)$$

* Notice that for $\alpha = 0$, $|g(x, y)|^2$ is proportional to the square of $\varphi(x, y)$.

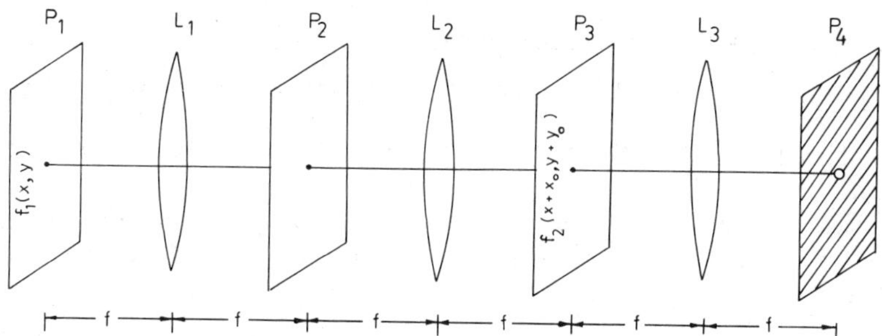

Fig. 6.16. The configuration for performing cross-correlation between two functions $f_1(x, y)$ and $f_2(x, y)$. A transparency whose transmittance is proportional to $f_1(x, y)$ is placed in plane P_1. The lens combination L_1 and L_2 produces its image in plane P_3. In plane P_3 is kept the transparency whose transmittance is proportional to $f_2(x, y)$ and displaced in the minus x and minus y directions by x_0 and y_0, respectively. In plane P_4, on the axis the cross-correlation between the two functions $f_1(x, y)$ and $f_2(x, y)$ is obtained.

where $f_1(x, y)$ and $f_2(x, y)$ are two real functions.* The configuration that performs the process of cross-correlation between two functions $f_1(x, y)$ and $f_2(x, y)$ is shown in Fig. 6.16. A transparency with transmittance proportional to $f_1(x, y)$ is placed in plane P_1, which is at a distance f from lens L_1: its Fourier spectrum is formed at plane P_2 at a distance f from L_1. Lens L_2 again Fourier transforms and the function $f_1(x, y)$ is displayed in plane P_3. We keep a transparency with transmittance proportional to $f_2(x, y)$ in plane P_3. The transparency $f_2(x, y)$ is displaced by amounts x_0 and y_0 along the minus x and minus y directions, respectively. Thus the field distribution in the front focal plane of L_3 becomes $f_1(x, y) f_2(x + x_0, y + y_0)$. In the back focal plane of L_3, namely, plane P_4, is formed the Fourier transform of this function:

$$\Phi_{12}(x_0, y_0, u, v) = \iint f_1(x, y) f_2(x + x_0, y + y_0)\, e^{2\pi i(ux + vy)}\, dx\, dy \qquad (6.5\text{-}12)$$

This is called the ambiguity function. If we restrict ourselves to $u = 0$, $v = 0$, in plane P_4, we obtain the amplitude at that point to be

$$\varphi_{12}(x_0, y_0) = \Phi_{12}\Big|_{\substack{u=0\\v=0}} = \iint f_1(x, y) f_2(x + x_0, y + y_0)\, dx\, dy \qquad (6.5\text{-}13)$$

which is the required cross-correlation. To obtain the cross-correlation, a

* The configuration to perform cross-correlation between two complex functions using holographic techniques is given in Section 6.5.3.

pinhole is placed on the axis in plane P_4 and the intensity behind the pinhole is measured for different values of x_0 and y_0, i.e., for different values of the displacement of $f_2(x, y)$ in plane P_3. Autocorrelation can also be performed by replacing $f_2(x, y)$ by $f_1(x, y)$ in plane P_3.

6.5.3. Character Recognition

It is sometimes necessary to detect the presence of particular characters in an optical image. This is known as a character recognition problem. In such problems it is necessary to make a cross-correlation between the required character and the image. At the positions of the required character in the image, one would obtain in the correlation (i.e., in plane P_4) bright spots. For example, if one wants to determine the positions of the letter A in a page, one has to cross-correlate the page with the letter A. The method described in Section 6.5.2 for cross-correlation cannot be applied here since it becomes cumbersome. We will describe a method using holographic techniques, to obtain the cross-correlation.*

Let $f(x, y)$ represent the function to be correlated with the character $h(x, y)$. Thus we have to determine the following integral:

$$\varphi_{fh} = \iint f(x', y')\, h^*(x + x', y + y')\, dx'\, dy' \tag{6.5-14}$$

Consider the function

$$\Phi_{fh} \equiv f(x, y) ⊛ h(x, y) = F(u, v)\, H^*(u, v) \tag{6.5-15}$$

If we take the Fourier transform of Φ_{fh}, we find

$$\begin{aligned}
\mathscr{F}[\Phi_{fh}] &= \iint F(u, v)\, H^*(u, v)\, e^{2\pi i(ux + vy)}\, du\, dv \\
&= \iint F(u, v)\left[\iint h^*(\xi, \eta)\, e^{-2\pi i(u\xi + v\eta)}\, d\xi\, d\eta\right] e^{2\pi i(ux + vy)}\, du\, dv \\
&= \iint h^*(\xi, \eta)\left[\iint F(u, v)\, e^{-2\pi i[u(\xi - x) + v(\eta - y)]}\, du\, dv\right] d\xi\, d\eta \\
&= \iint h^*(\xi, \eta)\, f(\xi - x, \eta - y)\, d\xi\, d\eta \\
&= \iint f(x', y')\, h^*(x + x', y + y')\, dx'\, dy'
\end{aligned} \tag{6.5-16}$$

* The principle of holography will be discussed in Chapter 8.

Thus the Fourier transform of $F(u, v)\,H^*(u, v)$ is the required cross-correlation between $h(x, y)$ and $f(x, y)$. Hence if it is possible to produce a transparency with transmittance proportional to $H^*(u, v)$, then all one needs to do is to place the transparency $f(x, y)$ in the front focal plane P_1 of a lens, as shown in Fig. 6.17, which will form its Fourier transform $F(u, v)$ in the back focal plane P_2. The transparency with transmittance proportional to $H^*(u, v)$ is kept in plane P_2. Consequently, the amplitude distribution immediately behind plane P_2 is proportional to $F(u, v)\,H^*(u, v)$. This is then Fourier transformed by lens L_2 and displayed in plane P_3. The Fourier transform of $F(u, v)\,H^*(u, v)$ is nothing but the cross-correlation of $f(x, y)$ and $h(x, y)$. Thus one is able to obtain the cross-correlation of the whole page without having physically to displace the functions as in the method discussed in Section 6.5.2. Corresponding to the values of (x, y) at which $f(x, y)$ corresponds to $h(x, y)$, one obtains bright spots in plane P_3. Thus the position of the bright spots gives directly the position of the required character in the image to be scanned. Hence the problem finally reduces to the production of filters having transmittance proportional to $H^*(u, v)$. $H(u, v)$ is in general a complex function of u and v. The production of complex filters had been a difficult problem until recently. But with the development of the holographic techniques, it has become possible to produce complex filters (Van der Lugt, 1964).

The function $h(x, y)$ is normally a real function of x and y. If we Fourier transform $h(x, y)$ to obtain $H(u, v)$, the recorded pattern will be the real part of $H(u, v)$, since it is the intensity that can be recorded; the phase information will be lost. Hence a holographic technique that helps in converting phase variations into intensity variations is required. This will be discussed below.

To obtain the required filter an optical configuration shown in Fig.

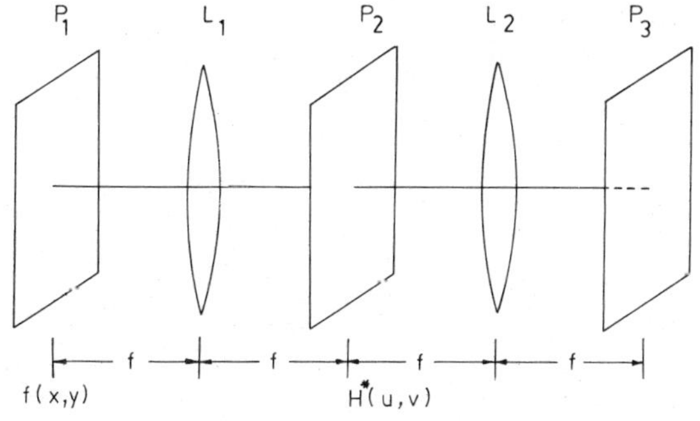

Fig. 6.17. Planes P_1 and P_2 contain transparencies with transmittances proportional to $f(x, y)$ and $H^*(x/\lambda f, y/\lambda f)$, respectively. In the above configuration, we obtain the cross-correlation of $f(x, y)$ and $h(x, y)$ in plane P_3.

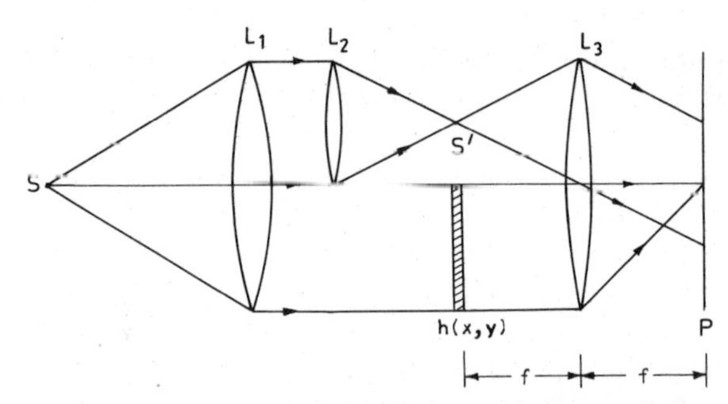

Fig. 6.18. Configuration for holographically producing complex filters.

6.18 is used. S is a point source placed at the focus of a lens L_1. A part of the emerging parallel beam illuminates the transparency with transmittance $h(x, y)$. A lens L_2 focuses the other part to a point S'. Lens L_3 Fourier transforms $h(x, y)$ and displays $H(u, v)$ in plane P. The wave emanating from S' becomes a plane wave after refraction through lens L_3; this plane wave is inclined to the axis at an angle determined by the displacement of S' from the axis of lens L_3. If a represents the displacement of the source S' from the axis of lens L_3, then in plane P the amplitude distribution due to S' is the Fourier transform of $P_0\,\delta(x - a)$, i.e.,

$$P_0 \int \delta(x - a)\, e^{2\pi i x u}\, dx = P_0 e^{2\pi i a u} \tag{6.5-17}$$

Hence the amplitude distribution in plane P is given by

$$\begin{aligned} G(u, v) &= H(u, v) + P_0 e^{2\pi i a u} \\ &= H_0(u, v)\, e^{-i\varphi(u, v)} + P_0 e^{2\pi i a u} \end{aligned} \tag{6.5-18}$$

where $H_0(u, v)$ and $\varphi(u, v)$ represent the amplitude and phase of $H(u, v)$. The pattern recorded in plane P is the intensity pattern, namely,

$$\begin{aligned} |G(u, v)|^2 &= |H(u, v)|^2 + P_0^2 + H(u, v)\, P_0 e^{-2\pi i a u} + H^*(u, v)\, P_0 e^{2\pi i a u} \\ &= H_0^2 + P_0^2 + 2H_0 P_0 \cos\left[2\pi a u + \varphi(u, v)\right] \end{aligned} \tag{6.5-19}$$

Thus the recorded pattern has also the phase of the filter.

Let $f(x, y)$ be the function to be filtered with $H(u, v)$. The transparency containing $f(x, y)$ is placed in the front focal plane P_1 of lens L_1 (see Fig. 6.17). The transparency containing $|G(u, v)|^2$ is placed in plane P_2. Lens L_1 forms the Fourier spectrum of $f(x, y)$ in plane P_2, and thus to the right of P_2 the field is

$$F(u, v)\, |G(u, v)|^2 = |H|^2 F + P_0^2 F + HFP_0 e^{-2\pi i a u} + P_0 H^* F e^{2\pi i a u} \tag{6.5-20}$$

Lens L_2 forms a further Fourier transform, and we obtain in plane P_3 the Fourier transform of the above function. In particular, the third term yields a field

$$\mathscr{F}\left[H(u, v) F(u, v) e^{-2\pi i a u}\right] = \iint h(x', y') f(a - x - x', -y - y')\, dx'\, dy' \tag{6.5-21}$$

which is the convolution between $h(x, y)$ and $f(x, y)$; the convolution is centered about $x = a$, $y = 0$, and is also inverted. Similarly the fourth term in Eq. (6.5-20) gives rise to a field proportional to

$$\mathscr{F}\left[H^*(u, v) F(u, v) e^{2\pi i a u}\right] = \iint f(x', y') h^*(x + x' + a, y + y')\, dx'\, dy' \tag{6.5-22}$$

which is nothing but the cross-correlation between $f(x, y)$ and $h(x, y)$; the cross-correlation can be seen to be centered at $x = -a$, $y = 0$. Thus, in the region of correlation, at different points in plane P_3 corresponding to the positions of the character to be recognized in plane P_1, one obtains bright dots, which is a consequence of high correlation. Figure 6.19 corresponds to the letters Q, W, and P in the front focal plane. When a holo-

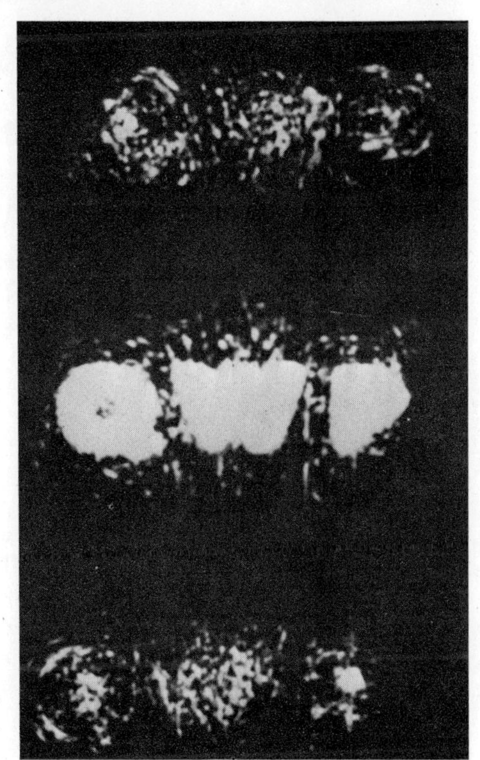

Fig. 6.19. Identification of the letter P by cross-correlation technique (after Goodman, 1968; used with permission).

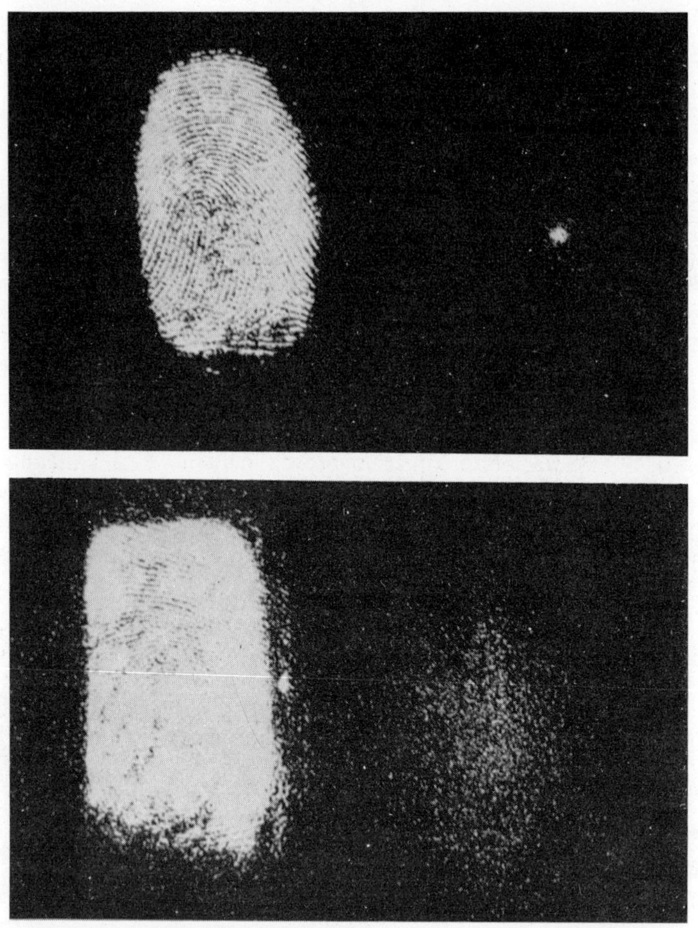

Fig. 6.20. Identification of fingerprints by cross-correlation technique (after Tsujiuchi *et al.*, 1971).

graphically prepared filter corresponding to the letter P is placed in plane P_2 (see Fig. 6.17), then in plane P_3 one obtains the intensity distribution shown in Fig. 6.19. The presence of a bright spot on the lower right corner (which represents a high correlation) indicates that the letter P appears in the corresponding position in plane P_1. The convolution (which appears in the upper portion of the figure) and the cross-correlation with the letters Q and W produce a smear. The central portion corresponds to the first two terms on the right-hand side of Eq. (6.5-20). In a similar manner, the correlation technique can be applied to fingerprint identification. Figure 6.20 shows the output of the correlator when the fingerprint obtained from the scene of the crime matches with that of the fingerprint under test and when they do not match (Tsujiuchi, 1971). Observe that when the fingerprints are matched a bright dot results.

6.5.4. Multichannel Operation

Optical systems operate in two dimensions, i.e., both x and y directions in the object plane are analyzed simultaneously, as compared to electrical systems, which operate in a single dimension, namely, time. This brings forward many advantages of speed and multichannel operation of computers using optical synthesis as compared to electrical computing techniques. By making use of cylindrical lenses (i.e., lenses that have finite radii of curvature along one of the transverse axes and infinite radius of curvature along a perpendicular axis), one can perform multichannel operations, i.e., many one-dimensional calculations can be done simultaneously. For example, if one wants to perform cross-correlation between different sets of functions, say $f_1(x)$ with $f_2(x)$, $g_1(x)$ with $g_2(x)$, $h_1(x)$ with $h_2(x)$, etc., then using the following procedure one can obtain all the required cross-correlations simultaneously. A transparency with the transmittance proportional to the functions $f_1(x, y_1)$, $g_1(x, y_2)$, $h_1(x, y_3)$, ..., is placed in the object plane P_1 (see Fig. 6.21). Here $y_1, y_2, y_3, \ldots$, are different values of the y

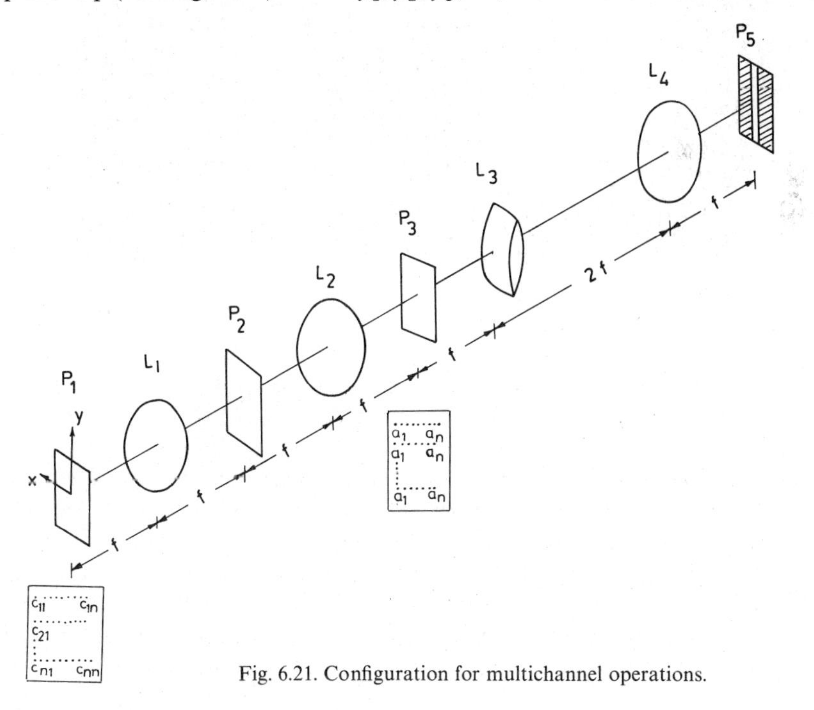

Fig. 6.21. Configuration for multichannel operations.

coordinate. In such a configuration, the transmittance along the line $y = y_1$ is proportional to $f_1(x)$, the transmittance along the line $y = y_2$ is proportional to $g_1(x)$, etc. Then in plane P_3 a transparency with transmittance proportional to $f_2(x, y_1), g_2(x, y_2), h_2(x, y_3), \ldots,$ is placed. Notice that the y coordinates of the functions f_1 and f_2, g_1 and g_2, etc., are the same. This is because the magnification of the system is unity in the present case; otherwise the magnification of the system must be taken into consideration while calculating the y coordinates of f_2, g_2, h_2, etc. At a distance f from plane P_3 is kept a cylindrical lens, which Fourier transforms in the y direction and leaves the x distribution undisturbed. At a distance $2f$ from the cylindrical lens a spherical lens is placed. At the back focal plane of spherical lens L_4, we find a field distribution that is the Fourier transform of the field distribution along the x direction in plane P_3 and is the image of the field distribution in the y direction in plane P_3. Thus by employing cylindrical lenses in conjunction with spherical lenses, one can perform multichannel operations.

Problem 6.5. Show that a filter of the form

$$T(u) = \alpha u \tag{6.5-23}$$

placed in the spatial frequency plane yields in the image plane a differential of the object amplitude function placed in the front focal plane. (We are considering a two-dimensional configuration for simplicity.)

Solution. Let $f(x)$ represent the amplitude distribution in the front focal plane of the lens. In the back focal plane appears the distribution $F(u)\,\{=\mathscr{F}[f(x)]\}$. The filtered spectrum is

$$G(u) = F(u)\,T(u) = \alpha u F(u) \tag{6.5-24}$$

Thus the image is

$$g(x) = \mathscr{F}[G(u)] = \alpha \int uF(u)\,e^{+2\pi iux}\,du$$

$$= \frac{\alpha}{2\pi i}\frac{d}{dx}\left[\int F(u)\,e^{+2\pi iux}\,du\right] = \frac{\alpha}{2\pi i}\frac{df(-x)}{dx} \tag{6.5-25}$$

Thus in the image plane is displayed the differential of the object distribution. Since the differential of a function is maximum near the edges, such a filtering process leads to edge enhancement.

Equation (6.5-23) implies that the transmittance increases with the value of $|x|$ (since $u = x/\lambda f$) and there is a phase change of π for negative values of x. Such a filter can be produced by superimposing an amplitude filter with a variation in transmittance that is proportional to $|x|$ and a phase filter that has a sudden phase change of π at $x = 0$.

Problem 6. 6. Show that a filter with a transmittance

$$T(u) = \alpha/u \tag{6.5-26}$$

displays, in the image plane, the integral of the function in the object plane. Notice that such a filter is unrealistic, since for sufficiently small values of u, $T(u) > 1$, i.e., an amplification of the incident signal is required. If the frequency spectrum of the object is spread out, one need not worry much about the value of $T(u)$ near $u = 0$.

6.5.5. *Matrix Multiplication**

We will now consider the optical configuration necessary for the following matrix operation:

$$B = \begin{pmatrix} b_1 \\ b_2 \\ \vdots \\ b_n \end{pmatrix} = \begin{pmatrix} c_{11} & c_{12} \cdots c_{1n} \\ c_{21} & c_{22} \cdots c_{2n} \\ \vdots & \\ c_{n1} & c_{n2} \cdots c_{nn} \end{pmatrix} \begin{pmatrix} a_1 \\ a_2 \\ \vdots \\ a_n \end{pmatrix} = CA \tag{6.5-27}$$

A and B are column matrices and C is a square matrix. Given C and A we want to evaluate B optically. The configuration is shown in Fig. 6.21. In plane P_1 we have a series of points with transmittance proportional to $c_{11}, c_{12}, \ldots, c_{nn}$ in matrix form. The lens combination L_1 and L_2 forms its image at plane P_3. If we put a screen at plane P_3, then we obtain the same intensity distribution as at P_1. As such, if we put a transparency with transmittance proportional to the coefficients $a_1, a_2, \ldots, a_n$ in n rows (see Fig. 6.21), then we will get the distribution, which will be the product of the two distributions, i.e., in plane P_3 we have a matrix of points with the intensity proportional to the various coefficients of the following matrix:

$$\begin{pmatrix} c_{11}a_1 & c_{12}a_2 \cdots c_{1n}a_n \\ c_{21}a_1 & c_{22}a_2 \cdots c_{2n}a_n \\ \vdots & \\ c_{n1}a_1 & c_{n2}a_2 \cdots c_{nn}a_n \end{pmatrix}$$

This acts as the object for lens L_3. The lens combination L_3, L_4 (as explained earlier), images along the y direction and Fourier transforms along the x direction. Thus in plane P_5, along $u = 0$ (i.e., $x = 0$), we have the values

* *Reference:* Cutrona (1965).

$$\begin{pmatrix} \sum_m c_{1m} a_m \\ \sum_m c_{2m} a_m \\ \vdots \\ \sum_m c_{nm} a_m \end{pmatrix}$$

where the convolution has been written in the form of a summation since the array is made up of discrete points. The above column matrix is the required matrix B.

7

Fourier Optics II. Optical Transfer Functions

7.1. Introduction

One of the most important developments in optics has been the realization that image formation by optical systems can be treated as a linear process and hence the general theory of linear systems (which are extensively used in electrical circuits) can be applied to optical systems. Consequently, as in electrical circuits, one can study the imaging quality of an optical system by applying frequency response techniques; this leads to the concept of transfer functions, which can be calculated from the design data of the optical system. In this chapter, we have developed the theory behind the optical transfer function (OTF). It may be mentioned that when an optical system has small aberrations, the geometrical optics analysis of the image is not even qualitatively correct and one must incorporate the effects of diffraction. The diffraction effects are implicitly contained in the optical transfer function. Consequently this gives the lens designer a method for accurately designing an optical system.

7.2. The Point-Spread Function

Let $f(x, y)$ represent the input to an optical system L; the input is assumed to be on plane P_1 (see Fig. 7.1). If $g(x, y)$ is the corresponding output on plane P_2, then one can represent the optical system by an operator P such that

$$g(x, y) = P[f(x, y)] \tag{7.2-1}$$

The optical system is said to be linear if it satisfies the following properties:

(1) If an input $f_1(x, y)$ produces an output $g_1(x, y)$ and another input

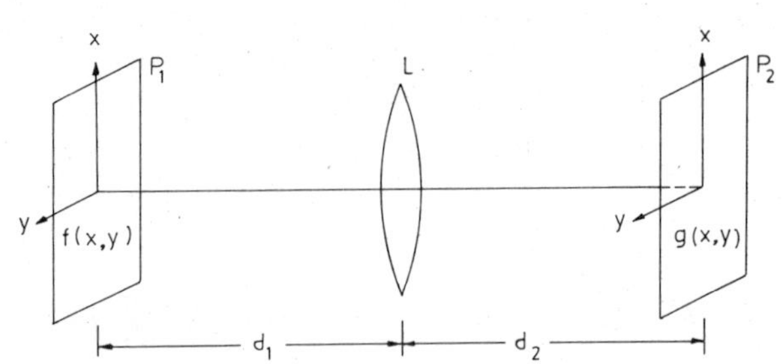

Fig. 7.1. L, in general, represents an optical system, which in this case is simply a thin lens.

$f_2(x, y)$ produces an output $g_2(x, y)$, then an input $f_1(x, y) + f_2(x, y)$ results in the output $g_1(x, y) + g_2(x, y)$. In mathematical terms, if

$$g_1(x, y) = P[f_1(x, y)], \qquad g_2(x, y) = P[f_2(x, y)] \tag{7.2-2}$$

then

$$g_1(x, y) + g_2(x, y) = P[f_1(x, y) + f_2(x, y)] \tag{7.2-3}$$

(2) Furthermore,

$$P[af(x, y)] = ag(x, y) \tag{7.2-4}$$

We have not yet mentioned the precise nature of the input to the optical system. Indeed, for the system to be linear, the nature of the input depends on the degree of coherence of the illumination.* For example, as will be shown, for perfectly coherent illumination the input and output are amplitude distributions; on the other hand, for incoherent illumination the functions $f(x, y)$ and $g(x, y)$ are intensity distributions.†

Let us consider a point object situated at (x_1, y_1) in the object plane; the corresponding input can be written as

$$f(x, y) = \delta(x - x_1)\,\delta(y - y_1) \tag{7.2-5}$$

where $\delta(x)$ is the Dirac delta function (see Appendix A). If this object distribution produces a distribution $h(x, y;\ x_1, y_1)$ in plane P_2, i.e., if

$$h(x, y;\ x_1, y_1) = P[\delta(x - x_1)\,\delta(y - y_1)] \tag{7.2-6}$$

then $h(x, y;\ x_1, y_1)$ is termed the point-spread function. In electrical engineering technology it is termed the impulse response function, and in

* See Chapter 5 for a detailed account of coherence.

† For a partially coherent system, the functions correspond to mutual intensity.

mathematical language it is the Green's function of the system. Thus for example, an ideal optical system is one that produces point images of point objects between two specific planes determined by the system. For such a system, the point-spread function is a delta function. In general, since no system is ideal, the point-spread function is not a delta function. For example, for an aberrationless lens, the point-spread function is simply the Airy pattern that is formed due to diffraction effects (see Section 4.7).

We write the input $f(x, y)$ of an optical system as

$$f(x, y) = \iint f(x_1, y_1)\,\delta(x - x_1)\,\delta(y - y_1)\,dx_1\,dy_1 \qquad (7.2\text{-}7)$$

which follows from the definition of the Dirac delta function. Equation (7.2-7) implies that an input can be considered a weighted sum of delta functions, the weight factors being $f(x_1, y_1)$. If $g(x, y)$ represents the output of the system for an input $f(x, y)$, then

$$g(x, y) = P[f(x, y)] = P\left[\iint f(x_1, y_1)\,\delta(x - x_1)\,\delta(y - y_1)\,dx_1\,dy_1\right]$$

or

$$\begin{aligned} g(x, y) &= \iint f(x_1, y_1)\,P[\delta(x - x_1)\,\delta(y - y_1)]\,dx_1\,dy_1 \\ &= \iint f(x_1, y_1)\,h(x, y, x_1, y_1)\,dx_1\,dy_1 \qquad (7.2\text{-}8) \end{aligned}$$

where we have used the condition that the system is linear. Thus, knowing the point-spread function of the system, one can determine the output of the system for a given input. Now, if $g(x, y)$ is the output for an input $f(x, y)$, then the system is termed space invariant,* if $g(x - \xi, y - \eta)$ is the output for an input $f(x - \xi, y - \eta)$, i.e., the shifting of the input in space results in a shift of the output in space without altering the output distribution.†

* It should be borne in mind that the coordinates in the output plane are scaled coordinates to take into account the magnification and possible inversion [see, e.g., Eq. (7.3-6)]. It must also be noted that real optical systems are never space invariant throughout the input plane. There are small regions over which the system is space invariant. These regions are called isoplanatic regions. In our analysis we assume that the object is small enough to be lying in one isoplanatic region. The condition of isoplanatism is satisfied sufficiently accurately for corrected optical systems.

† The corresponding property in electrical engineering is the time invariance property, which means that if an input $f(t)$ produces an output $g(t)$, then an input $f(t - t_0)$ produces an output $g(t - t_0)$; t represents the time coordinate.

Thus, the point-spread function $h(x, y;\ x_1, y_1)$ depends on x, y, x_1, and y_1 only through the combinations $(x - x_1)$ and $(y - y_1)$, i.e.,

$$h(x, y;\ x_1, y_1) \equiv h(x - x_1, y - y_1) \tag{7.2-9}$$

Thus, Eq. (7.2-8) becomes

$$g(x, y) = \iint f(x_1, y_1)\, h(x - x_1, y - y_1)\, dx_1\, dy_1 = f(x, y) * h(x, y) \tag{7.2-10}$$

i.e., the output field distribution is the convolution of the input field distribution with the point-spread function.

We will now show that a coherent optical system is linear in amplitude (and consequently nonlinear in intensity) and an incoherent optical system is linear in intensity. Further, if h is the point-spread function of the coherent system, then $|h|^2$ is the point-spread function of the incoherent system.

For an object illuminated by a quasi-monochromatic source, the field distribution is of the form

$$f(\xi, \eta, t) = \varphi(\xi, \eta, t) \exp(2\pi i \nu_0 t) \tag{7.2-11}$$

where $\varphi(\xi, \eta, t)$ is a slowly varying function of time and ν_0 is the carrier frequency of the source. If $F(\xi, \eta, \nu)$ represents the Fourier transform of $f(\xi, \eta, t)$, i.e.,

$$F(\xi, \eta, \nu) = \int_{-\infty}^{\infty} f(\xi, \eta, t)\, e^{-2\pi i \nu t}\, dt \tag{7.2-12}$$

then for the source to be quasi-monochromatic F must be sharply peaked around $\nu = \nu_0$ with a half-width $\Delta\nu \ll \nu_0$. In addition, we assume that the path differences involved are small in comparison to $c/\Delta\nu$ (see Section 5.4).

It must be noted that, in general, $h(x - \xi, y - \eta)$ depends on ν; however, since each frequency component propagates independently, the amplitude (corresponding to the frequency ν) in the image plane is given by

$$G(x, y, \nu) = \iint F(\xi, \eta, \nu)\, h(x - \xi, y - \eta, \nu)\, d\xi\, d\eta \tag{7.2-13}$$

Thus, the total field in the image plane is given by

$$\begin{aligned} g(x, y, t) &= \int G(x, y, \nu)\, e^{2\pi i \nu t}\, d\nu \\ &\simeq \iint d\xi\, d\eta\, h(x - \xi, y - \eta, \nu_0) \int F(\xi, \eta, \nu)\, e^{2\pi i \nu t}\, d\nu \\ &= \iint d\xi\, d\eta\, h(x - \xi, y - \eta, \nu_0)\, f(\xi, \eta, t) \end{aligned} \tag{7.2-14}$$

where we have made use of the fact that since $F(\xi, \eta, \nu)$ is a sharply peaked function around $\nu = \nu_0$, $h(x - \xi, y - \eta, \nu)$ may be replaced by its value at $\nu = \nu_0$. Thus the intensity in the image plane is given by

$$I(x, y) = \langle g^*(x, y, t)\, g(x, y, t)\rangle$$
$$= \iiiint h^*(x - \xi, y - \eta, \nu_0)\, h(x - \xi', y - \eta', \nu_0) \times \langle f^*(\xi, \eta, t)\, f(\xi', \eta', t)\rangle \, d\xi \, d\eta \, d\xi' \, d\eta' \tag{7.2-15}$$

where angular brackets denote time averaging (see Section 5.3). We may now consider some special cases.

(1) For coherent illumination, we may write

$$f(\xi, \eta, t) = f_0(\xi, \eta)\, f(0, 0, t)/[\langle |f(0, 0, t)|^2\rangle]^{1/2} \tag{7.2-16}$$

Thus

$$\langle f^*(\xi, \eta, t)\, f(\xi', \eta', t)\rangle = f_0^*(\xi, \eta)\, f_0(\xi', \eta') \tag{7.2-17}$$

Substituting the above equation into Eq. (7.2-15), we immediately obtain

$$I(x, y) = \left| \iint h(x - \xi, y - \eta, \nu_0)\, f_0(\xi, \eta)\, d\xi \, d\eta \right|^2 = g_0^*(x, y)\, g_0(x, y) \tag{7.2-18}$$

where

$$g_0(x, y) = \iint h(x - \xi, y - \eta, \nu_0)\, f_0(\xi, \eta)\, d\xi \, d\eta \tag{7.2-19}$$

Equation (7.2-19) tells us that a coherent optical system can be assumed to be linear in amplitude with a point-spread function $h(x - \xi, y - \eta)$; however, the system is nonlinear in intensity.

(2) When the object is illuminated by incoherent light, we may assume $\langle f^*(\xi, \eta, t)\, f(\xi', \eta', t)\rangle$ to vanish except in a small region around $\xi = \xi'$ and $\eta = \eta'$. We may, for example, assume

$$\begin{aligned} \langle f^*(\xi, \eta, t)\, f(\xi', \eta', t)\rangle &= O(\xi, \eta) \qquad \text{for } (\xi - \xi')^2 + (\eta - \eta')^2 < a^2 \\ &= 0 \qquad \text{otherwise} \end{aligned} \tag{7.2-20}$$

where $O(\xi, \eta)$ represents the object intensity distribution. For small values of a,* we may in the first approximation replace $h(x - \xi', y - \eta', \nu_0)$ in

* We must mention here that we cannot make the area πa^2 go to zero. Indeed, for incoherent illumination a must be $\gtrsim \lambda$. For a more thorough discussion on this point see Beran and Parrent (1964).

Eq. (7.2-15) by its value at (ξ, η) to obtain

$$I(x, y) = \pi a^2 \iint O(\xi, \eta) |h(x - \xi, y - \eta)|^2 \, d\xi \, d\eta \qquad (7.2\text{-}21)$$

Thus, an incoherent optical system is linear in intensity and the point-spread function is proportional to the modulus square of the point-spread function of the same system under coherent illumination.

7.3. Point-Spread Function of a Thin Lens

We now obtain an explicit expression for the point-spread function of a thin lens of radius a. Let P_1 be the input plane at a distance d_1 from the thin lens of focal length f and let P_2 be the output plane at a distance d_2 from the lens (see Fig. 7.1). Consider a point object situated at the point (x_1, y_1) in the input plane. The field in the output plane (which is just the point-spread function) is given by [see Eq. (6.4-2)]

$$\begin{aligned} h(x_2, y_2;\, x_1, y_1) = \frac{1}{\lambda^2 d_1 d_2} \iiiint_{-\infty}^{\infty} & \delta(\xi - x_1)\,\delta(\eta - y_1) \\ & \times \exp\left\{ -\frac{ik}{2d_1}[(x - \xi)^2 + (y - \eta)^2] \right\} p(x, y) \\ & \times \exp\left[\frac{ik}{2f}(x^2 + y^2) \right] \\ & \times \exp\left\{ -\frac{ik}{2d_2}[(x - x_2)^2 + (y - y_2)^2] \right\} d\xi \, d\eta \, dx \, dy \end{aligned}$$

$$(7.3\text{-}1)$$

where $p(x, y)$ is the pupil function of the lens and is defined, for an aberrationless lens of circular aperture, by Eq. (6.3-6). The integration over ξ and η in Eq. (7.3-1) can be immediately performed, yielding

$$\begin{aligned} h(x_2, y_2;\, x_1, y_1) = \frac{1}{\lambda^2 d_1 d_2} \iint_{-\infty}^{\infty} p(x, y) \exp\left\{ -\frac{ik}{2}\left[\frac{(x - x_1)^2 + (y - y_1)^2}{d_1} \right.\right. \\ \left.\left. - \frac{x^2 + y^2}{f} + \frac{(x - x_2)^2 + (y - y_2)^2}{d_2} \right] \right\} dx \, dy \qquad (7.3\text{-}2) \end{aligned}$$

We are interested in the output plane that corresponds to the geometrical image plane, i.e., the value of d_2 that satisfies the lens law [see Eq. (6.3-4)].

Thus

$$h(x_2, y_2;\ x_1, y_1) = \frac{1}{\lambda^2 d_1 d_2} \exp\left[-\frac{ik}{2d_1}(x_1^2 + y_1^2)\right] \exp\left[-\frac{ik}{2d_2}(x_2^2 + y_2^2)\right]$$

$$\times \iint p(x, y) \exp\left\{\frac{2\pi i}{\lambda d_2}[x(x_2 + Mx_1) + y(y_2 + My_1)]\right\} dx\, dy \quad (7.3\text{-}3)$$

where, as before, $M = d_2/d_1$. The above integral is the Fourier transform of $p(x, y)$.

Since it is the intensity that will be observed, the factor

$$\exp\left[-\frac{ik}{2d_2}(x_2^2 + y_2^2)\right]$$

in $h(x_2, y_2;\ x_1, y_1)$ can be omitted. At the same time, the factor

$$\exp\left[-\frac{ik}{2d_1}(x_1^2 + y_1^2)\right]$$

cannot be neglected on the same grounds, because the final image pattern is obtained by carrying out an integration over x_1 and y_1. Since in most practical cases, the main contribution to an image point comes from points lying near the corresponding ideal object point, we may in the first approximation neglect the terms depending quadratically on x_1 and y_1. Under this approximation, the first exponential factor in Eq. (7.3-3) can also be neglected, and one obtains

$$h(x_2, y_2;\ x_1, y_1) = \frac{1}{\lambda^2 d_1 d_2} \iint_{-\infty}^{\infty} p(x, y)$$

$$\times \exp\left\{\frac{2\pi i}{\lambda d_2}[x(x_2 + Mx_1) + y(y_2 + My_1)]\right\} dx\, dy \quad (7.3\text{-}4)$$

Thus the point-spread function is proportional to the Fourier transform of the pupil function $p(x, y)$ centered at the ideal image point. Defining

$$u' = x/\lambda d_2, \qquad v' = y/\lambda d_2, \qquad x_1' = -Mx_1, \qquad y_1' = -My_1 \quad (7.3\text{-}5)$$

we obtain

$$h(x_2, y_2;\ x_1', y_1') = M \iint p(\lambda d_2 u', \lambda d_2 v')$$

$$\times \exp\{2\pi i[u'(x_2 - x_1') + v'(y_2 - y_1')]\}\, du'\, dv' \quad (7.3\text{-}6)$$

Thus, h considered as a function of x_2, y_2, x'_1, and y'_1 is space invariant and can be written $h(x_2 - x'_1, y_2 - y'_1)$. If we put $x_2 - x'_1 = \tilde{x}$ and $y_2 - y'_1 = \tilde{y}$, then Eq. (7.3-6) can be rewritten

$$h(\tilde{x}, \tilde{y}) = M \iint p(\lambda d_2 u', \lambda d_2 v') \exp[2\pi i(u'\tilde{x} + v'\tilde{y})]\, du'\, dv' \qquad (7.3\text{-}7)$$

For a circular pupil of radius a, the point-spread function can easily be shown to be proportional to $J_1(\tau)/\tau$, where

$$\tau = \frac{2\pi a}{\lambda d_2}(\tilde{x}^2 + \tilde{y}^2)^{1/2}$$

In the above analysis, we assumed the lens to be aberrationless. In general, lenses suffer from aberrations and the origin of aberrations lies in the fact that for an incident spherical wavefront, the wavefront coming out of the optical system may not be spherical. The pupil function of an aberrated lens may be redefined by the relation

$$p(x, y) = \begin{cases} \exp[-ikW(x, y)], & x^2 + y^2 \leq a^2 \\ 0, & x^2 + y^2 > a^2 \end{cases} \qquad (7.3\text{-}8)$$

where $W(x, y)$ defines the wavefront aberration, because it gives the deviation of the actual wavefront from the spherical shape. In Section 7.4 we obtain an explicit expression for $W(x, y)$ for a system suffering from defocusing.

It may be mentioned that for a general (rotationally symmetric) optical system, Eq. (7.3-7) would remain valid with the pupil function given by Eq. (7.3-8), a representing the radius of the exit pupil* and d_2 representing the distance of the image plane from the exit pupil.

7.4. Frequency Analysis

In Section 7.2, we showed that the output to an optical system is obtained by convoluting the input with the point-spread function [see Eq. (7.2-10)]. If we take the Fourier transform of Eq. (7.2-10) and use the convolution theorem (see Appendix B), we obtain

$$G(u, v) = F(u, v)\, H(u, v) \qquad (7.4\text{-}1)$$

where $F(u, v)$, $G(u, v)$, and $H(u, v)$ are the Fourier transforms of $f(x, y)$,

* The entrance pupil is a pupil (either real or imaginary) that restricts the size of the cone of rays entering the optical system and the exit pupil is a pupil (the geometrical image of the entrance pupil) that restricts the cone of rays leaving the optical system.

$g(x, y)$, and $h(x, y)$, respectively [see Eq. (6.2-7)]. As can be seen from Eq. (7.4-1), the frequency components in the image plane can be obtained by simply multiplying the object frequency components with $H(u, v)$; the function $H(u, v)$ is known as the coherent transfer function of the optical system and is given by

$$H(u, v) = \iint h(\tilde{x}, \tilde{y})\, e^{2\pi i(u\tilde{x} + v\tilde{y})}\, d\tilde{x}\, d\tilde{y}$$

Using Eq. (7.3-7), we obtain, within a proportionality constant,

$$H(u, v) = \iint du'\, dv'\, p(\lambda d_2 u', \lambda d_2 v') \int d\tilde{x}\, e^{2\pi i\tilde{x}(u' + u)} \int d\tilde{y}\, e^{2\pi i\tilde{y}(v' + v)} \quad (7.4\text{-}2)$$

The last two integrals are simply delta functions [see Eq. (A-8)] and hence

$$H(u, v) = p(-\lambda d_2 u, -\lambda d_2 v) \quad (7.4\text{-}3)$$

Thus, the transfer function of a coherent optical system is simply the pupil function of the system in a reflected frame of reference. For a diffraction-limited system defined by Eq. (6.3-6), the transfer function is of the form

$$H(u, v) = \begin{cases} 1 & \text{for } \lambda^2 d_2^2(u^2 + v^2) \le a^2 \\ 0 & \text{for } \lambda^2 d_2^2(u^2 + v^2) > a^2 \end{cases} \quad (7.4\text{-}4)$$

(see Fig. 7.2). Thus the system passes (without distortion) all frequency components satisfying

$$u^2 + v^2 \le a^2/\lambda^2 d_2^2 \quad (7.4\text{-}5)$$

For a system afflicted with aberrations, one simply has to use Eq. (7.3-8) for the pupil function and once again the system passes all frequency components satisfying Eq. (7.4-5), however, with a distortion of phase.

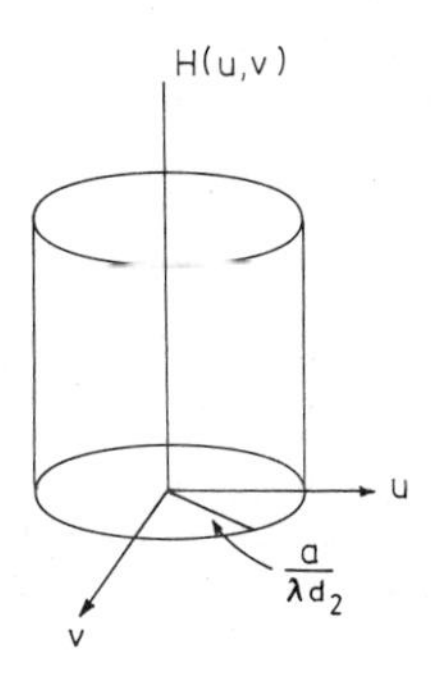

Fig. 7.2. The coherent transfer function of a diffraction-limited optical system with a circular aperture of radius a.

Problem 7.1. Determine $H(u, v)$ for a rectangular pupil defined by

$$p(x, y) = \begin{cases} 1 & |x| \le a, |y| \le b \\ 0 & \text{otherwise} \end{cases} \tag{7.4-6}$$

For an incoherent system, the point-spread function is proportional to $|h(x, y)|^2$, and as such the frequency response can be written as

$$\mathscr{H}(u, v) = \frac{\iint h^*(x, y)\, h(x, y)\, e^{2\pi i(ux+vy)}\, dx\, dy}{\iint h^*(x, y)\, h(x, y)\, dx\, dy} \tag{7.4-7}$$

where we have normalized the function such that $\mathscr{H}(u = 0, v = 0) = 1$. Now since

$$\int f(x)\, g^*(x)\, e^{2\pi iux}\, dx = \int\left[\int F(u')\, e^{-2\pi iu'x}\, du'\right]\left[\int G^*(u'')\, e^{2\pi iu''x}\, du''\right] e^{2\pi iux}\, dx$$

$$= \iint du'\, du'' F(u')\, G^*(u'')\, \delta(u' - u'' - u) \quad = \int F(u')\, G^*(u' - u)\, du'$$

Thus

$$\iint h(x, y)\, h^*(x, y)\, e^{2\pi i(ux+vy)} dx\, dy = \iint H^*\left(\xi' - \frac{1}{2}u, \eta' - \frac{1}{2}v\right)$$

$$\times H\left(\xi' + \frac{1}{2}u, \eta' + \frac{1}{2}v\right) d\xi'\, d\eta' \tag{7.4-8}$$

$$\mathscr{H}(u, v) = \frac{\iint H^*(\xi' - \frac{1}{2}u, \eta' - \frac{1}{2}v)\, H(\xi' + \frac{1}{2}u, \eta' + \frac{1}{2}v)\, d\xi'\, d\eta'}{\iint |H(\xi', \eta')|^2\, d\xi'\, d\eta'} \tag{7.4-9}$$

where the denominator is obtained by evaluating the numerator at $u = 0$, $v = 0$. The function $\mathscr{H}(u, v)$ is called the optical transfer function (OTF) of the system and the modulus of $\mathscr{H}(u, v)$ is known as the modulation transfer function (MTF). Since according to Eq. (7.4-3), $\mathscr{H}(u, v)$ is the pupil function, we can also write

$$\mathscr{H}(u, v) = \frac{\begin{array}{c}\iint p^*(-\lambda d_2\xi' + \frac{1}{2}\lambda d_2 u, -\lambda d_2\eta' + \frac{1}{2}\lambda d_2 v) \\ \times\, p(-\lambda d_2\xi' - \frac{1}{2}\lambda d_2 u, -\lambda d_2\eta' - \frac{1}{2}\lambda d_2 v)\, d\xi'\, d\eta'\end{array}}{\iint |p(-\lambda d_2\xi', -\lambda d_2\eta')|^2\, d\xi'\, d\eta'}$$

$$= \frac{\iint p^*(\xi + \frac{1}{2}\lambda d_2 u, \eta + \frac{1}{2}\lambda d_2 v)\, p(\xi - \frac{1}{2}\lambda d_2 u, \eta - \frac{1}{2}\lambda d_2 v)\, d\xi\, d\eta}{\iint |p(\xi, \eta)|^2\, d\xi\, d\eta} \tag{7.4-10}$$

where $\xi = -\lambda d_2\xi'$ and $\eta = -\lambda d_2\eta'$.

Problem 7.2. Calculate the OTF of an aberration-free system whose pupil function is of the form given by Eq. (6.3-6).

Solution. In order to calculate the OTF of the optical system we have to calculate the area of overlap between two pupil functions displaced by $(\frac{1}{2}\lambda d_2 u, \frac{1}{2}\lambda d_2 v)$ and $(-\frac{1}{2}\lambda d_2 u, -\frac{1}{2}\lambda d_2 v)$ as shown in Fig. 7.3. The equation of circle 1 is

$$(\xi' + \alpha)^2 + \eta'^2 = a^2$$

where $\alpha = \frac{1}{2}\lambda d_2 (u^2 + v^2)^{1/2}$. Thus, the area of overlap (which is twice the area of of the shaded region) is

$$2\int_0^{a-\alpha} 2\eta' \, d\xi' = 4\int_0^{a-\alpha} [a^2 - (\xi' + \alpha)^2]^{1/2} \, d\xi'$$

Carrying out the above integration, we finally obtain

$$\mathscr{H}(u, v) = \begin{cases} \dfrac{2}{\pi}\left\{\cos^{-1}\left(\dfrac{\sigma}{2\sigma_0}\right) - \dfrac{\sigma}{2\sigma_0}\left[1 - \dfrac{\sigma^2}{(2\sigma_0)^2}\right]^{1/2}\right\} & \text{for } \sigma[=(u^2+v^2)^{1/2}] < 2\sigma_0 \\ 0 & \text{for } \sigma > 2\sigma_0 \end{cases} \tag{7.4-11}$$

where $\sigma_0 = a/\lambda d_2$ represents the cutoff frequency of the system under coherent illumination. For a lens having $f/2a = 5$ (commonly referred to as an $f/5$ lens) and operating in d light (5875 Å) the cutoff frequency under incoherent illumination is $2\sigma_0 = 1/(5 \times 5.875 \times 10^{-4}) \simeq 340$ lines/mm. Figure 7.4a shows a plot of OTF as a function of u and v. If we compare the results obtained by using the coherent and incoherent illuminations, we find that whereas the cutoff frequency of the coherent system was $a/\lambda d_2$, the cutoff frequency of the same system working in incoherent illumination is $2a/\lambda d_2$. There seems to be an apparent increase in resolving power by changing from coherent to incoherent illumination. This point is discussed in greater detail in Section 7.5.

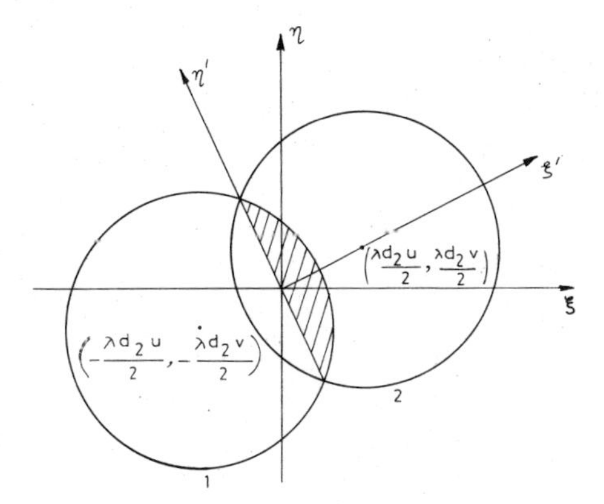

Fig. 7.3. OTF of a diffraction-limited system is the area of overlap of two displaced pupil functions.

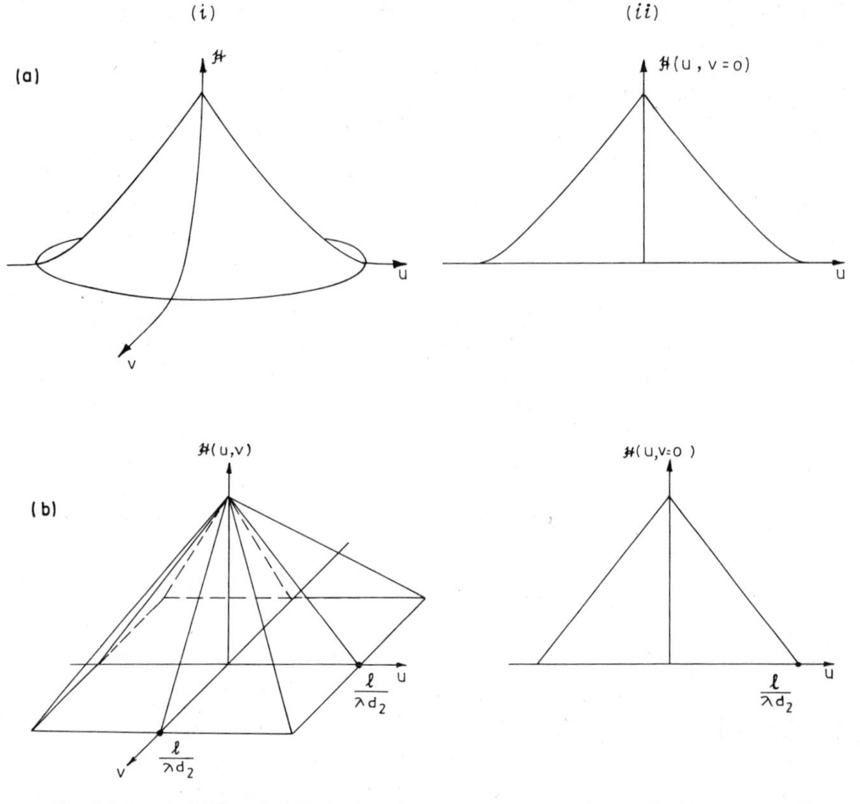

Fig. 7.4. OTF of diffraction-limited optical systems: (a) circular pupil, (b) square pupil.

Problem 7.3. Calculate the OTF of an aberrationless system with square aperture, the pupil function of which is given by

$$p(x, y) = \begin{cases} 1, & |x|, |y| \leq l/2 \\ 0, & |x|, |y| > l/2 \end{cases} \tag{7.4-12}$$

Solution. Once again we have to find the overlap area of two displaced rectangular functions, and the OTF can be shown to be given by

$$\mathscr{H}(u, v) = \Lambda\left(\frac{u\lambda d_2}{l}\right)\Lambda\left(\frac{v\lambda d_2}{l}\right) \tag{7.4-13}$$

where $\Lambda(x)$ is the triangle function defined by the relation

$$\Lambda(x) = \begin{cases} 1 - |x|, & |x| \leq 1 \\ 0, & |x| > 1 \end{cases} \tag{7.4-14}$$

Figure 7.4b shows the OTF of the square aperture as a function of u and v.

Problem 7.4. Consider an object with intensity distribution

$$f(x) = \alpha + \beta \cos 2\pi u x \tag{7.4-15}$$

being imaged through an optical system. The object represented by Eq. (7.4-15) is known as a sinusoidal object of spatial frequency u; notice that the intensity distribution is independent of y coordinate; such patterns are known as one-dimensional patterns. If C_i and C_o represent the contrasts in the image and the object, respectively, then show that the MTF is simply C_i/C_o; the contrast is defined as

$$C = \frac{I_{\max} - I_{\min}}{I_{\max} + I_{\min}} \tag{7.4-16}$$

where $I_{\max}$ and $I_{\min}$ represent the maximum and minimum values of the intensity. This property is used in measuring the MTF of optical systems. [*Note:* In carrying out the convolution, the magnification factor of the system has to be taken into account; see Eq. (7.3-5). This factor involves appropriate scaling of the coordinate system in the image plane and for the sake of simplicity we assume unit magnification.]

Solution. Let $l(x)$ be the intensity line-spread function of the optical system.* Then the intensity distribution in the image of an object of the form given by Eq. (7.4-15) is given by

$$\begin{aligned} g(x) &= \int f(x')\, l(x - x')\, dx' = \int l(x')\, f(x - x')\, dx' \\ &= \int l(x')\{\alpha + \beta \cos[2\pi u(x - x')]\}\, dx' \\ &= \alpha + \tfrac{1}{2}\beta[e^{2\pi i u x} L^*(u) + e^{-2\pi i u x} L(u)] \\ &= \alpha + \beta|L(u)| \cos[2\pi u x + \theta(u)] \end{aligned} \tag{7.4-17}$$

where we have assumed $\int l(x')\, dx' = 1$ and

$$L(u) = \frac{\int l(x')\, e^{2\pi i u x'}\, dx'}{\int l(x')\, dx'} = |L(u)|\, e^{-i\theta(u)} \tag{7.4-18}$$

[cf. Eq. (7.4-7)]. Thus the image of a sinusoidal pattern is also sinusoidal. Furthermore,

$$\frac{C_i}{C_o} = \left\{ \frac{[\alpha + \beta|L(u)|] - [\alpha - \beta|L(u)|]}{[\alpha + \beta|L(u)|] + [\alpha - \beta|L(u)|]} \right\} \left\{ \frac{[\alpha + \beta] - [\alpha - \beta]}{[\alpha + \beta] + [\alpha + \beta]} \right\}^{-1} = |L(u)| \tag{7.4-19}$$

The phase $\theta(u)$ is determined by the displacement of the image distribution from the axis.

* Since the intensity distribution is independent of the y coordinate, it is more convenient to deal with the line-spread function $l(x)$, which is the field distribution produced by a line object.

Problem 7.5. (a) Show that the intensity line-spread function of a slit aperture is proportional to $\operatorname{sinc}^2\zeta$, where $\zeta = (2/\lambda)\, ax/d_2$, $2a$ being the width of the slit. Hence show that for a sinusoidal object of the form given by Eq. (7.4-15),* the light intensity in the image is given by

$$g(x) = \begin{cases} \alpha + (1 - \tfrac{1}{2}\omega)\,\beta \cos \omega\xi & \text{for } 0 < \omega < 2 \\ 0 & \text{for } \omega > 2 \end{cases}$$

where $\omega = \lambda d_2 u/a$ and $\xi = ax/\lambda d_2$.

(b) Show that the intensity point-spread function for a circular aperture is simply the Airy pattern. Thus the corresponding line-spread function will be a line integral through the Airy pattern. The result is (for small ζ)

$$L(\zeta) = 3\left[\frac{1}{1^2 \cdot 3} - \frac{\zeta^2}{1^2 \cdot 3^2 \cdot 5} + \frac{\zeta^4}{1^2 \cdot 3^2 \cdot 5^2 \cdot 7} + \cdots\right]$$

where $\zeta = 2\pi x/F\lambda$. *Reference:* Kottler and Perrin (1966).

Problem 7.6. Consider a periodic bar pattern of the form

$$f(\xi) = \begin{cases} \alpha + \beta & \text{for } -\dfrac{\pi}{2\omega} < \xi < \dfrac{\pi}{2\omega} \\ \alpha - \beta & \text{for } \dfrac{\pi}{2\omega} < \xi < \dfrac{3\pi}{2\omega} \end{cases}$$

with $f(\xi + 2\pi/\omega) = f(\xi)$, where $\omega = \lambda d_2 u/a$ and $\xi = ax/\lambda d_2$. Show that the image function is proportional to

$$\alpha + \frac{4\beta}{\pi} \sum_{n=1}^{n'} \frac{\sin(\tfrac{1}{2}n\pi)}{n} A_s(n) \cos n\omega\xi$$

where $A_s(n) = 1 - \tfrac{1}{2}n\omega$ and n' is the maximum integer less than $2/\omega$. [*Hint:* Make a Fourier expansion of $f(\xi)$.] *Reference:* Kottler and Perrin (1966).

As seen earlier, when the optical system suffers from aberrations, the pupil function is given by Eq. (7.3-8) and in such a case the OTF becomes

$$\mathscr{H}(u, v) = \frac{\iint \exp\{ik[W(\xi + \tfrac{1}{2}\lambda d_2 u, \eta + \tfrac{1}{2}\lambda d_2 v) - W(\xi - \tfrac{1}{2}\lambda d_2 u, \eta - \tfrac{1}{2}\lambda d_2 v)]\}\, d\xi\, d\eta}{\iint d\xi\, d\eta} \tag{7.4-20}$$

where the numerator is the overlap integral of the two displaced pupil

* In many investigations, gratings having sinusoidally varying transmission [see Eq. (7.4-11)] have been taken as the object; however, since it is difficult to obtain a test object with unit contrast and sinusoidal transmission, various other alternatives to the sinusoidal gratings have been proposed. For example, Washer and Rosberry (1951) have used periodic bar patterns with a rectangular waveform, and Katti *et al.* (1969) have carried out investigations using triangular wave objects.

functions and the denominator denotes the area of the pupil. Thus, if the form of wavefront aberration $W(x, y)$ for different kinds of aberrations (such as spherical aberration, defocusing) is determined, then the corresponding OTF can be calculated. We will now discuss the effect of defocusing on the OTF of an otherwise aberration-free system. To determine the OTF we have first to determine the expression for $W(x, y)$. For a slightly defocused system

$$d_2 = \frac{fd_1}{d_1 - f} + \delta l \tag{7.4-21}$$

or

$$\gamma \simeq \beta - \alpha - \frac{2}{k}\delta l(\beta - \alpha)^2$$

where $\delta l = 0$ corresponds to the focused system, and α, β, and γ have been defined in Section 6.4. In deriving Eq. (6.4-2) we had not taken into account the pupil function. If we do so we obtain

$$g(x, y) = \frac{1}{\lambda^2 d_1 d_2} \iiiint f(\xi, \eta) p(\xi', \eta') \exp\{-i\alpha[(\xi' - \xi)^2 + (\eta' - \eta)^2]\}$$
$$\times \exp[i\beta(\xi'^2 + \eta'^2)] \exp\{-i\gamma[(x - \xi')^2 + (y - \eta')^2]\}\, d\xi\, d\eta\, d\xi'\, d\eta' \tag{7.4-22}$$

where it is assumed that the lens is free from aberrations. Using Eq. (7.4-21), Eq. (7.4-22) becomes

$$g(x, y) = \frac{1}{\lambda^2 d_1 d_2} \exp[-i\gamma(x^2 + y^2)] \iiiint f(\xi, \eta)\, P(\xi', \eta')$$
$$\times \exp[-i\alpha(\xi^2 + \eta^2)] \exp\left\{2\pi i\left[\xi'\left(\frac{\xi}{\lambda d_1} + \frac{x}{\lambda d_2}\right)\right.\right.$$
$$\left.\left. + \eta'\left(\frac{\eta}{\lambda d_1} + \frac{y}{\lambda d_2}\right)\right]\right\} d\xi\, d\eta\, d\xi'\, d\eta' \tag{7.4-23}$$

where

$$P(x, y) = p(x, y) \exp[-ikW(x, y)], \qquad W(x, y) = 2\Delta(x^2 + y^2) \tag{7.4-24}$$

with $\Delta = \delta l(\beta - \alpha)^2/k^2$. If we compare Eq. (7.4-24) with Eq. (6.4-4), we immediately see that the defocusing corresponds to a wavefront aberration*

* In the presence of primary and secondary spherical aberration and defocusing, the aberration function $W(x, y)$ is of the form

$$W(x, y) = \omega_{20}(x^2 + y^2) + \omega_{40}(x^2 + y^2)^2 + \omega_{60}(x^2 + y^2)^3$$

where the coefficients ω_{20}, ω_{40}, and ω_{60} denote, respectively, defocusing, primary spherical aberration, and secondary spherical aberration. The corresponding calculation of frequency response has been reported by Goodbody (1958). The influence of astigmatism and of primary coma on the response function of an optical system has been studied by De (1955) and De and Nath (1958).

of $2\Delta(x^2 + y^2)$ and the effective pupil function is $P(x, y)$. Hence the OTF becomes

$$\mathscr{H}(u, v) = \frac{1}{\iint d\xi\, d\eta}\left[\iint \exp\left\{\frac{ik\Delta}{2}\left[\left(\xi + \frac{\lambda d_2 u}{2}\right)^2 + \left(\eta + \frac{\lambda d_2 v}{2}\right)^2 - \left(\xi - \frac{\lambda d_2 u}{2}\right)^2 - \left(\eta - \frac{\lambda d_2 v}{2}\right)^2\right]\right\} d\xi\, d\eta\right]$$

$$= \frac{1}{l^2}\int_{-\frac{1}{2}(l-\lambda d_2|u|)}^{\frac{1}{2}(l-\lambda d_2|u|)}\int_{-\frac{1}{2}(l-\lambda d_2|v|)}^{\frac{1}{2}(l-\lambda d_2|v|)} \exp[2\pi i\Delta d_2(\xi u + \eta v)]\, d\xi\, d\eta \qquad (7.4\text{-}25)$$

where in writing the last expression we have assumed the pupil to be a square of side l. Carrying out the integration, we obtain

$$\mathscr{H}(u, v) = \Lambda\left(\frac{u}{2\omega_0}\right)\Lambda\left(\frac{v}{2\omega_0}\right)\operatorname{sinc}\left[\frac{\Delta l^2}{\lambda}\frac{u}{2\omega_0}\left(1 - \frac{|u|}{2\omega_0}\right)\right] \times \operatorname{sinc}\left[\frac{\Delta l^2}{\lambda}\frac{v}{2\omega_0}\left(1 - \frac{|v|}{2\omega_0}\right)\right] \qquad (7.4\text{-}26)$$

where $\omega_0 = l/2\lambda d_2$ is the cutoff frequency of the coherent system and $\Lambda(x)$ is the triangle function defined by Eq. (7.4-14). In Fig. 7.5 we have plotted the measured and theoretical response curves for a defocused lens for various values of the defocusing parameter. We can show that for large enough values of δl, the OTF can become negative at some values of frequency (see Fig. 7.5a). It was shown in Problem 7.4 that the OTF at a particular frequency is the ratio of the contrast in the image to that in the object of that particular frequency. Thus, a negative value of OTF implies a contrast reversal, i.e., at those frequencies at which the OTF is negative there is a reversal of contrast in the image as compared to that in the object (see Fig. 7.5b).

We have shown above that if the pupil function of an optical system is known, then the transfer function can be obtained by carrying out an autocorrelation [see Eq. (7.4-10)]. We have also made an explicit calculation of the pupil function of a defocused optical system free from other aberrations. In the presence of aberrations, the pupil function is a function of the design data of the system (i.e., refractive indices, radii of curvatures, etc.) and its computation is in general quite cumbersome. A considerable amount of work has been reported on the calculation of the aberration

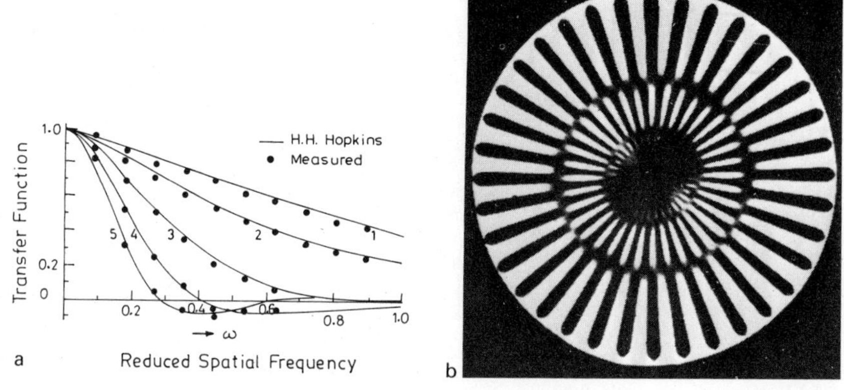

Fig. 7.5. Measured (●) and theoretical (–) response curves for a defocused lens for various values of the defocusing parameter (after Hopkins, 1955). (b) Contrast reversal (after Lindberg, 1954).

function and the subsequent evaluation of the transfer function (see, e.g., Hopkins, 1957, 1962; Goodbody, 1958; Barakat, 1962; Barakat and Morello, 1962; Miyamoto, 1966). Figure 7.6 is the result of such a numerical calculation and shows the effect of defocusing and of primary spherical aberration on the MTF of an $f/4$ optical system operating at $\lambda = 5 \times 10^{-4}$ mm. One can immediately see that in the presence of spherical aberration, the contrast (at least for small values of frequency) becomes poorer. Thus the aberrations in general reduce the MTF and may even cause contrast reversal as shown in Fig. 7.5. The resolution by the slightly defocused system above the frequency 117 lines/mm (see curve b in Fig. 7.6) is usually referred to as spurious. In Fig. 7.7 we have plotted Barakat and Morello's calculations of MTF for rotationally symmetric aberrations for an $f/5$, 66-in.

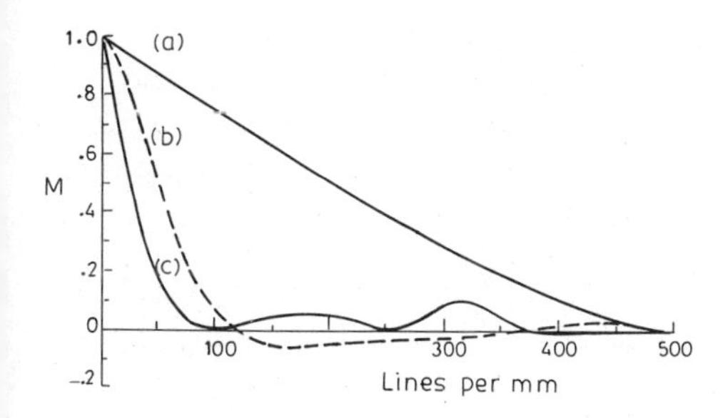

Fig. 7.6. (a) The OTF of an aberrationless system working at $f/4$ at $\lambda = 5 \times 10^{-4}$ mm. (b, c) The same system suffering from defocusing and primary spherical aberration, respectively (after Linfoot, 1964).

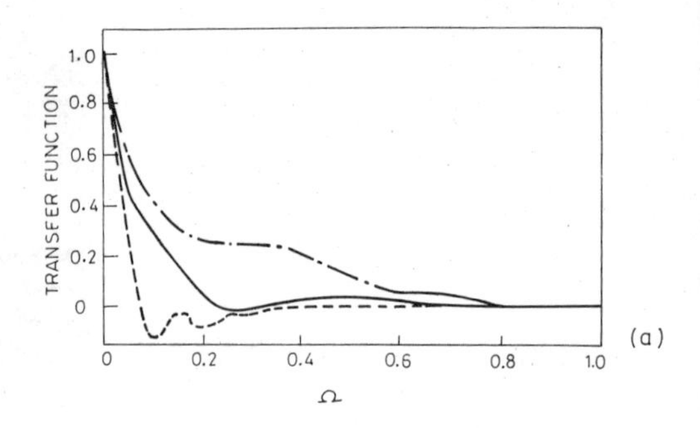

Fig. 7.7. (a) MTF of an $f/5$, 66-in. spherical doublet ($\delta l = -0.018$ in.), and (b) MTF of an $f/5$, 66-in. aspheric doublet ($\delta l = -0.003$ in.). Notice that the aspheric doublet is very well corrected for d light. (– –, c light; - - -, d light; —, e light; after Barakat and Morello, 1962.)

spherical doublet and for an $f/5$, 66-in. aspheric doublet, respectively. The paraxial focus corresponds to $\delta l = 0$, and a negative value of δl implies that we are defocusing toward the marginal focus. The c, d, and e light correspond to 6562.8, 5875.5, and 5460.7 Å, respectively. As can be seen from the figures the aspheric doublet is unquestionably superior over the spherical doublet. One can also see that the aspheric doublet is very well corrected for the d light.

It is clear from the above discussion that the transfer function of an optical system gives the complete imaging qualities of the system. In Problem 7.4 we saw that the OTF can be measured using a sinusoidal object and determining the contrast in the image and in the object. For detailed descriptions of the methods of the determination of the OTF, the reader is referred to Murata (1966) and Rosenhauer (1967).

7.5. Coherence and Resolution

We saw in Problems 7.2 and 7.3 that the cutoff frequency of an incoherent system is twice the cutoff frequency of the same system working in coherent illumination, from which one may deduce that a system working in incoherent illumination is better than the same system working in coherent illumination. However, this is not always true. We will show through examples that one cannot uniquely specify whether a particular system is better for coherent or incoherent illumination; in fact, the resolution of the system depends very much on the phase distribution in the object. For example, let us consider two-point resolution in incoherent and coherent

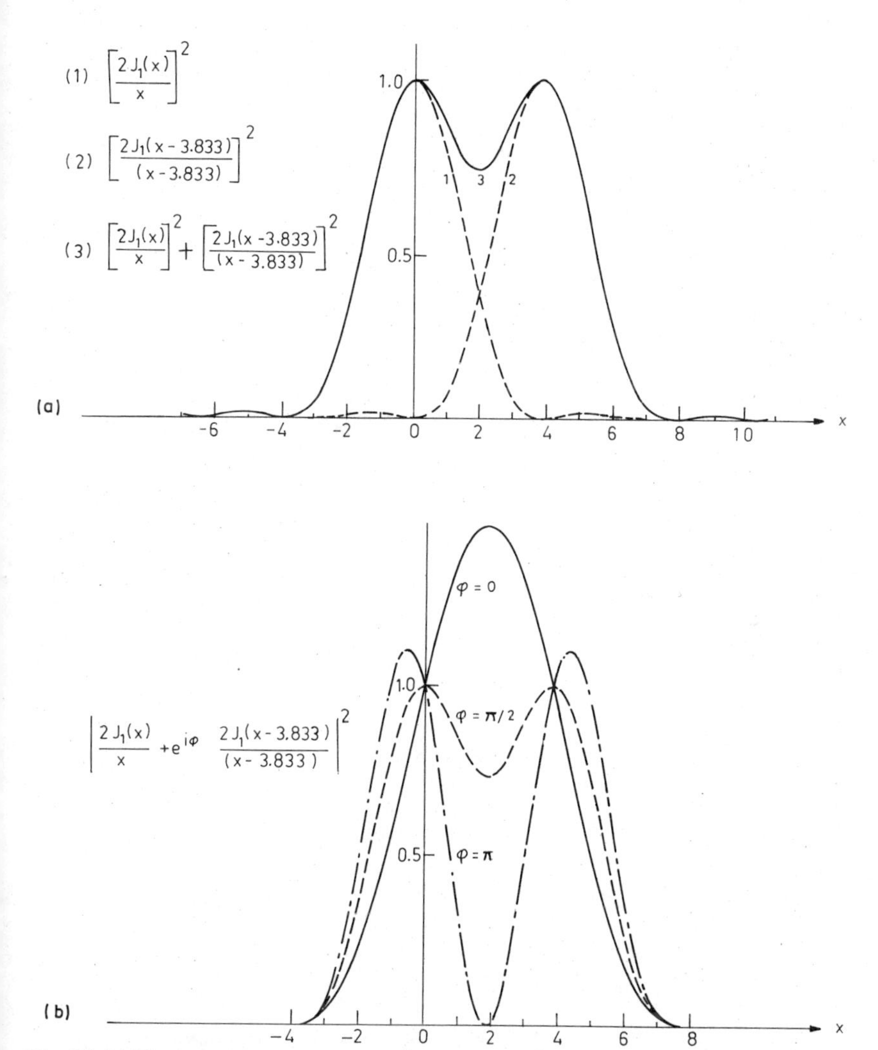

Fig. 7.8. (a) The intensity pattern of two incoherent point sources at the Rayleigh resolution limit. (b) The intensity pattern of two coherent point sources. —, the two points in phase; – –, the two points differ in phase by $\pi/2$; — · —, the two points are out of phase.

illumination. In the incoherent case, under the Rayleigh criterion, the Airy patterns of the two points add up to give a dip of about 27% at the center (see Fig. 7.8a). For the coherent case, we have first to add the amplitudes and then take the modulus square to obtain the resultant intensity pattern.

Since one is adding amplitudes, the relative phases of the two points are very important. For example, if the phase difference between the two points were zero, then the resultant intensity pattern would have no dip in the center corresponding to the separation determined by the Rayleigh limit (see Fig. 7.8b). If the two points are $\pi/2$ out of phase, then the resultant intensity pattern is the same as in incoherent illumination. If the phase difference happens to be π, then the dip at the center is greater than that for the incoherent case (see Fig. 7.8b). Thus, comparing the incoherent and coherent illumination, one observes:

(a) If the two points are in phase, then the resolution in incoherent illumination is higher than in coherent illumination.

(b) If the two points are $\pi/2$ out of phase, then the system behaves equally well under coherent and incoherent illumination.

(c) When the two points are out of phase, then the system working in coherent illumination has a higher resolution than in incoherent illumination.

Thus, there is no single criterion by which one can definitely say whether coherent or incoherent illumination is better for two-point resolution.

Problem 7.7. Consider the imaging of two objects with transmittances

$$f_1(x) = |\cos(2\pi x/b)| \tag{7.5-1}$$

$$f_2(x) = \cos(2\pi x/b) \tag{7.5-2}$$

by a system with a circular aperture of radius a satisfying $\lambda d_2/b < a < 2\lambda d_2/b$. Compare the imaging of the two objects under coherent and incoherent illuminations.

Solution. For the object specified by Eq. (7.5-1), the spatial frequency is $2/b$. Since this frequency is greater than the cutoff frequency of the system in coherent illumination ($= a/\lambda d_2$) this object will not be imaged by the system under coherent illumination. At the same time since the frequency of the object lies below the cutoff frequency under incoherent illumination ($= 2a/\lambda d_2$), this object will be imaged under incoherent illumination.

For the object specified by a transmittance of the form Eq. (7.5-2), the spatial frequency is $1/b$. Hence it will be imaged both under coherent and incoherent illuminations. At the same time, since the transfer function of an incoherent system falls off with increasing spatial frequency [see Eq. (7.4-11)] and that of the coherent system is constant within the cutoff value [see Eq. (7.4-4)], the contrast in coherent illumination will be much better than in incoherent illumination.

8

Holography

8.1. *Introduction*

In taking a photograph, an image of the object is formed on a plane where some photosensitive material (e.g., a photographic plate) is kept. The photosensitive material reacts to the *intensity* of light falling on it and consequently, after developing and fixing of the photographic plate, one obtains a permanent record of the intensity distribution that existed at the plane occupied by the photographic plate when it was exposed. In this process, the phase distribution that existed in the plane of the photographic plate is completely lost. Thus, if the amplitude distribution at the plane of the photographic plate was, say, $A(x, y)\exp[i\varphi(x, y)]$, where $A(x, y)$ and $\varphi(x, y)$ are real functions of x and y (the x-y plane being the plane of the photograph), then the recorded pattern is proportional to $|A(x, y)\, e^{i\varphi(x,y)}|^2 = A^2(x, y)$. Thus, the phase information, which is contained in $\varphi(x, y)$, is lost. In contrast, the principle of holography, which we are going to develop in this chapter, is one in which not only the amplitude distribution but also the phase distribution can be recorded.

The principle of holography was introduced by Gabor in 1948 in his attempt to increase the resolving power of electron microscopes. He showed that one can indeed record both the amplitude and the phase of a wave by making use of an interferometric principle. In such a method, a coherent reference wave is added to the wave emanating from the object and the resulting interference pattern is recorded. This interference pattern is characteristic of the object and is called the hologram (hologram means total recording, i.e., a record of both amplitude and phase). Although when we look at a hologram it does not resemble the object, when this recorded pattern is illuminated by a suitably chosen reconstruction wave, then out of the many component waves emerging from the hologram, one wave completely resembles the object wave in both the amplitude and phase structure. Thus, when this wave is viewed, the effect is as if the object were still in position even though the object may not be present there. Since during reconstruction the object wave is itself emerging from the hologram, one has all the effects of three dimensionality while viewing such a wave. One can indeed move and "look behind" the objects present and one can focus at different distances and thus feel the depth in the image.

Although holography was discovered by Gabor in 1948, not much work was reported on this subject until the advent of the laser. However, since laser sources became available, there has been a tremendous amount of work in this field. There have been also numerous applications in many diverse areas (see, e.g., Thompson, 1971; Nalimov, 1969; Stroke, 1969; Shulman, 1970; Cathey, 1974; Collier *et al.*, 1971). In Sections 8.2–8.6, we discuss the basic principles of holography. In Section 8.6, we discuss Fourier transform holograms, which find application in character recognition problems. In Section 8.7, we give an elementary theory of volume holograms. Finally, in Section 8.8, we briefly discuss the various applications of holography.

8.2. The Underlying Principle

In this section, we will consider a simple configuration of taking a hologram and see how the object wave may be reconstructed. The experi-

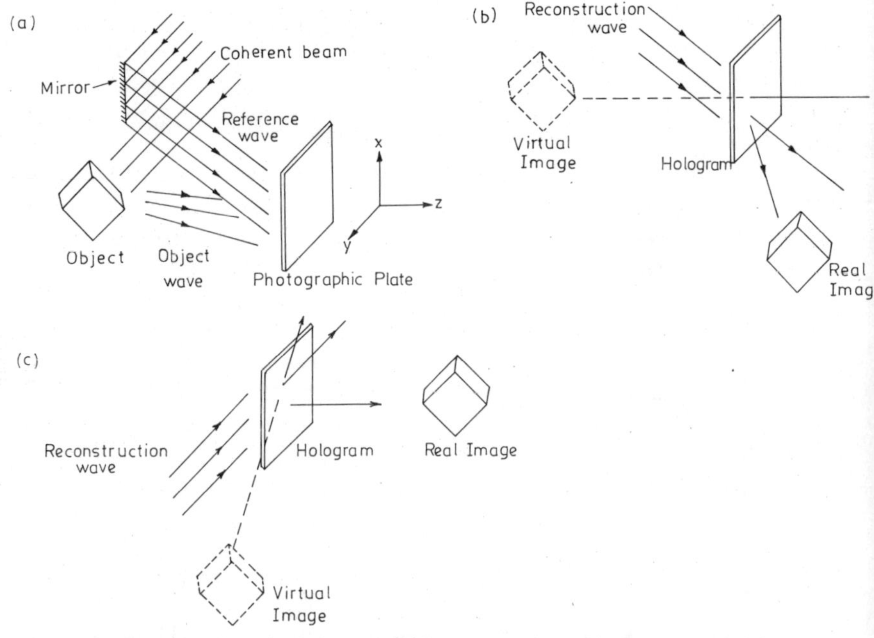

Fig. 8.1. (a) The recording of a hologram. (b) Reconstruction with a wave that is identical to the reference wave. (c) Reconstruction by the conjugate of the reference wave.

mental set-up is shown in Fig. 8.1a, where part of a coherent beam of light is allowed to fall on a plane mirror and the remaining part is made to illuminate the object whose hologram is to be recorded. The light scattered from the object is made to fall on a photographic plate along with the reference wave, as shown in Fig. 8.1a. The photographic plate records the resulting interference pattern. This recorded interference pattern is developed suitably and forms the hologram. We assume that the photographic plate corresponds to the x-y plane. Let $O(x, y)$ represent the field at the photographic plate due to the object wave and $R(x, y)$ the field due to the reference wave (the reference wave is usually a plane wave).† Thus the resultant field on the photographic plate is given by

$$U(x, y) = O(x, y) + R(x, y) \tag{8.2-1}$$

The photographic plate responds only to the intensity variation, which will be given by

$$I(x, y) = |U(x, y)|^2 = |O(x, y)|^2 + |R(x, y)|^2 + O(x, y)\,R^*(x, y) + O^*(x, y)\,R(x, y) \tag{8.2-2}$$

where, in writing Eq. (8.2-2), we have carried out a time averaging (see Section 5.3). When this photographic plate is developed suitably, it will represent the hologram of the object.

In order to reconstruct the image we illuminate the photographic plate by a reconstruction wave that has a field distribution $R(x, y)$ on the plane of the hologram (see Fig. 8.1b). The wave transmitted by the hologram will depend on its transmittance, which is defined as the ratio of the amplitude of the emergent wave to that of the incident wave. The dependence of the amplitude transmittance on the intensity (incident during the exposure) is, in general, nonlinear. Figure 8.2 gives the variation of the amplitude transmittance with the exposure of the photographic plate. The exposure is defined as the total energy that has fallen on the photographic plate; it is equal to the product of the intensity falling on the plate and the exposure time. The curve shows a nonlinear dependence, but if we restrict ourselves to the region AB shown in Fig. 8.2 then the dependence is almost linear.

† We are assuming here that the fields are monochromatic so that the time dependence is of the form $\exp(i\omega t)$. Now, if we have two functions f and g whose time variations are of the form $\exp(i\omega t)$, then

$$\langle \operatorname{Re} f \operatorname{Re} g\rangle = \tfrac{1}{2}\langle \operatorname{Re} f^* g\rangle$$

[see the discussion following Eq. (5.3-5)]. The actual field on the plane of the photographic plate is $\operatorname{Re}\{[O(x, y) + R(x, y)]\,e^{i\omega t}\}$ and the intensity will therefore be proportional to

$$[O(x, y) + R(x, y)]^*\,[O(x, y) + R(x, y)]$$

which gives Eq. (8.2-2).

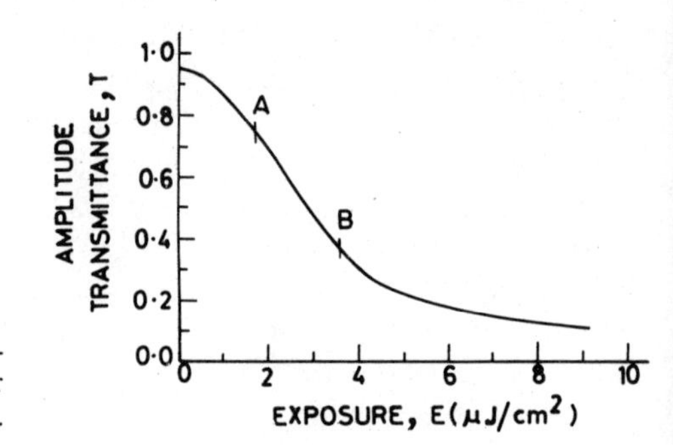

Fig. 8.2. The *T–E* curve for Agfa-Gevaert 10E70 emulsion (from Collier *et al.*, 1971; reprinted with permission).

In the present and future analysis, we will assume that the amplitude transmittance T is linearly related to the intensity incident during exposure I, so that we may write

$$T = I \tag{8.2-3}$$

where we have omitted a constant of proportionality.†

Thus, the amplitude transmittance T of the developed transparency will be

$$T(x, y) = |O(x, y)|^2 + |R(x, y)|^2 + O(x, y)\, R^*(x, y) + O^*(x, y)\, R(x, y) \tag{8.2-4}$$

If we now illuminate this transparency with the reference wave $R(x, y)$, then the field emerging from the hologram is given by

$$\begin{aligned}\Phi(x, y) = T(x, y)\, R(x, y) &= [|O(x, y)|^2 + |R(x, y)|^2]\, R(x, y) \\ &+ O(x, y) |R(x, y)|^2 + O^*(x, y)\, R(x, y)\, R(x, y)\end{aligned} \tag{8.2-5}$$

If we now assume that the reference wave is a plane wave with its propagation vector lying in the x-z plane, then

$$R(x, y) = R_0 \exp(-ikx \sin\theta) \tag{8.2-6}$$

where $k \sin\theta$ and $k \cos\theta$ represent the x and z components of $\mathbf{k}$. Thus,

$$\Phi(x, y) = [|O(x, y)|^2 + R_0^2]\, R_0 e^{-ikx \sin\theta} + O(x, y)\, R_0^2 + R_0^2 O^*(x, y)\, e^{-2ikx \sin\theta} \tag{8.2-7}$$

The first term on the right-hand side corresponds to a wave propagating in the direction of the reference wave with an amplitude distortion. The second

† We will assume that the film characteristics are such that even the largest spatial frequency present in the object is recordable. A thorough treatment taking into account the transfer function of the film has been given by Collier *et al.* (1971).

term is proportional to the object wave $O(x, y)$, and the effect of viewing this wave will be the same as the effect of viewing the object itself: all the essential features of the object (such as depth) will be reproduced (see Fig. 8.1b). The last term on the right-hand side of Eq. (8.2-7) is proportional to $O^*(x, y)$, which is the conjugate of the object wave. This wave, in general, produces a real image lying on the opposite side of the hologram as shown in Fig. 8.1b. However, the additional phase term not only tilts the wave but also introduces distortion in the image. On the other hand, if the hologram is illuminated by a wave such that the field distribution (on the hologram plane) is $R^*(x, y)$, then

$$\Phi(x, y) = [|O(x, y)|^2 + R_0^2]\, R^*(x, y) + O(x, y)\, R^*(x, y)\, R^*(x, y) + O^*(x, y)\, R_0^2 \tag{8.2-8}$$

In this case, the virtual image will be distorted and the real image produced by $O^*(x, y)$ will be free from distortion (see Fig. 8.1c).

It may be mentioned that since holography is essentially an interference phenomenon, the radiation should satisfy certain coherence requirements

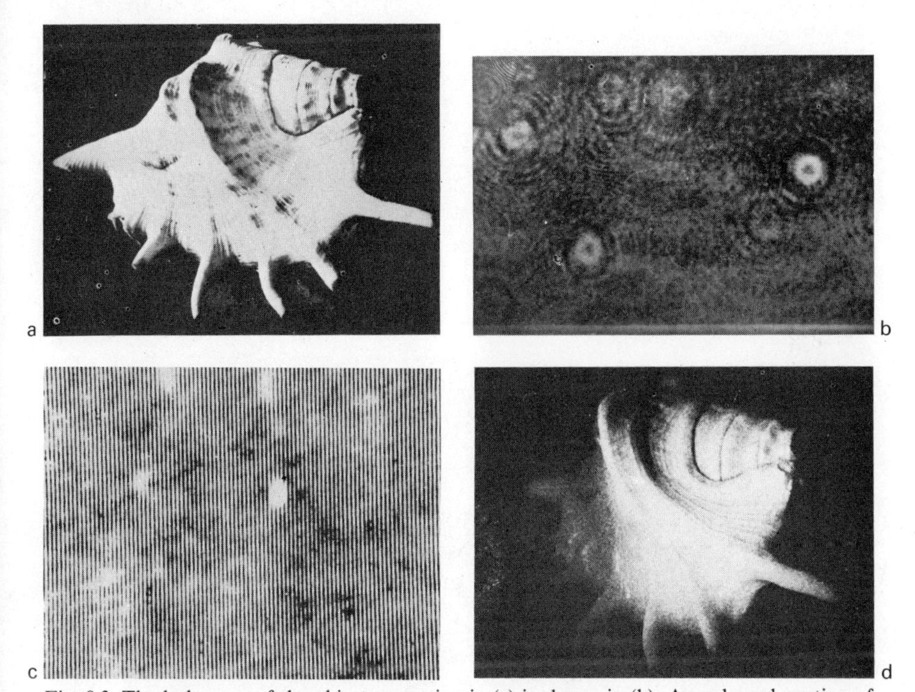

Fig. 8.3. The hologram of the object appearing in (a) is shown in (b). An enlarged portion of the hologram is shown in (c). (d) gives the reconstructed image from the hologram (photographs courtesy of Dr. R. S. Sirohi).

to form good holograms. A critical requirement for the recording of interference fringes is that they be stable over the recording time. It is also required that the maximum path difference be less than the coherence length of the radiation. In fact, each object point must be capable of forming a stable interference pattern with the reference beam. In reconstruction, since the wave from each point of the hologram has to interfere to produce the object wave, the reconstruction wave must again be spatially coherent.

Figure 8.3a shows an object whose hologram is shown in Fig. 8.3b. Notice the little resemblance between the object and the hologram. Figure 8.3c shows an enlarged portion of the hologram and Fig. 8.3d shows the reconstructed image.

8.3. Interference between Two Plane Waves

In this section, we will consider the interference between two inclined plane waves and show explicitly the reconstruction of the object and its conjugate waves. Let us assume that the photographic plate (which is to record the interference pattern) lies in the plane $z = 0$. The fields due to the two plane waves at the plane $z = 0$ are given by

$$u_1 = A \exp(-ikx \sin \theta_1) \tag{8.3-1a}$$

and

$$u_2 = B \exp(-ikx \sin \theta_2) \tag{8.3-1b}$$

where A and B represent the amplitudes of the two plane waves, and θ_1 and θ_2 represent the angles subtended by the propagation directions of the two plane waves with the z axis (see Fig. 8.4). The resultant field at the photographic plane will be $u_1 + u_2$ and the intensity recorded by the photographic plate will be given by

$$\begin{aligned} I = |u_1 + u_2|^2 &= A^2 + B^2 + AB \exp[-ikx(\sin \theta_1 - \sin \theta_2)] \\ &\quad + AB \exp[ikx(\sin \theta_1 - \sin \theta_2)] \\ &= A^2 + B^2 + 2AB \cos[kx(\sin \theta_1 - \sin \theta_2)] \end{aligned} \tag{8.3-2}$$

Thus, the photographic plate has recorded the phase distribution also. The developed plate (which is the hologram) represents essentially a sinusoidal grating. We now illuminate the hologram by a plane wave whose direction of propagation makes an angle θ_3 with the z axis (see Fig. 8.4b). Thus, the wave emerging from the hologram is given by

$$\begin{aligned} \Phi = u_3 I &= (A^2 + B^2) C \exp(-ikx \sin \theta_3) + \\ &\quad + ABC \exp[-ikx(\sin \theta_1 - \sin \theta_2 + \sin \theta_3)] \\ &\quad + ABC \exp[-ikx(\sin \theta_2 - \sin \theta_1 + \sin \theta_3)] \end{aligned} \tag{8.3-3}$$

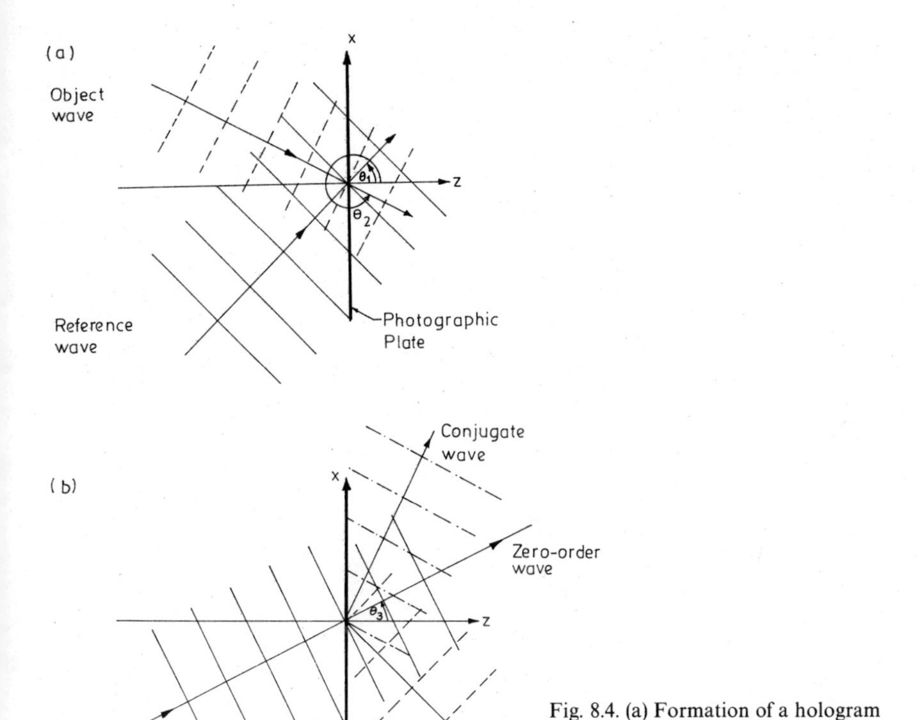

Fig. 8.4. (a) Formation of a hologram when the object and the reference waves are plane waves. (b) Reconstruction by a plane wave.

Let us consider the simple case when $\theta_2 = \theta_3 = 0$. Then

$$\Phi = (A^2 + B^2)\,C + ABC\exp(-ikx\sin\theta_1) + ABC\exp(ikx\sin\theta_1) \qquad (8.3\text{-}4)$$

The first term in Eq. (8.3-4) corresponds to a plane wave traveling along the z direction. [Notice that Eq. (8.3-4) gives the field in the plane $z = 0$.] The second term gives a plane wave traveling at an angle θ_1, which is now the reconstruction of the original wave u_1, and the third term represents another plane wave traveling along the direction $-\theta_1$, which is the conjugate of the original wave u_1. These represent the different orders of diffraction. Since the hologram is simply a sinusoidal grating, only two first orders are present. Thus, it is possible to reconstruct the original wavefront from the recording of the intensity pattern.

In Section 8.4, we consider the interference between waves emanating from two point sources and show the formation of real and virtual images.

8.4. Point Source Holograms

Let $S_1(x_1, y_1, z_1)$ and $S_2(x_2, y_2, z_2)$ represent two point sources emitting radiation of wavelength λ (see Fig. 8.5a). The plane of the photographic film is represented by $z = 0$. Since z_1 and z_2 are negative quantities, we introduce

$$\zeta_1 = |z_1|, \qquad \zeta_2 = |z_2|$$

We assume that

$$|x_1|, |y_1|, |x_2|, |y_2| \ll \zeta_1, \zeta_2 \tag{8.4-1}$$

The field due to the point source S_1 on the plane of the plate is

$$u_1 = A' e^{-ikr}/r \tag{8.4-2}$$

where $k = 2\pi/\lambda$, and

$$\begin{aligned} r_1 = S_1 P &= [(x_1 - \xi)^2 + (y_1 - \eta)^2 + z_1^2]^{1/2} \\ &= \zeta_1 \left[1 + \frac{x_1^2 + y_1^2}{\zeta_1^2} + \frac{\xi^2 + \eta^2}{\zeta_1^2} - \frac{2(x_1\xi + y_1\eta)}{\zeta_1^2} \right]^{1/2} \\ &\simeq \zeta_1 + \frac{x_1^2 + y_1^2}{2\zeta_1} + \frac{\xi^2 + \eta^2}{2\zeta_1} - \frac{(x_1\xi + y_1\eta)}{\zeta_1} \end{aligned} \tag{8.4-3}$$

The point $P(\xi, \eta, 0)$ represents an arbitrary point on the photographic plate. Thus

$$u_1 \simeq \frac{A_1}{\zeta_1} \exp\left\{ -\frac{ik}{2\zeta_1} [\xi^2 + \eta^2 - 2(x_1\xi + y_1\eta)] \right\} \tag{8.4-4}$$

where we have absorbed the constant phase factor

$$\exp\left\{ -ik \left[\zeta_1 + \frac{x_1^2 + y_1^2}{2\zeta_1} \right] \right\}$$

in the coefficient A_1 and have put $r_1 = \zeta_1$ in the denominator. Equation (8.4-4) represents the quadratic approximation of a spherical wave.

In an exactly similar manner, if u_2 represents the field at the plane $z = 0$ due to the point source S_2, then

$$u_2 \simeq \frac{A_2}{\zeta_2} \exp\left\{ -\frac{ik}{2\zeta_2} [\xi^2 + \eta^2 - 2x_2\xi - 2y_2\eta] \right\} \tag{8.4-5}$$

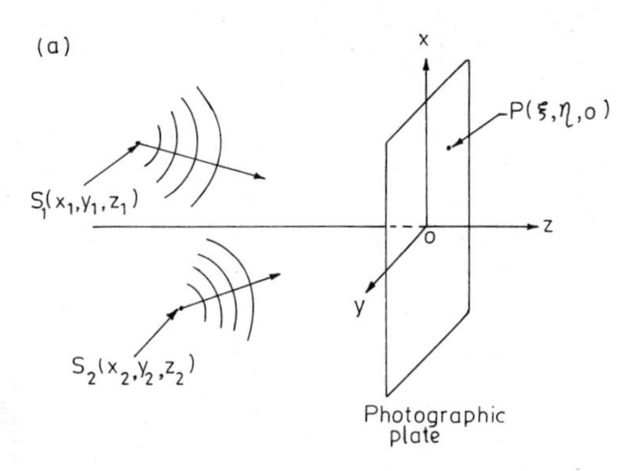

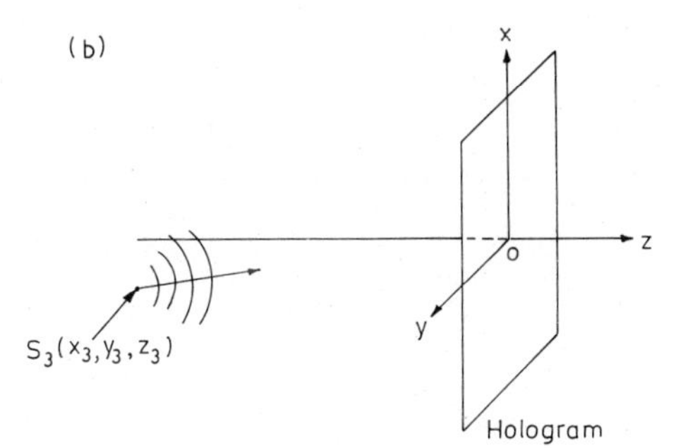

Fig. 8.5. (a) Hologram formed by two spherical waves emanating from points S_1 and S_2. (b) Reconstruction of the hologram.

Since the resultant field on the plane $z = 0$ is $u_1 + u_2$, the intensity pattern recorded will be

$$I = |u_1 + u_2|^2 = \frac{|A_1|^2}{\zeta_1^2} + \frac{|A_2|^2}{\zeta_2^2} + \frac{A_1 A_2^*}{\zeta_1 \zeta_2} e^{-i\varphi} + \frac{A_1^* A_2}{\zeta_1 \zeta_2} e^{i\varphi} \qquad (8.4\text{-}6)$$

where

$$\varphi = \frac{k}{2}\left[(\xi^2 + \eta^2)\left(\frac{1}{\zeta_1} - \frac{1}{\zeta_2}\right) - 2\xi\left(\frac{x_1}{\zeta_1} - \frac{x_2}{\zeta_2}\right) - 2\eta\left(\frac{y_1}{\zeta_1} - \frac{y_2}{\zeta_2}\right)\right] \qquad (8.4\text{-}7)$$

We now illuminate the hologram by a point source placed at the point $S_3(x_3, y_3, z_3)$ (see Fig. 8.5b) emitting a radiation of wavelength λ'. Under the quadratic approximation, the field produced on the hologram plane is given by

$$u_3 \simeq \frac{A_3}{\zeta_3} \exp\left\{ -\frac{ik'}{2\zeta_3}[\xi^2 + \eta^2 - 2x_3\xi - 2y_3\eta] \right\} \qquad (8.4\text{-}8)$$

where $k' = 2\pi/\lambda'$ and $\zeta_3 = |z_3|$. When this wave illuminates the hologram, the emergent wave is given by

$$\Phi = u_3 T = u_3 I = \left(\frac{|A_1|^2}{\zeta_1^2} + \frac{|A_2|^2}{\zeta_2^2}\right)\frac{A_3}{\zeta_3}\exp\left[-\frac{ik'}{2\zeta_3}(\xi^2 + \eta^2 - 2x_3\xi - 2y_3\eta)\right]$$
$$+ \frac{A_1 A_2^* A_3}{\zeta_1\zeta_2\zeta_3}\exp\left[-\frac{ik'}{2z_p}(\xi^2 + \eta^2 - 2\xi x_p - 2\eta y_p)\right]$$
$$+ \frac{A_1^* A_2 A_3}{\zeta_1\zeta_2\zeta_3}\exp\left[-\frac{ik'}{2z_c}(\xi^2 + \eta^2 - 2\xi x_c - 2\eta y_c)\right] \tag{8.4-9}$$

where

$$z_p = \left(\frac{\mu}{\zeta_1} - \frac{\mu}{\zeta_2} + \frac{1}{\zeta_3}\right)^{-1} = \frac{\zeta_1\zeta_2\zeta_3}{\zeta_1\zeta_2 + \mu\zeta_2\zeta_3 - \mu\zeta_1\zeta_3} \tag{8.4-10a}$$

$$x_p = z_p\left[\mu\left(\frac{x_1}{\zeta_1} - \frac{x_2}{\zeta_2}\right) + \frac{x_3}{\zeta_3}\right] = \frac{x_3\zeta_1\zeta_2 + \mu x_1\zeta_2\zeta_3 - \mu x_2\zeta_1\zeta_3}{\zeta_1\zeta_2 + \mu\zeta_2\zeta_3 - \mu\zeta_1\zeta_3} \tag{8.4-10b}$$

$$y_p = z_p\left[\mu\left(\frac{y_1}{\zeta_1} - \frac{y_2}{\zeta_2}\right) + \frac{y_3}{\zeta_3}\right] = \frac{y_3\zeta_1\zeta_2 + \mu y_1\zeta_2\zeta_3 - \mu y_2\zeta_1\zeta_3}{\zeta_1\zeta_2 + \mu\zeta_2\zeta_3 - \mu\zeta_1\zeta_3} \tag{8.4-10c}$$

$$z_c = \left(\frac{\mu}{\zeta_2} - \frac{\mu}{\zeta_1} + \frac{1}{\zeta_3}\right)^{-1} = \frac{\zeta_1\zeta_2\zeta_3}{\zeta_2\zeta_1 + \mu\zeta_1\zeta_3 - \mu\zeta_2\zeta_3} \tag{8.4-10d}$$

$$x_c = z_c\left[\mu\left(\frac{x_2}{\zeta_2} - \frac{x_1}{\zeta_1}\right) + \frac{x_3}{\zeta_3}\right] = \frac{\zeta_1\zeta_2 x_3 + \mu x_2\zeta_1\zeta_3 - \mu x_1\zeta_2\zeta_3}{\zeta_2\zeta_1 + \mu\zeta_1\zeta_3 - \mu\zeta_2\zeta_3} \tag{8.4-10e}$$

$$y_c = z_c\left[\mu\left(\frac{y_2}{\zeta_2} - \frac{y_1}{\zeta_1}\right) + \frac{y_3}{\zeta_3}\right] = \frac{\zeta_1\zeta_2 y_3 + \mu y_2\zeta_1\zeta_3 - \mu y_1\zeta_2\zeta_3}{\zeta_2\zeta_1 + \mu\zeta_1\zeta_3 - \mu\zeta_2\zeta_3} \tag{8.4-10f}$$

and $\mu = \lambda'/\lambda$.

If we compare the first term on the right-hand side of Eq. (8.4-9) with Eq. (8.4-4), we see that it represents a spherical wave emanating from the point (x_3, y_3, z_3). Thus the first term is the reconstruction wave itself (with amplitude modification). The second term represents a spherical wave, but the sign of z_p will determine whether the spherical wave is diverging or converging. For $z_p > 0$ it will be a diverging spherical wave emanating from the point $(x_p, y_p, -z_p)$. Similarly, if $z_p < 0$, it will be a converging spherical wave, converging to the point $(x_p, y_p, |z_p|)$. Thus, a positive z_p corresponds to a diverging spherical wave; hence a virtual image and a negative z_p corresponds to a converging spherical wave and therefore a real image. The last term on the right-hand side of Eq. (8.4-9) can be interpreted similarly. It should be mentioned that in the quadratic approximation, i.e., when Eq. (8.4-4) is valid there is no distortion in the two images.* Furthermore, since the phase dependence of the second term in Eq. (8.4-9) has the same form as the object wave, it is said to give rise to the primary image; the last term, which has a phase dependence conjugate to that of the object wave gives rise to the conjugate image: hence the subscripts p and c.

The magnification in the primary image is dx_p/dx_1, where dx_p represents the change in x_p when x_1 is changed by dx_1. Thus,

$$M_p = \frac{dx_p}{dx_1} = \frac{dy_p}{dy_1} = \mu \frac{\zeta_2\zeta_3}{\zeta_1\zeta_2 + \mu\zeta_2\zeta_3 - \mu\zeta_1\zeta_3} \tag{8.4-11}$$

$$M_c = \frac{dx_c}{dx_1} = \frac{dy_c}{dy_1} = -\mu \frac{\zeta_2\zeta_3}{\zeta_2\zeta_1 + \mu\zeta_1\zeta_3 - \mu\zeta_2\zeta_3} \tag{8.4-12}$$

where M_p and M_c represent the magnifications in the primary and conjugate images, respectively, and use has been made of Eq. (8.4-10).

We will now consider some special cases:

(1). When the source corresponding to the reconstruction wave is placed at the same point as the source corresponding to the reference wave and it emits radiation of the same wavelength as the reference source, then we have

$$x_3 = x_2, \qquad y_3 = y_2, \qquad z_3 = z_2, \qquad \mu = 1 \tag{8.4-13}$$

Thus Eqs. (8.4-10a)–(8.4-10f) simplify to

$$\begin{aligned} z_p &= \zeta_1, & x_p &= x_1, & y_p &= y_1 \\ z_c &= \frac{\zeta_3\zeta_1}{2\zeta_1 - \zeta_3}, & x_c &= \frac{2x_3\zeta_1 - x_1\zeta_3}{2\zeta_1 - \zeta_3}, & y_c &= \frac{2y_3\zeta_1 - y_1\zeta_3}{2\zeta_1 - \zeta_3} \end{aligned} \tag{8.4-14}$$

Thus, the second term on the right-hand side of Eq. (8.4-9) represents a virtual image of the object point. The last term of Eq. (8.4-9) represents a real or a virtual image depending on whether z_c is negative or positive, i.e., whether ζ_3 is less than or greater than $2\zeta_1$. The magnification corresponding to the virtual image is unity. However, the conjugate image does not, in general, correspond to unit magnification.

(2). For $\zeta_1 = \zeta_2$, the expression for the magnification [see Eq. (8.4-11)]

* An analysis of hologram aberrations has been given by Meier (1965); the final results can also be found in Smith (1975). It can be shown that if the reconstruction wave is identical to the reference wave, then one of the images is free from all aberrations. In particular, if one uses plane waves, then the primary image will be free from aberrations. On the other hand, if the plane wave falls on the hologram as shown in Fig. 8.1c, then the conjugate image is free from aberrations. Furthermore, whenever one uses a different λ for reconstruction, i.e., when $\lambda' \neq \lambda$, then one invariably introduces aberrations unless the hologram is also scaled accordingly.

becomes

$$M_p = \mu\zeta_3/\zeta_1 \tag{8.4-15}$$

which shows that by increasing the wavelength of the reconstruction wave, we can increase the magnification. Further, if ζ_1 is also equal to ζ_3, then the magnification is simply μ. For such a case $z_p = z_c = \zeta_3$ and both primary and conjugate images are virtual and are formed in the original object plane. For such a configuration the terms quadratic in ξ and η in the expression for φ vanish; this corresponds to the lensless Fourier transform hologram configuration (see Section 8.6).

(3). Let us consider the case when

$$x_1 = y_1 = x_2 = y_2 = x_3 = y_3 = 0 \tag{8.4-16}$$

i.e., the sources for the reference and reconstruction waves are situated on the z axis and the object point also lies on the axis. For such a case Eq. (8.4-10) tells us that $x_p = y_p = x_c = y_c = 0$. Also, if we assume that the object is close to the source, then we may write

$$\zeta_2 = \zeta_1 + \varepsilon \tag{8.4-17}$$

where ε is assumed to be small. If $\zeta_3 = \zeta_2$ and $\mu = 1$, Eqs. (8.4-10a) and (8.4-10d) give

$$z_p = \zeta_1 = \zeta_2 - \varepsilon, \qquad z_c \simeq \zeta_2 + \varepsilon \tag{8.4-18}$$

Thus the two images are virtual and situated symmetrically about the reference source.

(4). When the reference and reconstruction waves are plane waves propagating in the $+z$ direction, then in addition to Eq. (8.4-16) we have $\zeta_2 = \zeta_3 = \infty$, and for $\mu = 1$, we get

$$z_p = \zeta_1, \qquad z_c = -\zeta_1 \tag{8.4-19}$$

Thus one of the images is virtual and the other real, and the images are situated symmetrically about the hologram plane.

Figure 8.6 shows an arrangement similar to that used by Gabor (1948), in which a plane wave falls on a highly transparent object and the transmitted wave interferes with the scattered wave to form a hologram. During reconstruction a plane wave is allowed to fall on the hologram (Fig. 8.6b), and as discussed above, one obtains a virtual image on the left of the hologram and a real image on the right. Thus while viewing either of the images, one also views a defocused image. This is one of the drawbacks of such an arrangement. Also, the object has to be highly transparent so that one has a strong coherent background. To overcome this difficulty, Leith and Upatneiks (1962) have suggested a modified technique in which the reference beam instead of being incident parallel to the object beam is

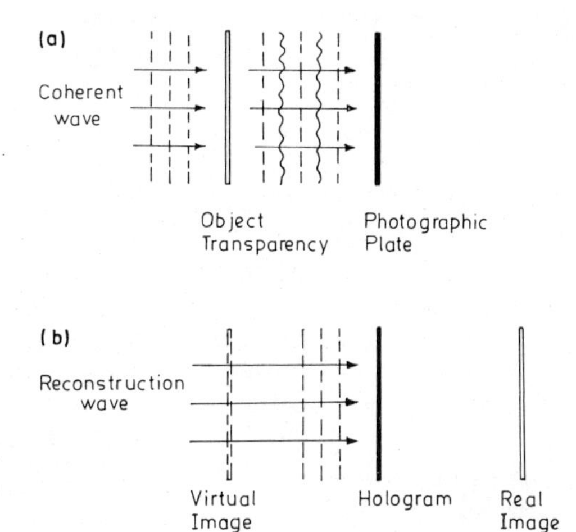

Fig. 8.6. (a) Recording of a hologram of a transparency by a method similar to that used originally by Gabor. (b) Reconstruction.

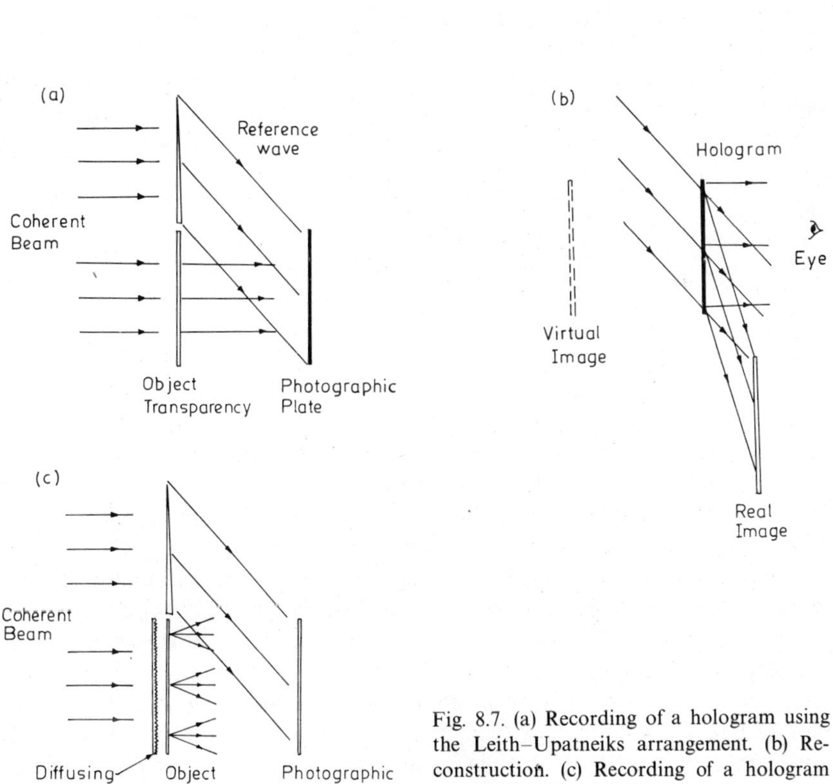

Fig. 8.7. (a) Recording of a hologram using the Leith–Upatneiks arrangement. (b) Reconstruction. (c) Recording of a hologram using diffuse illumination.

incident at an angle (Fig. 8.7a). This results in a separation of the primary and conjugate images (see Fig. 8.7b) and extends the principle to objects that are not highly transparent and also to continuous-tone objects.

8.5. Diffuse Illumination of the Object

Let us consider the recording of the hologram of a transparency illuminated by a plane wave as shown in Fig. 8.7a. It can be seen at once that there is almost a one to one correspondence between the object and the hologram and the object detail is recorded at small areas. Thus, corresponding to transparent portions in the object, there are portions in the hologram plane with a high density of the incident radiation, and corresponding to opaque portions, there are very low intensity levels. In fact, the process results in a shadowgram of the object. Since high-intensity ratios are involved, it is difficult everywhere to restrict to the linear region of the emulsion of the photographic plate (see Fig. 8.2). This will lead, in general, to aberrations in the image. Also when such a hologram is used for reconstruction, then to view the complete virtual image one has to shift one's position because each small region of the hologram corresponds to a small region in the object. In fact, viewing such images requires optical aids since only that portion of the image that is between the observer and the source is visible at one time.

The preceding problems may be overcome by inserting a diffusing screen (like a piece of ground glass) between the illuminating beam and the transparency (see Fig. 8.7c). The diffusing screen scatters the illuminating wave over a wide angle. Consequently, information from each point of the object transparency is spread over the hologram plane and each point of the hologram plane contains information about the complete object transparency. Thus even when highly transparent and opaque regions are present, high- and low-intensity regions do not result in the hologram as in the case of nondiffuse illumination. In fact, one can operate in the linear range of the emulsion. From holograms formed in such a manner, the original object transparency can be viewed from a single position of the observer without having to change position or use optical aids. When the viewing position is changed, the perspective in the image is changed. Since each object point illuminates the complete hologram, even if the holographic plate is broken into different fragments, each fragment is capable of reconstructing the entire object; of course, the resolution in the image decreases as the size of the hologram plate decreases. In fact, the hologram formed by diffuse illumination is more immune to imperfections of the recording plate than a picture recorded by conventional photographic techniques.

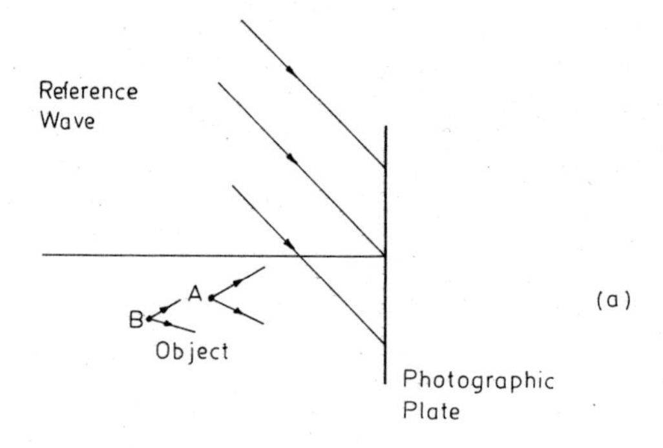

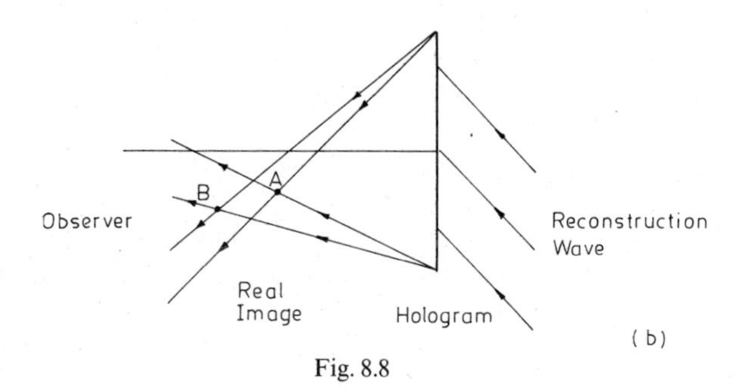

Fig. 8.8

It should be mentioned here that object points closest to the hologram plane are reproduced closest to the hologram plane in the real image. Thus, if A and B represent two object points as shown in Fig. 8.8a, when they are reconstructed they appear as shown in Fig. 8.8b in the real image. When the object is viewed, then to the observer, point A appears nearer than point B, while when the reconstructed real image is viewed, point B appears nearer than point A. Thus while observing the objects (as in Fig. 8.8a), if a portion of object B is obstructed by A, then in the reconstructed image a portion of B will also be obstructed by A, although A will be behind B. Such an image is called pseudoscopic. For plane objects, this effect is not important.

8.6. Fourier Transform Holograms

In the recording configurations discussed so far, the photographic plate used to make a hologram was assumed to be at a finite distance from the object. In such an arrangement, the object diffracts the illuminating wave and forms the Fresnel diffraction pattern at the photographic plate. This diffraction pattern, which is characteristic of the object, is then made to interfere with a reference wave and the resulting interference pattern is recorded on the photographic plate. Such holograms are called Fresnel holograms.

In Chapter 6, we showed that the field distribution at the back focal plane of a lens is the Fourier transform of the field distribution at the front focal plane of the lens. Thus if $f(x, y)$ represents the amplitude distribution at the front focal plane of the lens, then the field in the back focal plane is given by (apart from a proportionality constant)

$$F(\xi, \eta) = \iint f(x, y) \exp[2\pi i(x\xi + y\eta)]\, dx\, dy \tag{8.6-1}$$

where $\xi = x/\lambda f$ and $\eta = y/\lambda f$ represent spatial frequencies and λ and f represent the wavelength of the radiation and the focal length of the lens, respectively. Since the Fourier transform of the transparency is characteristic of the transparency, a recording of the Fourier transform will also contain all the information about the object. Thus a hologram recorded at the back focal plane of the lens contains information about the object and is termed a Fourier transform hologram. In order to record a Fourier transform hologram, the object transparency is placed on the front focal plane of a lens and is illuminated by a plane wave (see Fig. 8.9a). A photographic plate is placed at the back focal plane and is illuminated by a plane wave that forms the reference wave; this reference wave may be produced by a point source placed at the front focal plane as shown in Fig. 8.9a. The lens forms the Fourier transform of the object transparency on the plane of the photographic plate and the photographic plate records the resulting interference pattern formed between the plane reference wave and the Fourier transform of the object transparency. The recorded pattern is known as the Fourier transform hologram. The reconstruction of the object transparency can be carried out by placing a lens in front of the hologram and illuminating it by a plane wave as shown in Fig. 8.9b; the two reconstructed images lie on the back focal plane of the lens.

Let $f(x, y)$ represent the transmittance of the object placed in the front focal plane (see Fig. 8.9a). To provide the reference wave, we locate a point source of light, also in the front focal plane, but displaced from the axis by a distance $-a$ along the x direction. At the back focal plane of the lens we find the Fourier transform of $f(x, y)$, namely, $F(x/\lambda f, y/\lambda f)$. The point source at $x = -a$, $y = 0$, can be represented by $A\,\delta(x + a)\,\delta(y)$, where $\delta(x)$ represents the Dirac delta function. The field due to the point source at the back focal plane is given by

$$R = A\iint \delta(x' + a)\,\delta(y')\exp[2\pi i(x'\xi + y'\eta)]\, dx'\, dy' = A\exp(-2\pi i a\xi) \tag{8.6-2}$$

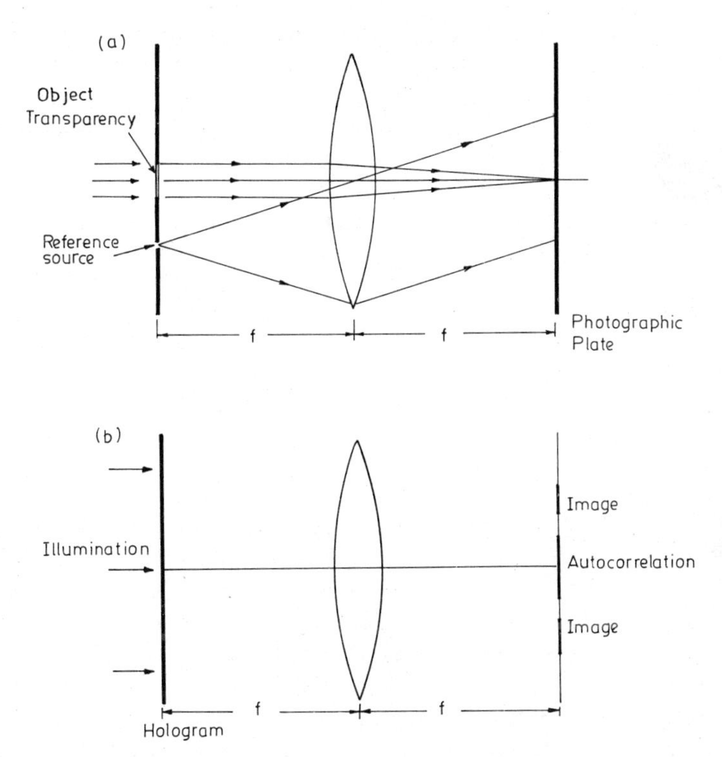

Fig. 8.9. (a) Recording of a Fourier transform hologram. (b) Reconstruction.

where $\xi = x/\lambda f$ and $\eta = y/\lambda f$. Thus, the total amplitude at the back focal plane is given by $F(\xi, \eta) + A\exp(-2\pi i a\xi)$ and the intensity pattern recorded is given by

$$I(\xi, \eta) = F(\xi, \eta)\,F^*(\xi, \eta) + A^2 + AF(\xi, \eta)\,e^{2\pi i a\xi} + AF^*(\xi, \eta)\,e^{-2\pi i a\xi} \quad (8.6\text{-}3)$$

The photographic plate that records the above intensity pattern forms the hologram. The amplitude transmittance of the hologram is given by

$$T(\xi, \eta) = I(\xi, \eta) = F(\xi, \eta)\,F^*(\xi, \eta) + A^2 + AF(\xi, \eta)\,e^{2\pi i a\xi} + AF^*(\xi, \eta)\,e^{-2\pi i a\xi} \quad (8.6\text{-}4)$$

where we have assumed the validity of Eq. (8.2-3). When such a hologram is placed in the front focal plane of another lens and illuminated by a plane wave, then in the back focal plane we find the Fourier transform of the distribution given by Eq. (8.6-4). Thus, in the back focal plane of the lens, we find the field distribution

$$\begin{aligned}\mathscr{F}[T(\xi, \eta)] = {} & f(x, y) \circledast f(x, y) + A^2\,\delta(x)\,\delta(y) + Af(-x - a, -y) \\ & + Af^*(x - a, y) \quad (8.6\text{-}5)\end{aligned}$$

where ⊛ represents correlation [see Eq. (6.5-14)] and we have used the fact that

$$\mathscr{F}\{\mathscr{F}[f(x, y)]\} = f(-x, -y)$$

$$\mathscr{F}[|F(\xi, \eta)|^2] = \mathscr{F}[F(\xi, \eta)] * \mathscr{F}[F^*(\xi, \eta)] = f(x, y) \circledast f(x, y) \quad (8.6\text{-}6)$$

and

$$\mathscr{F}[F(\xi, \eta)\, e^{2\pi i a \xi}] = f(-x-a, -y) \quad (8.6\text{-}7)$$

The first term in Eq. (8.6-5) represents the autocorrelation of the function $f(x, y)$ and is centered on the axis. The second term also lies on the axis. The third and fourth terms represent the object distribution and its conjugate, which are reconstructed around $x = -a$ and $x = a$, respectively. Thus, both the primary and conjugate images are situated on the back focal plane. We will now show that by choosing large enough values of a, the primary and conjugate images given in Eq. (8.6-5) can be spatially separated.

To determine the minimum offset distance a of the point reference source, let us consider an object transparency of width b. In the image plane we have an autocorrelation centered on the axis; the autocorrelation of the transparency of width b has dimension $2b$.† Also, the two images are situated at a distance a on either side of the axis (see Fig. 8.9b). Thus, if the three terms have to remain separated, then the condition $2a \geq 3b$ must be satisfied. Alternatively, if θ is the angle at which the reference plane wave falls on the hologram plane, then since $a \simeq \theta f$, the condition to be satisfied becomes $\theta \geq 3b/2f$. Thus, if the above condition is satisfied, two clear images can be obtained.

The Fourier transform arrangement of recording holograms finds application in character recognition problems. As discussed in Section 6.5, the character recognition problem involves performing a cross-correlation of the character to be recognized with that of the object. The cross-correlation is obtained by recording holographically the Fourier transform of the character and then placing it in the back focal plane of a lens in whose front focal plane lies the function with which the cross-correlation has to be performed. Another lens that performs a further transform is placed such that its front focal plane coincides with the back focal plane of the first lens (see Fig. 6.17). In the back focal plane of the second lens, one obtains both a cross-correlation and a convolution. Corresponding to the positions of the character to be recognized, one obtains bright dots in the back focal plane of the second lens. The character recognition process finds application in problems of detection of isolated signals in random noise, of alphanumerics, etc.

† For example, the autocorrelation of a rectangular function of width b is triangular with base length $2b$.

8.6.1. Resolution in Fresnel and Fourier Transform Holograms

We will now consider the resolution properties of Fresnel and Fourier transform holograms. Let us first consider a Fresnel hologram recording arrangement, and for simplicity we consider a point object (see Fig. 8.10a). We consider the reference wave to be a plane wave incident at an angle θ_r on the photographic plate as shown in Fig. 8.10a. The photographic plate records the intensity distribution given by

$$I(x, y) = \left| \frac{A}{d} \exp\left[-\frac{ik}{2d}(x - x_0)^2 \right] + B \exp(-ikx \sin \theta_r) \right|^2$$

$$= \frac{A^2}{d^2} + B^2 + \frac{2AB}{d} \cos\left[\frac{k}{2d}(x - x_0)^2 - kx \sin \theta_r \right] \quad (8.6\text{-}8)$$

where we have considered a two-dimensional configuration for simplicity. Thus the spatial frequency of the recorded intensity pattern along the

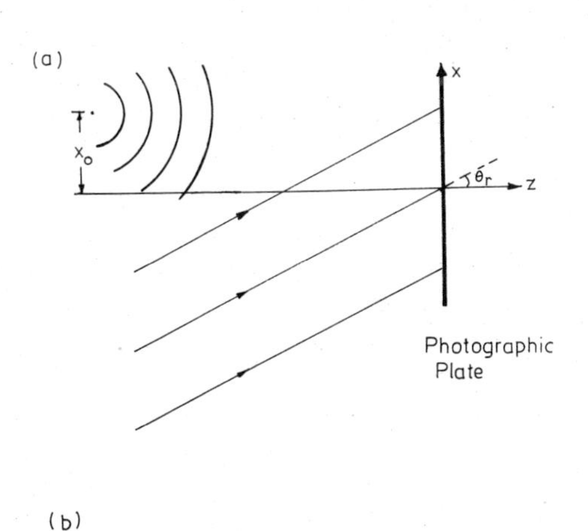

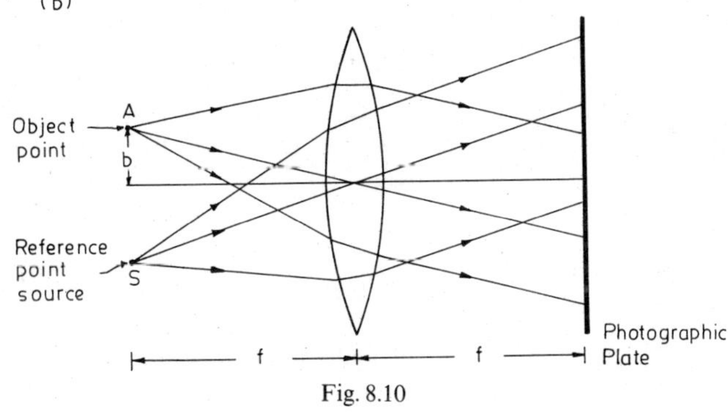

Fig. 8.10

x direction is given by

$$\frac{d}{dx}\left[\frac{(x-x_0)^2}{2\lambda d} - \frac{x}{\lambda}\sin\theta_r\right] = \frac{x-x_0}{\lambda d} - \frac{\sin\theta_r}{\lambda} \tag{8.6-9}$$

which is dependent on x, the distance along the hologram plane. Thus, if ν_c represents the cutoff frequency beyond which the photographic plate fails to record, then the hologram records only those frequencies that satisfy the relation

$$\left|\frac{x-x_0}{\lambda d} - \frac{\sin\theta_r}{\lambda}\right| < \nu_c \quad \text{or} \quad -\nu_c + \frac{x_0}{\lambda d} + \frac{\sin\theta_r}{\lambda} < \frac{x}{\lambda d} < \nu_c + \frac{x_0}{\lambda d} + \frac{\sin\theta_r}{\lambda} \tag{8.6-10}$$

Thus the given object point is recorded in a region $2\lambda d\nu_c$ of the hologram. While reconstructing, the finite size of the hologram results in a resolution limit of $\lambda d/2\lambda d\nu_c = 1/2\nu_c$, where we have considered the hologram to be bounded by a finite aperture of width $2\lambda d\nu_c$ (see Section 7.4).

In contrast, let us consider the Fourier transform hologram configuration as shown in Fig. 8.10b, where an object point A and a reference point source S are placed at the front focal plane of a lens. The waves emanating from A and S become plane waves as they pass through the lens. Hence at the back focal plane (in which the photographic plate lies), the plane waves intersect to produce fringes of constant spatial frequency $(\sin\theta_r + \sin\theta)/\lambda$. Notice that this frequency is independent of the position on the photographic plate but is dependent on the distance b of the object point from the axis. Thus by placing the reference source close to the object [within the limits obtained earlier for spatial separation of the various terms in Eq. (8.6-4)] the spatial frequencies of the resulting fringes may be kept low. Thus for every point in the object, the hologram records a constant spatial frequency fringe system throughout its plane. Object points that lie so far off as to produce a fringe system with spatial frequency $> \nu_c$ will not be imaged by the hologram. Thus the finite value of ν_c has the effect of restricting the field of view of the image. For small objects, one can see that the Fourier transform hologram configuration is capable of producing a low-frequency fringe system over a larger aperture, and thus the images produced using this configuration have higher resolution.

Problem 8.1. Show that if the wavelength of the reconstruction source is different from that used for forming the hologram, then one indeed obtains a magnification of λ'/λ, where λ' and λ are the wavelengths of the reconstruction and reference sources, respectively.

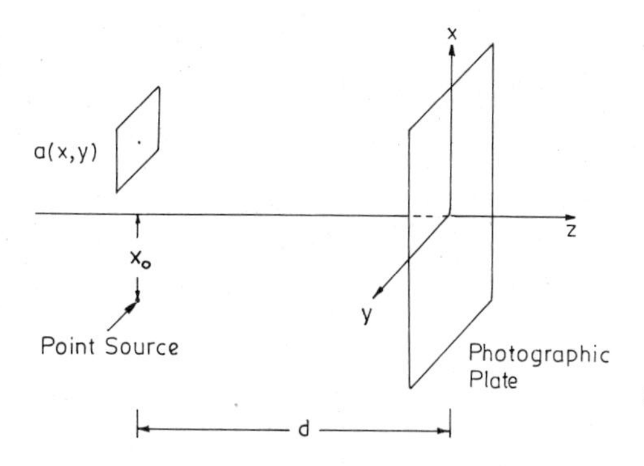

Fig. 8.11. Recording of a lensless Fourier transform hologram.

8.6.2. Lensless Fourier Transform Holograms

We have seen that Fourier transform holograms can be formed by locating the object in the front focal plane of a lens and recording the resultant spatial frequency spectrum formed at the back focal plane. As already discussed, the advantage of the Fourier transform configuration is that one can indeed obtain high resolutions. Stroke (1965) has developed another method, which does not involve lenses in the recording process, to obtain Fourier transform holograms.

To understand the theory behind the lensless Fourier transform hologram, we refer to Fig. 8.11, where $a(x, y)$ represents an object and $r(x, y)$ the reference source, which is assumed to be a point source situated at, say $x = -x_0, y = 0$. The object and the reference source are situated at the same distance d from the recording plate. We have already seen in Section 6.2 that if $a(x, y)$ is the field distribution in any plane, then the field distribution at a distance d from it is

$$a(x, y) * \exp\left[-\frac{ik}{2d}(x^2 + y^2)\right]$$

where $*$ represents convolution. Thus the fields at the plane of the photographic plate due to the object and reference waves, respectively, are given by (apart from some constant factors)

$$A(x, y) = \iint a(x', y') \exp\left\{-\frac{ik}{2d}[(x - x')^2 + (y - y')^2]\right\} dx'\, dy' \quad (8.6\text{-}11)$$

$$R(x, y) = \exp\left\{-\frac{ik}{2d}[(x + x_0)^2 + y^2]\right\} \quad (8.6\text{-}12)$$

Thus the total field is $A(x, y) + R(x, y)$, and the intensity pattern recorded is

$$|A(x, y)|^2 + |R(x, y)|^2 + A(x, y) R^*(x, y) + A^*(x, y) R(x, y)$$

Let us look at the third term. We have, using Eqs. (8.6-11) and (8.6-12),

$$A(x, y) R^*(x, y) = \iint a(x', y') \exp\left\{ -\frac{ik}{2d}[(x - x')^2 + (y - y')^2 - (x + x_0)^2 - y^2] \right\} dx'\, dy'$$

$$= \exp\left[\frac{\pi i}{\lambda d}(x_0^2 + 2x_0 x)\right] \iint dx'\, dy' a(x', y') \times \exp\left[-\frac{ik}{2d}(x'^2 + y'^2)\right] \exp\left[\frac{2\pi i}{\lambda d}(xx' + yy')\right] \quad (8.6\text{-}13)$$

which, apart from a constant factor, is the Fourier transform of

$$a(x, y) \exp\left[-\frac{ik}{2d}(x^2 + y^2)\right]$$

If we write

$$g(x, y) = a(x, y) \exp\left[-\frac{ik}{2d}(x^2 + y^2)\right]$$

and if G represents the Fourier transform of g, we have

$$A(x, y) R^*(x, y) = \exp\left(\frac{\pi i x_0^2}{\lambda d}\right) G\left(\frac{x}{\lambda d}, \frac{y}{\lambda d}\right) \exp\left(\frac{2\pi i x x_0}{\lambda d}\right) \quad (8.6\text{-}14)$$

Similarly,

$$A^*(x, y) R(x, y) = \exp\left(-\frac{\pi i x_0^2}{\lambda d}\right) G^*\left(\frac{x}{\lambda d}, \frac{y}{\lambda d}\right) \exp\left(-\frac{2\pi i x x_0}{\lambda d}\right) \quad (8.6\text{-}15)$$

When the developed hologram is placed in the front focal plane of a lens (see Fig. 8.9b) and illuminated with parallel light, then in its back focal plane we obtain the Fourier transform of the quantities given in Eqs. (8.6-14) and (8.6-15) in addition to that of $|A|^2 + |R|^2$, which is centered about the origin. Thus, the image formed due to the term $A(x, y) R^*(x, y)$ is

$$\mathscr{F}[A(x, y) R^*(x, y)] = \exp\left(\frac{\pi i x_0^2}{\lambda d}\right) \iint G\left(\frac{x'}{\lambda d}, \frac{y'}{\lambda d}\right) \exp\left(\frac{2\pi i x_0 x'}{\lambda d}\right) \times \exp\left[2\pi i\left(x'\frac{x}{\lambda f} + y'\frac{y}{\lambda f}\right)\right] dx'\, dy' \quad (8.6\text{-}16)$$

where f represents the focal length of the lens. On using the explicit expression for $G(x'/\lambda d, y'/\lambda d)$ in terms of $g(x, y)$, we can simplify Eq. (8.6-16) to obtain

$$\lambda^2 d^2 \exp\left(\frac{\pi i x_0^2}{\lambda d}\right) g\left(-\frac{xd}{f} - x_0 - \frac{yd}{f}\right)$$

Since only the intensity will be observed, we see that the term given by Eq. (8.6-14) produces an intensity distribution that is proportional to

$$\left|g\left(-x\frac{d}{f} - x_0, -y\frac{d}{f}\right)\right|^2 = \left|a\left(-x\frac{d}{f} - x_0, -y\frac{d}{f}\right)\right|^2 \qquad (8.6\text{-}17)$$

Thus the object distribution is reproduced in the back focal plane with a magnification f/d and is centered at $x = -x_0 f/d$. In an exactly similar manner it can be shown that the term given by Eq. (8.6-15) also gives rise to the object intensity distribution in the back focal plane.

8.7. Volume Holograms

Until now, we have assumed in our analysis that the photographic plate has, in fact, a negligible thickness compared to the fringe spacing. Such holograms are called plane holograms. In general, one cannot neglect the thickness of the emulsion as compared to the fringe spacing. For example, typical emulsion thicknesses range from 5 to 20 μm, while fringe spacing may be as small as 2 to 3 μm. Thus the thickness of the emulsion could be many times the fringe spacing. Holograms in which the whole volume of the emulsion takes part in reconstruction are termed volume holograms.

For an elementary understanding of the formation and reconstruction with volume holograms we consider the interference between two inclined plane waves (see Fig. 8.12a). We assume that the propagation vectors of the two plane waves lie in the x-z plane. The field due to the two plane waves can be written in the form

$$u_1(x, z) = A \exp[-2\pi i(x\xi_1 + z\eta_1)] \qquad (8.7\text{-}1)$$

$$u_2(x, z) = B \exp[-2\pi i(x\xi_2 + z\eta_2)] \qquad (8.7\text{-}2)$$

where $\xi_1 = (\sin\theta)/\lambda$, $\eta_1 = (\cos\theta)/\lambda$, $\xi_2 = (\sin\theta')/\lambda$, and $\eta_2 = (\cos\theta')/\lambda$; θ and θ' are the angles made by the two waves with the z axis when they are traveling inside the medium and λ is the wavelength of the radiation in the medium. The resultant amplitude is given by

$$u = u_1 + u_2 = A \exp[-2\pi i(x\xi_1 + z\eta_1)] + B \exp[-2\pi i(x\xi_2 + z\eta_2)]$$

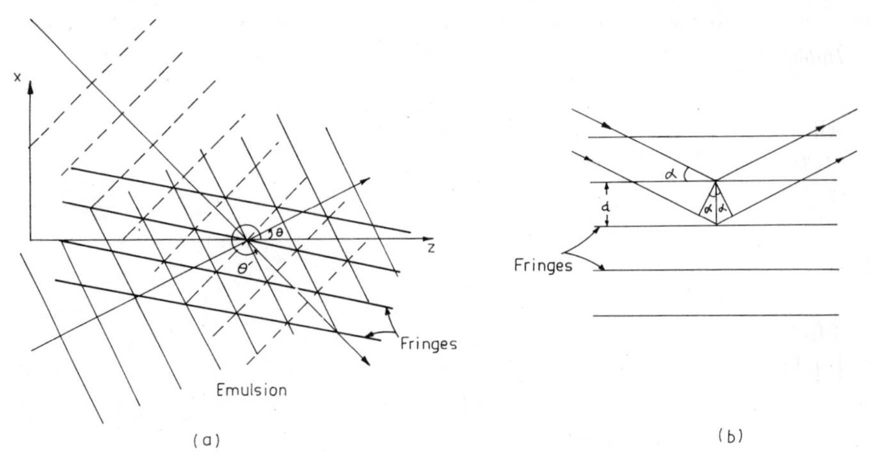

Fig. 8.12. (a) Formation of volume hologram by plane waves. (b) The reconstruction process.

Hence, we get for the recorded intensity variation

$$\begin{aligned} I &= A^2 + B^2 + AB \exp\{-2\pi i[x(\xi_1 - \xi_2) + z(\eta_1 - \eta_2)]\} \\ &\quad + AB \exp\{2\pi i[x(\xi_1 - \xi_2) + z(\eta_1 - \eta_2)]\} \\ &= A^2 + B^2 + 2AB \cos\{2\pi[x(\xi_1 - \xi_2) + z(\eta_1 - \eta_2)]\} \end{aligned} \tag{8.7-3}$$

The minima and maxima are planes given by

$$x(\xi_1 - \xi_2) + z(\eta_1 - \eta_2) = m + \tfrac{1}{2} \tag{8.7-4}$$

$$x(\xi_1 - \xi_2) + z(\eta_1 - \eta_2) = m \tag{8.7-5}$$

respectively. The fringes are planes normal to the x-z plane. The angle ψ that the planes make with the z axis is given by

$$\tan\psi = \frac{dx}{dz} = -\frac{\eta_1 - \eta_2}{\xi_1 - \xi_2} = -\frac{\cos\theta - \cos\theta'}{\sin\theta - \sin\theta'} = \tan\left(\frac{\theta + \theta'}{2}\right) \tag{8.7-6}$$

i.e., $\psi = (\theta + \theta')/2$, which implies that the planes of the fringes bisect the propagation directions of the two plane waves. Thus, when the photographic emulsion is developed, silver atoms are formed along planes of the maxima. Let us consider the specific case when $\theta' = 2\pi - \theta$ (see Fig. 8.12a). Then $\xi_1 = (\sin\theta)/\lambda$, $\xi_2 = -(\sin\theta)/\lambda$, $\eta_1 = (\cos\theta)/\lambda$, $\eta_2 = (\cos\theta)/\lambda$. Thus, the planes of maxima will be given by

$$2x \sin\theta = m\lambda \tag{8.7-7}$$

and separation between the fringes d is

$$d = \lambda/(2 \sin\theta) \tag{8.7-8}$$

For reconstruction, we illuminate the fringe system by another plane wave. The planes of silver atoms behave like plane partially reflecting mirrors. The waves reflected from different planes differ in phase and interfere constructively or destructively depending on the relative phase differences. Referring to Fig. 8.12b, we see that the path difference between the waves reflected from adjacent planes for an illumination at an angle α is $2d \sin \alpha$. When

$$2d \sin \alpha = \lambda \tag{8.7-9}$$

the reflected waves will add in phase and produce a bright reconstructed image. This is the Bragg condition. Comparing Eqs. (8.7-8) and (8.7-9), we find that for maximum intensity, the illumination angle α must be either θ or $\pi - \theta$. Thus maximum intensity of the reconstructed image occurs when the reconstructing wave is incident at the same angle as the original reference wave or is supplementary to the original reference wave. At all other angles, the reconstructed image is very weak. When the reconstruction is made at an angle θ, one generally obtains the reconstruction of the virtual image, and when it is illuminated at an angle $\pi - \theta$, then the real image is reconstructed. Because of this directional sensitivity, volume holograms find applications in information storage. For example, the various pages of a book can be stored throughout the volume of the emulsion. Each page is recorded after changing the orientation of the reference beam. Since the information is stored throughout the volume in a nonlocalized manner, the hologram will be insensitive to dust, scratches, etc.

We now consider another particular case when the object beam and the reference beam are traveling in approximately opposite directions (i.e., $\theta \simeq \pi/2$) (see Fig. 8.13). Thus on developing the emulsion, we obtain planes of silver atoms that are approximately parallel to the emulsion surface and are separated by a distance $\sim\lambda/2$. The developed plate behaves like an interference filter, i.e., the reflectivity is high only for that wavelength with which it was formed. Since the hologram becomes highly wavelength selective, the image can be reconstructed even with white light, and in such a case the reconstructed image appears colored; the color of the image is the same as that used for forming the hologram unless the emulsion shrinks during development and fixing. Such holograms are called reflection holograms because the reconstructed image can be viewed by reflection.

In the above analysis, we have discussed the formation of volume holograms when the object and the reference waves are plane waves. A general theory for volume holograms is rather involved, and interested readers are referred to Kogelnik (1969) and Collier *et al.* (1971). We would, however, like to mention that in addition to information storage, volume holograms also find application in the recording of colored objects. To record colored holograms one uses three recording wavelengths in the red,

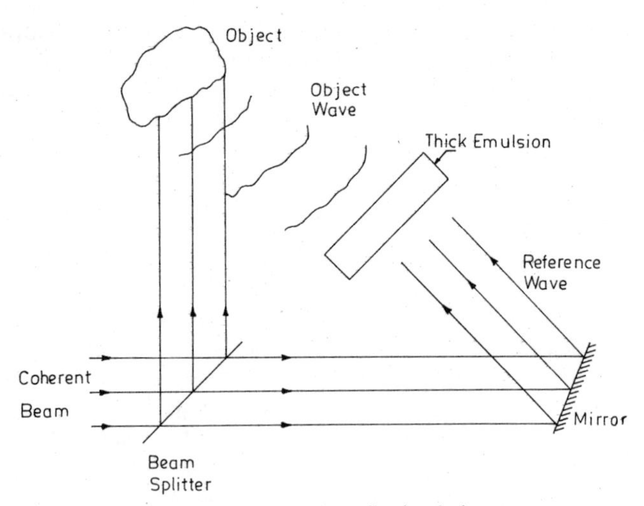

Fig. 8.13. Recording of a reflection hologram.

green, and blue regions and records the holograms in the same emulsion. The Bragg reflection planes formed in the emulsion are separated by different distances depending on the wavelength and one obtains a set of planes for each wavelength used in the recording. Since each set of planes behaves as an interference filter for the particular wavelength that formed it, when such a hologram is illuminated with white light, the three colors are reproduced and one obtains the reconstruction of the colored object. The distinct advantage of volume holograms over plane holograms in color recording lies in the fact that there is very little cross-talk between different wavelengths, especially in the case of reflection hologram configurations.

8.8. *Applications of Holography*

Although holography was invented in an effort to increase the resolving power of microscopes, with the advent of the laser many diverse applications have emerged. We will discuss some of these in this section.

8.8.1. *Three-Dimensional Reconstruction*

As we have already stated, the hologram is capable of displaying the image of the object in its full three-dimensional perspective. The depth of the object is also recorded. This implies that a hologram has much more information stored in it than a photograph. In fact, it is this ability to display the image in its true three-dimensional form that has resulted in its popularity. This property of holograms can indeed be used for display purposes. Also, when one has to view different depths in an object sample whose properties are changing with time, this technique offers a unique advantage in that one can record, once and for all, the whole volume of the object and view each section leisurely. This has indeed been used by Thompson *et al.* (1967) for observation of aerosol particles.

8.8.2. Interferometry

Whenever measurements of great precision are required, interferometry techniques may be used with much advantage. In the conventional interferometric methods, the optical components must be made with great precision. Thus, for example, the mirrors in the Michelson interferometer arrangement must be well polished and well ground for the observation and subsequent measurements of interference fringes. In this branch of optics, holography offers unique advantages and has extended the applicability of interferometric principle to diffusely reflecting objects.

There are many variations of the holographic interferometry principle and we will discuss some of them here. The applicability of the holographic principle can be understood from the fact that since the hologram stores the object wave itself, one can compare the object wave with another wave from the object, which might have been distorted. Thus in the method called real-time interferometry, the object wave is recorded and the hologram is replaced in exactly the original position. Now when the reference wave illuminates the hologram, a virtual image is formed at exactly the same location as the object. If the object is still in its original position and is now stressed, then the wave emerging from the stressed object will be of a different form than the wave from the object before stressing. An observer viewing through the hologram receives both the original object wave and the wave from the stressed object. These two waves interfere and produce interference fringes. As the object surface changes, the interference pattern changes correspondingly and thus one can indeed study the changes produced as a function of time. Some of the difficulties encountered with this technique include the fact that it is difficult to place the hologram back into exactly the position it occupied during formation. Also when the emulsion is processed, shrinkage may occur, thus distorting the emergent wavefront.

A variant of the above technique is double-exposure interferometry, where two exposures of the same emulsion are made prior to processing. Thus a hologram of the original object is taken, and in the same emulsion, a hologram of the object with which it is to be compared is again taken with

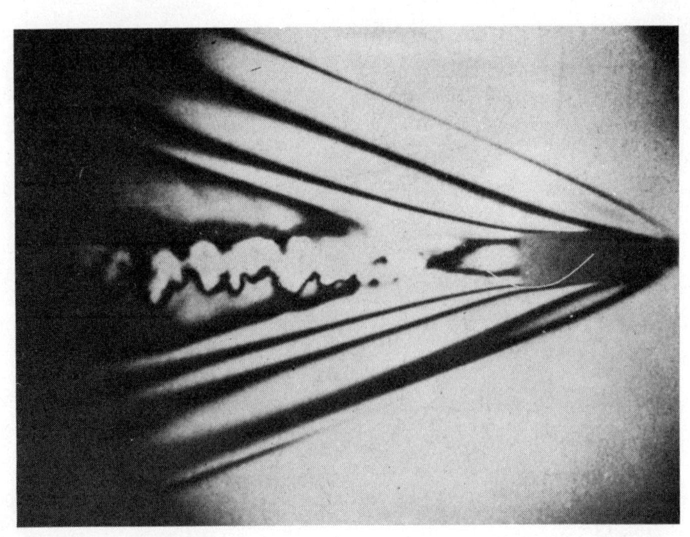

Fig. 8.14. Double-exposure hologram of a bullet in flight (after Wuerker, 1971).

the same reference beam. If the emulsion is now processed and illuminated with the reference wave, each of the two component object waves is reconstructed, and they interfere to produce the interference pattern from which one can make a comparison of the two objects. Usually the two objects correspond to the same object under different conditions of strain. This method does not require any exact replacement of the hologram as in the last method, and has been studied using pulsed-lasers transient events. Figure 8.14 shows a pulsed-laser double-exposure hologram of a bullet in flight. Such a hologram is made in the following way. First, a partial exposure of the holographic plate is made of the chamber through which the bullet is to pass. The hologram is made using a pulse from the laser. Then, the passage of the bullet through the volume triggers another pulse from the laser and a second recording on the same plate is made. Since the bullet produces shock waves, the density of the gas changes at some points. Thus, a reconstruction yields interference fringes as shown in Fig. 8.14. Note that the walls of the chamber need not be optically flat because both the interfering waves pass through the same region, and unless the glass walls themselves have changed their form, they will have no effect. For further details see Brooks *et al.* (1966) and Heflinger *et al.* (1966).

To understand the principle of double-exposure holographic interferometry, we consider the simple case when, between the two exposures, the object is disturbed in a direction normal to the plane of the hologram plate. Let $O(x, y)$ be the object wave in the original position and $O'(x, y)$ the object wave when the object is disturbed by a distance $\delta(x, y)$ along the z direction (see Fig. 8.15). If we assume that the amplitudes of $O'(x, y)$ and $O(x, y)$ are same but there is only a change of phase due to the displacement, then it can be seen that

$$O'(x, y) = O(x, y) \exp[4\pi i \delta(x, y)/\lambda] \tag{8.8-1}$$

where we have assumed an approximately normal illumination of the object surface. The object waves $O(x, y)$ and $O'(x, y)$ are both recorded on the same hologram plate. Upon reconstruction, we obtain the resultant wave $O(x, y)+ O'(x, y)$, and hence the observed intensity is given by

$$|O(x, y) + O'(x, y)|^2 = 2|O(x, y)|^2 [1 + \cos(4\pi\, \delta/\lambda)] \tag{8.8-2}$$

Thus the reconstructed image shows the object superimposed with a fringe pattern corresponding to contours of equal displacement δ. Thus from the fringe pattern one can indeed obtain the amplitudes of displacement at each point on the object surface.

In general, the displacement of the object points may be along a general direction, and in such a case the fringe interpretation is quite involved; for further details see Collier *et al.* (1971) and Hecht *et al.* (1973).

The method of holographic interferometry also finds application in vibration analysis. Suppose we take the hologram of a vibrating object; then the hologram will record, in the limit, the waves emanating from the surface for each position of the vibrating surface between the two extreme positions of the amplitude. Thus, if we are considering a surface vibrating in its normal mode, then at the nodes, the surface will be stationary, with

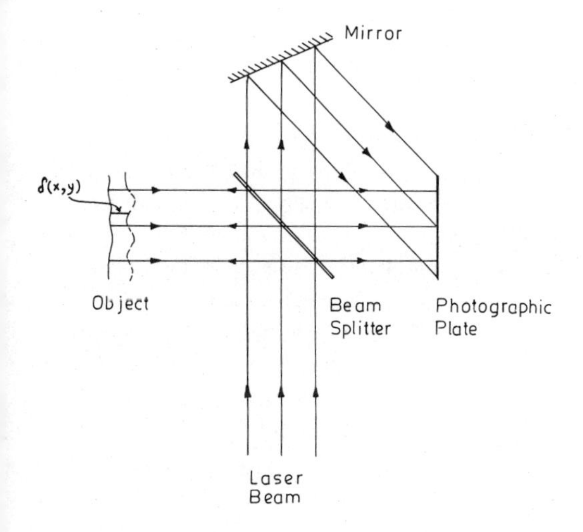

Fig. 8.15. Double-exposure interferometry when the object undergoes a displacement.

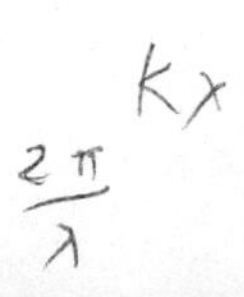

a b

Fig. 8.16. Reconstruction from the time-average hologram of a can top vibrating in (a) the lowest resonance frequency and (b) the second resonance. The rings are loci of equal amplitude (after Powell and Stetson, 1965).

the maximum amount of vibration amplitude at the antinodes. Figure 8.16 shows the holographic images of a diaphragm vibrating in two different modes.

To understand how such a fringe pattern can be produced by a vibrating object, we consider a surface vibrating simple harmonically as shown in Fig. 8.17. Thus, the displacement (which is assumed to be along the z axis) of any point $P(x)$ will be given by

$$z(x, t) = A(x) \cos \omega t \tag{8.8-3}$$

where $A(x)$ is the amplitude of vibration and is a function of position. From the figure it can be seen that the phase shift when the surface moves to a position different from the rest position is

$$\varphi(x, t) = \frac{2\pi}{\lambda} A(x)(\cos \alpha_1 + \cos \alpha_2) \cos \omega t \tag{8.8-4}$$

where α_1 and α_2 are the angles that the direction of observation and the direction of propagation of the incident light make with the z axis.

Now we expose the holographic plate to radiation from the vibrating surface along with a reference wave. Thus, if $O(x, t)$ represents the instantaneous field at the hologram plate due to the vibrating surface, then assuming that the recording time is much larger than the time period of vibration, the hologram will record the field that is the average of the intensity falling on it. The intensity falling on the plate will be

$$I(x) = \frac{1}{T} \int_0^T [R^2 + O^2(x, t) + R^* O(x, t) + RO^*(x, t)]\, dt \tag{8.8-5}$$

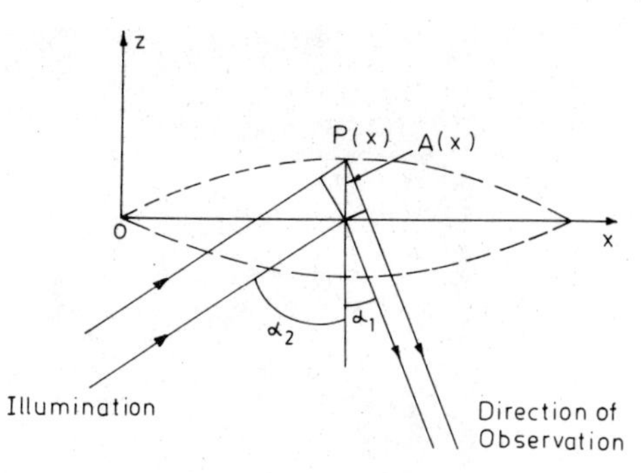

Fig. 8.17. Schematic for calculation of phase shift.

where R represents the reference wave and we have averaged over one period of vibration. On reconstruction, the term of interest, namely, the reconstructed object wave, is given by

$$\frac{R^*R}{T}\int_0^T O(x,t)\,dt = \frac{|R|^2}{2\pi}\int_0^{2\pi} O(x,t)\,d\,(\omega t) \tag{8.8-6}$$

The object wave $O(x, t)$ will be given by

$$O(x,t) = u(x)\exp[i\varphi(x,t)] = u(x)\exp\left[\frac{2\pi i}{\lambda}A(x)\cos\omega t(\cos\alpha_1 + \cos\alpha_2)\right] \tag{8.8-7}$$

where we have used Eq. (8.8-4) for $\varphi(x, t)$. Thus the reconstructed object wave will be proportional to

$$\frac{u(x)}{2\pi}\int_0^{2\pi} \exp\left[\frac{2\pi i}{\lambda}A(x)(\cos\alpha_1 + \cos\alpha_2)\cos\omega t\right]d(\omega t) \tag{8.8-8}$$

Since

$$\frac{1}{2\pi}\int_0^{2\pi} e^{ip\cos\theta}\,d\theta = J_0(p) \tag{8.8-9}$$

the amplitude at the observation point is

$$u(x)\,J_0(\overline{w})$$

where

$$\overline{w} = \frac{2\pi}{\lambda}A(x)(\cos\alpha_1 + \cos\alpha_2)$$

and hence the intensity at the observation point is proportional to

$$u(x)\,u^*(x)\,[J_0(\overline{w})]^2$$

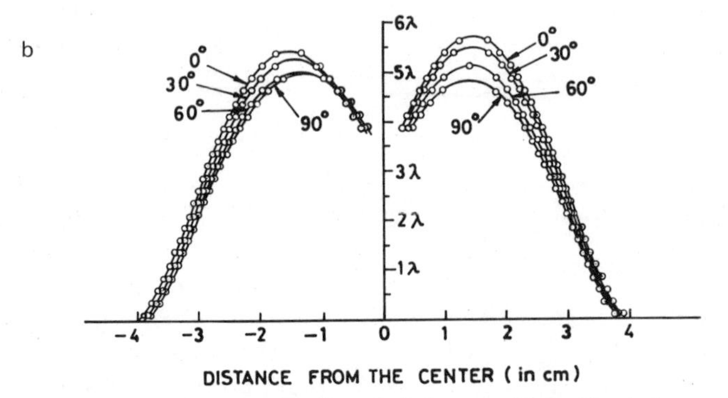

Fig. 8.18. (a) Reconstruction from the hologram obtained by stroboscopically illuminating a vibrating disk at the two extreme positions of vibrations. (b) The vibration amplitude distribution across the face of the plate (after Saito *et al.*, 1971).

Thus we will observe bright and dark fringes on the object, where the fringes represent contours of equal amplitude; the minima in the reconstruction correspond to the zeroes of $[J_0(\overline{w})]^2$ and the maxima correspond to the maxima of $[J_0(\overline{w})]^2$. Since the values of the maxima of $J_0^2(x)$ decrease with increasing values of the argument, the intensity of the fringes decreases with increasing $A(x)$, i.e., with increasing amplitude of vibration.

The main advantage of the holographic interferometry technique lies in the fact that the surfaces of the vibrating objects need not be made optically flat and complex vibrating objects can also be studied.

Another technique is stroboscopic holographic interferometry. In this technique, the object under vibration is stroboscopically illuminated, i.e., the object is illuminated by pulses of light of a short duration (short compared to the time period of vibration) and the hologram is exposed. By illuminating the object at proper times one can obtain interference between object waves emerging from the object corresponding to different phases of vibration of the object. Thus one could provide a chopping technique such that the object is illuminated when it is at the two extreme positions of displacement. This method has the advantage that uniform fringe visibility is obtained, and one can also compare the state of vibration of the object between any two times. Figure 8.18a shows such an interferogram obtained by Saito *et al.* (1971) of a circular aluminum plate of diameter 8 cm and thickness 0.1 mm supported on the circumference and excited at the center. The frequency of vibration of the plate was 200 Hz and the pulse width of the illuminating beam was about 1/20 the period of vibration. Figure 8.18b shows the amplitude distribution in the plate. For further details and an extensive bibliography on vibration analysis, see Singh (1973).

8.8.3. *Microscopy*

We saw in Section 8.3 that the magnification associated with the reconstruction (when a radiation of different wavelength is employed) is proportional to the ratio of wavelengths used for reconstruction to that used during formation. This gives us a method of observing magnified images of objects by increasing the wavelength of the radiation while reconstructing. This was the original idea proposed by Gabor, but because of the lack of sufficiently coherent sources of radiation of shorter wavelengths (say X-rays), this method has still not been put to significant use. At the same time, magnifications of about 100 have been attained with a resolution of several microns. It should be mentioned that a wavelength change during reconstruction always introduces aberrations unless the hologram is also scaled accordingly (see, e.g., Meier, 1965).

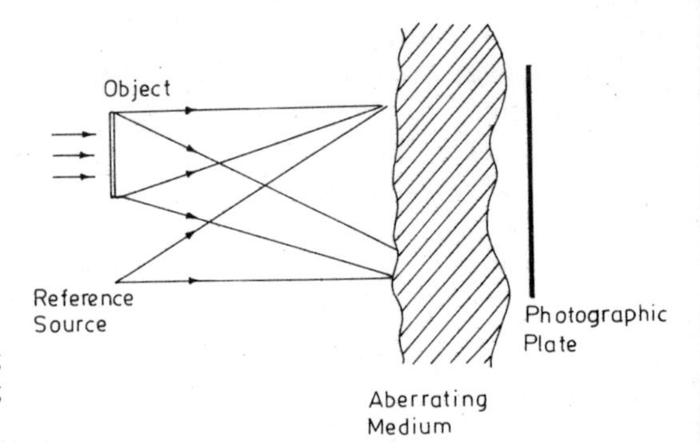

Fig. 8.19. A hologram recording in the presence of an aberrating medium.

8.8.4. Imaging through Aberrating Media

The principle of holography can also be employed for imaging objects through media that aberrate any wave passing through them. Thus, for example, the optical components (or the intervening medium) may introduce aberrations in the wave. We will consider the case when both the object wave and the reference wave pass through the same region of the aberrating medium as shown in Fig. 8.19. We assume that the photographic plate is situated just behind the aberrating medium. Let the effect of the aberrating medium be represented by a phase factor $\exp[iW(x, y)]$. If $O(x, y)$ and $R(x, y)$ represent the unaberrated object and reference waves at the film in the absence of the aberrating medium, then in the presence

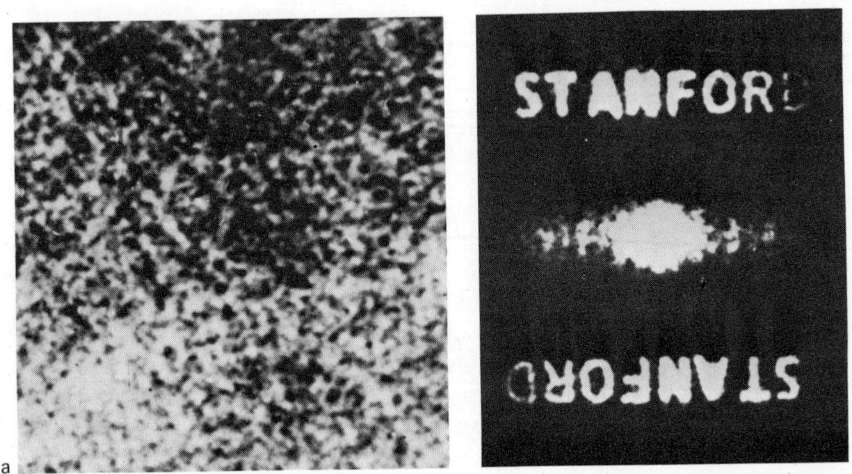

Fig. 8.20. Holographic imaging through an aberrating medium (after Goodman *et al.*, 1966).

of the aberrating medium, the waves get modified to $O(x, y) \exp[iW(x, y)]$ and $R(x, y) \exp[iW(x, y)]$. Since both waves have been assumed to pass through the same region of the aberrating medium, the effect on both is the same. Thus the recorded intensity pattern is given by

$$|O(x, y)\, e^{iW(x,y)} + R(x, y)\, e^{iW(x,y)}|^2 = |O(x, y)|^2 + |R(x, y)|^2 + O^*(x, y)\, R(x, y) + O(x, y)\, R^*(x, y)$$

One can see from this equation that the effect of the aberrating medium has been removed. The recording geometry corresponds to the lensless Fourier transform recording geometry. Thus reconstruction may be effected by the arrangement shown in Fig. 8.9b. Figure 8.20 shows the experimental results; notice that Fig. 8.20a, which corresponds to conventional photography through the aberrating medium, is unintelligible.

9

Self-Focusing

9.1. Introduction

In the last decade, with the availability of high-power laser beams, a large number of interesting nonlinear optical phenomena have been studied, among which the self-focusing of intense laser beams occupies an important place because of its relevance to other nonlinear effects. The self-focusing effect arises primarily due to the dependence of the refractive index of a material on the intensity of the propagating electromagnetic wave. This nonlinear dependence may arise, among others, from the following mechanisms:

1. Electrostriction. In the presence of an inhomogeneous electric field, dielectrics are subject to a volume force; this phenomenon is known as electrostriction. Because of this volume force, the material tends to be drawn into the high-field region, which affects the density of the material. The refractive index changes due to density variations and one obtains a dependence of the refractive index on the intensity of the beam. For a quantitative analysis of the phenomenon, see, e.g., Panofsky and Phillips (1962) and Sodha *et al.* (1974).

2. Thermal effects. When an intense electromagnetic wave (having an intensity distribution along its wavefront) propagates through an absorbing medium, a transverse temperature gradient is set up. This temperature gradient leads to a variation of the refractive index that causes either focusing or defocusing of the beam.

3. Kerr effect. If a liquid molecule (like CS_2) possesses anisotropic polarizability, then the electric field of an intense laser beam will tend to orient these anisotropically polarized molecules such that the direction of maximum polarizability is along the electric vector. This results in a nonlinear dependence of the dielectric constant on the electric field. For weak fields the dielectric constant varies linearly with the intensity; however, for strong fields all the molecules will get aligned and the dielectric constant attains a saturation value (see Problem 9.7).

For weak fields, each of the above mechanisms leads to the following nonlinear dependence of the refractive index on the electric field of the electromagnetic wave:

$$n = n_0 + \tfrac{1}{2}n_2 E_0^2 \tag{9.1-1}$$

where n_0 is the refractive index of the medium (in the absence of the electromagnetic wave), n_2 is a constant, and E_0 represents the amplitude of the electric field. If we write

$$E = E_0 \exp[i(\omega t - kz)] \tag{9.1-2}$$

then $E_0^2 = E^*E$, where E^* is the complex conjugate of E.

In this chapter, we will be studying the effect of such a nonlinearity on the propagation of electromagnetic waves. Since the electromagnetic beam, because of its high power, produces a refractive-index variation, which acts back on the beam to produce a focusing (and in some cases defocusing) the phenomenon is called self-focusing. Section 9.2 gives an elementary treatment of the self-focusing phenomenon, while Section 9.3 combined with Section 9.5 gives a more rigorous treatment. Section 9.4 deals with defocusing due to thermal effects.

9.2. *Elementary Theory of Self-Focusing**

The basic physics of self-focusing of an intense electromagnetic beam can be understood by considering the propagation of a parallel cylindrical beam of uniform intensity and having a circular cross section of radius a.† Clearly, the refractive index in the illuminated region is given by Eq. (9.1-1) and the refractive index in the nonilluminated region is simply n_0 (see Fig. 9.1). Thus we will have a discontinuity in the refractive-index variation at $r = a$. Consider the propagation of a ray as shown in Fig. 9.1. If n_2 is positive, the ray will suffer total internal reflection if the angle of incidence i is greater than

$$\sin^{-1}\left(\frac{n_0}{n_0 + \tfrac{1}{2}n_2 E_0^2}\right)$$

Thus if the ray makes an angle θ with the z axis (which represents the direction of the propagation of the wave) then for total internal reflection to

* The treatment is based on the review by Akhmanov *et al.* (1968*b*) and a pedagogically simple version by Sodha (1973). [See also Svelto (1974).]

† Actually most laser beams have a gradual variation of intensity along the wavefront, e.g., the intensity may vary as $\exp(-r^2/r_0^2)$ (see Section 9.3); the present model is useful for a qualitative understanding of the physics of self-focusing.

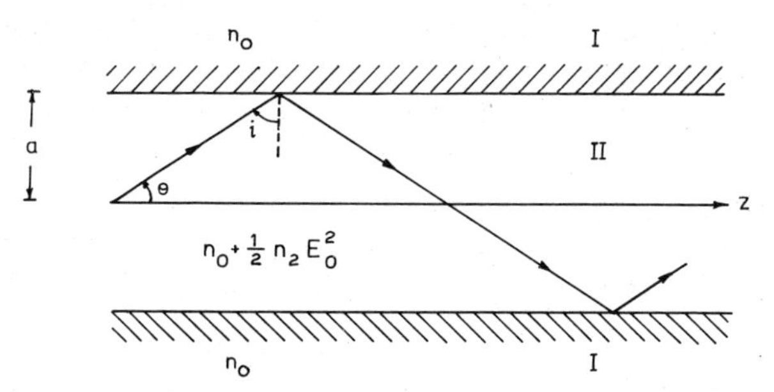

Fig. 9.1. Total internal reflection of a ray.

occur θ should be less than θ_c where

$$\theta_c = \frac{\pi}{2} - \sin^{-1}\left(\frac{n_0}{n_0 + \frac{1}{2}n_2E_0^2}\right) \tag{9.2-1}$$

or

$$\cos\theta_c = \frac{n_0}{n_0 + \frac{1}{2}n_2E_0^2} \tag{9.2-2}$$

We assume that the nonlinearity is weak, i.e., $\frac{1}{2}n_2E_0^2 \ll n_0$, which is indeed true for most systems; consequently, θ_c will also be small and we will have

$$1 - \frac{\theta_c^2}{2} \simeq 1 - \frac{1}{2}\frac{n_2}{n_0}E_0^2$$

or

$$\theta_c \approx \left[\frac{n_2}{n_0}E_0^2\right]^{1/2} \tag{9.2-3}$$

Now, a beam that is limited by an aperture of radius a will undergo diffraction, and from diffraction theory we know that a large fraction of the power will be carried by rays that make an angle less than θ_d with the axis, where

$$\theta_d \simeq \frac{0.61\lambda}{2a} = \frac{0.61}{2a}\frac{\lambda_0}{n_0} \tag{9.2-4}$$

where λ $(=\lambda_0/n_0)$ represents the wavelength in the medium.* Clearly, if $\theta_d < \theta_c$, then the diffracted rays will make an angle less than θ_c with the

* The angle $2\theta_d$ $(= 1.22\lambda/2a)$ corresponds to the first minimum of the Fraunhofer diffraction pattern produced by a circular aperture of radius a (see Chapter 4).

axis; hence the rays will suffer total internal reflection at the boundary and will return to the beam. When $\theta_d > \theta_c$, the beam will spread by diffraction. The critical power for uniform waveguide-like propagation will correspond to $\theta_c \simeq \theta_d$ or

$$E_0^2 \simeq \frac{1}{n_2 n_0}\left(\frac{0.61}{2a}\lambda_0\right)^2 \tag{9.2-5}$$

Now, the energy per unit volume associated with the electromagnetic wave is

$$\tfrac{1}{2}\langle(\mathbf{E}\cdot\mathbf{D} + \mathbf{B}\cdot\mathbf{H})\rangle = \langle\mathbf{E}\cdot\mathbf{D}\rangle = \varepsilon_0 K_0 \langle E^2\rangle = \tfrac{1}{2}n_0^2\varepsilon_0 E_0^2 \tag{9.2-6}$$

where we have assumed the energy associated with the magnetic field to be the same as the energy associated with the electric field and have used the fact that

$$\langle E^2\rangle = \tfrac{1}{2}E_0^2 \tag{9.2-7}$$

where $\langle\ \rangle$ denotes time average. Thus the power of the beam is given by

$$P = \pi a^2 \frac{c}{n_0}\cdot\frac{1}{2}n_0^2\varepsilon_0 E_0^2 \tag{9.2-8}$$

If we substitute the expression for E_0^2 from Eq. (9.2-5) we will get the critical power of the beam:

$$P_{\text{cr}} = (0.61)^2 \frac{\pi c}{8}\frac{\varepsilon_0\lambda_0^2}{n_2} \simeq 0.15\frac{c\varepsilon_0\lambda_0^2}{n_2} \tag{9.2-9}$$

Garmire *et al.* (1966) carried out experiments on the self-focusing of a ruby laser beam ($\lambda_0 = 0.6943\ \mu$m) in CS_2 and found that the critical power was 25 ± 5 kW. On the other hand, if we use Eq. (9.2-9) we obtain

$$P_{\text{cr}} = \frac{0.15 \times 3 \times 10^8 \times 8.854 \times 10^{-12} \times (0.6943 \times 10^{-6})^2}{2 \times 10^{-11}} \simeq 100\ \text{kW} \tag{9.2-10}$$

where we have used the following parameters for CS_2 (Garmire *et al.*, 1966):

$$n_0 = 1.6276, \qquad n_2 = 1.8 \times 10^{-10}\ \text{cgs units} = 2 \times 10^{-19}\ \text{mks units} \tag{9.2-11}$$

The mks unit for n_2 is (meter/volt)^2. It should be mentioned that the nonlinear coefficient n_2 (in CS_2) is predominantly due to the Kerr effect.

It should also be noted that the result expressed by Eq. (9.2-9) is through qualitative arguments and is therefore expected to give results that are correct within a factor of about 10. If we use the more accurate theoretical

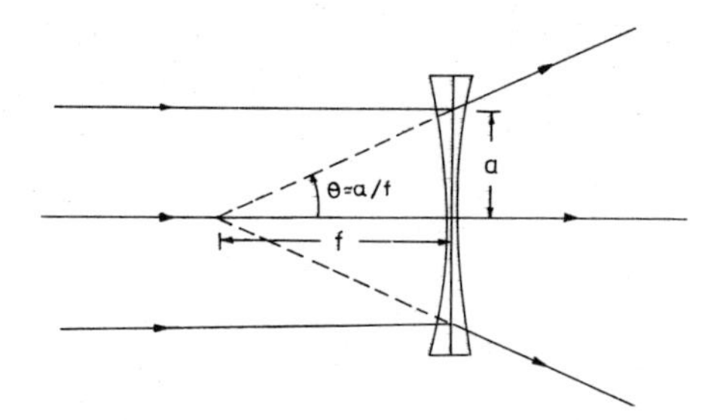

Fig. 9.2. If a parallel beam of light is incident on a diverging lens (of focal length f) the beam, after passing through the lens, will diverge at an angle $\theta = a/f$, where a is the radius of the aperture.

result [see Eq. (9.3-17)] we obtain 25 kW, which compares well with the experimental data.

We may conclude the following:

1. When $P < P_{cr}$, $\theta_d > \theta_c$, and the beam will diverge due to diffraction.

2. When $P = P_{cr}$, $\theta_d = \theta_c$, and the beam will propagate without divergence or convergence. This is often known as the condition for uniform waveguide-like propagation.

3. When $P > P_{cr}$, we may extrapolate that the beam will converge, which is indeed borne out by more rigorous analysis. This is known as self-focusing of the beam.

In order to have an alternative figurative description we note that if a parallel beam of light is incident on a concave lens, we obtain a diverging spherical wave with $\theta \simeq a/f$ (see Fig. 9.2). Now, Eq. (9.2-4) can be rewritten

$$\theta_d = a/f_d \tag{9.2-12}$$

where

$$\begin{aligned} f_d &= \frac{a}{\theta_d} \simeq \frac{2a^2 n_0}{0.61\lambda_0} \simeq \frac{2\pi}{\lambda_0} n_0 \frac{a^2}{2} = \frac{ka^2}{2} \\ k &= \frac{2\pi}{\lambda} = \frac{2\pi}{\lambda_0} n_0 \end{aligned} \tag{9.2-13}$$

Thus the phenomenon of diffraction can be approximated by a diverging lens of focal length $ka^2/2$.

Now we know that if a converging lens of a certain focal length is placed adjacent to a diverging lens of the same focal length, then a parallel beam incident on such a combination will emerge as a parallel beam, i.e., neither any focusing nor any defocusing will occur. In the present case we have just seen that an incident beam propagates without any divergence or convergence when

$$E_0^2 = \frac{1}{n_2 n_0}\left(\frac{0.61\lambda_0}{2a}\right)^2 \tag{9.2-14}$$

which may be rewritten in the form

$$f_{\rm d} = \frac{1}{2}ka^2 \simeq a\left(\frac{n_0}{2n_2E_0^2}\right)^{1/2}$$

Thus the nonlinear medium can be thought of as a converging lens of focal length

$$f_{\rm nl} \simeq a\left(\frac{n_0}{2n_2E_0^2}\right)^{1/2} \tag{9.2-15}$$

When $f_{\rm d} < f_{\rm nl}$, the diffraction divergence will dominate, and when $f_{\rm d} > f_{\rm nl}$, the focusing action will dominate, giving rise to a converging wave in the nonlinear medium.

9.3. More Rigorous Theory for Self-Focusing

In the previous section, we gave an approximate theory for qualitative understanding of the self-focusing phenomenon. In this section, we will present a more rigorous theory, the starting point of which is the scalar wave equation

$$\nabla^2 E = \frac{K}{c^2}\frac{\partial^2 E}{\partial t^2} \tag{9.3-1}$$

where E represents the electric field (which is assumed to be transverse to the direction of propagation) and K is the dielectric constant. The use of the scalar wave equation is justified when the characteristic length for the inhomogeneity in the medium (introduced by the nonlinear refractive-index variation) is large in comparison to the wavelength* and therefore the transverse character of the waves is approximately maintained. Since the dielectric constant is the square of the refractive index, we may write

$$K = n^2 = (n_0 + \tfrac{1}{2}n_2E_0^2)^2 \simeq K_0 + \tfrac{1}{2}K_2E_0^2 \tag{9.3-2}$$

where

$$K_0 = n_0^2, \qquad K_2 = 2n_0n_2 \tag{9.3-3}$$

and we have assumed $\frac{1}{2}n_2E_0^2 \ll n_0$. Assuming a time dependence of the form $\exp(i\omega t)$, we obtain

$$\nabla^2 E + (K_0 + \tfrac{1}{2}K_2E_0^2)\frac{\omega^2}{c^2}E = 0 \tag{9.3-4}$$

* This will indeed be shown in Problem 9.1, where the characteristic length has been shown to be $\simeq 5$ cm (see also Problem 10.2).

The solution of the above equation for azimuthally symmetric beams (i.e., the field being independent of the φ coordinate) is given in Section 9.5. In particular, if we assume a Gaussian beam at $z = 0$, i.e.,

$$E_0^2(r, z = 0) = A_0^2 \exp(-r^2/r_0^2) \tag{9.3-5}$$

then in the paraxial approximation, it will be shown that

$$E_0^2(r, z) = \frac{A_0^2}{f^2(z)} \exp\left[-\frac{r^2}{r_0^2 f^2(z)} \right] \tag{9.3-6}$$

where $f(z)$ (known as the dimensionless beam width parameter) satisfies the following differential equation:

$$\frac{1}{f}\frac{d^2 f}{dz^2} = \left(\frac{1}{k^2 r_0^4 f^4}\right) - \left(\frac{1}{2}\frac{K_2}{K_0}\frac{A_0^2}{r_0^2}\frac{1}{f^4}\right) \tag{9.3-7}$$

and

$$f(z = 0) = 1 \tag{9.3-8}$$

Furthermore,

$$\frac{1}{f}\frac{df}{dz} = \frac{1}{R} \tag{9.3-9}$$

where R represents the radius of curvature of the wave. Thus for an incident plane wave

$$\left.\frac{1}{f}\frac{df}{dz}\right|_{z=0} = 0 \tag{9.3-10}$$

If we substitute the expression for E_0^2 from Eq. (9.3-6) into Eq. (9.3-2), we obtain

$$\begin{aligned} K &= K_0 + \frac{1}{2} K_2 \left\{ \frac{A_0^2}{f^2(z)} \left[1 - \frac{r^2}{r_0^2 f^2(z)} + \frac{r^4}{2 r_0^4 f^4(z)} + \cdots \right] \right\} \\ &\simeq K_0 + \frac{1}{2} K_2 \frac{A_0^2}{f^2(z)} - \frac{1}{2}\frac{K_2 A_0^2}{r_0^2 f^4(z)} r^2 \\ &= K_0 \left\{ 1 + \frac{1}{2}\frac{K_2}{K_0}\frac{A_0^2}{f^2(z)} - \left[\frac{1}{2}\frac{K_2}{K_0}\frac{A_0^2}{r_0^2 f^4(z)} \right] r^2 \right\} \end{aligned} \tag{9.3-11}$$

where we have assumed $r/[r_0 f(z)] \ll 1$, which is the paraxial approximation. It can easily be seen from Eq. (9.3-11) that in the paraxial approximation one obtains a parabolic variation of the dielectric constant (and hence of the refractive index).* It should be pointed out that the first term on the right-hand side of Eq. (9.3-7) is due to diffraction and the second term [which is essentially the term in square brackets in Eq. (9.3-11)] is due to nonlinear effects.

The power of the beam is given by

$$P = \int_0^\infty \left(n^2 \varepsilon_0 \frac{E_0^2}{2} \right) \frac{c}{n} 2\pi r\, dr \tag{9.3-12}$$

where the quantity inside the parentheses represents the energy per unit volume. Substituting for E_0^2 from Eq. (9.3-6), we get

$$P = \frac{A_0^2}{f^2(z)} \varepsilon_0 n_0 \pi c \int_0^\infty \exp\left[-\frac{r^2}{r_0^2 f^2(z)} \right] r\, dr = \frac{\pi c}{2} \varepsilon_0 n_0 A_0^2 r_0^2 \tag{9.3-13}$$

Thus the power of the beam is independent of z, which it indeed ought to be, because we have neglected absorption. Note that when

$$\frac{1}{k^2 r_0^4} = \frac{1}{2} \frac{K_2}{K_0} \frac{A_0^2}{r_0^2} \tag{9.3-14}$$

Eq. (9.3-7) becomes

$$d^2 f / dz^2 = 0 \tag{9.3-15}$$

Thus df/dz is a constant and for an incident plane wave, this constant will be zero [see Eq. (9.3-9)]. Thus

$$f(z) = 1 \tag{9.3-16}$$

where we have used Eq. (9.3-8). Since $f(z)$ is unity for all values of z, Eq. (9.3-6) predicts no change in the transverse intensity distribution, as the beam propagates through the nonlinear medium. This is indeed the condition for uniform waveguide-like propagation as discussed in Section 9.2. Thus the critical power of the beam should correspond to Eq. (9.3-12), and if we substitute the expression for A_0^2 from Eq. (9.3-14) into Eq. (9.3-13), we should get the critical power:

$$P_{\text{cr}} = \frac{\pi c}{2} \varepsilon_0 n_0 r_0^2 \times \frac{2K_0}{K_2} \frac{1}{r_0^2 k^2} \simeq 0.039 \frac{c \varepsilon_0 \lambda_0^2}{n_2} \tag{9.3-17}$$

The above equation may be compared with Eq. (9.2-9). Substituting the values corresponding to the experiments carried out by Garmire *et al.* (1966) [see Eq. (9.2-11)], we obtain $P_{\text{cr}} \simeq 25$ kW, which compares well with the experimental value, which is 25 ± 5 kW.

* In Chapter 10 we discuss the propagation of electromagnetic waves through inhomogeneous media characterized by a parabolic refractive-index variation.

In order to solve Eq. (9.3-7) we multiply it by $2f\,(df/dz)\,dz$ and integrate:

$$\int 2f \frac{df}{dz}\frac{d^2f}{dz^2}\,dz = (R_d^{-2} - R_n^{-2}) \int 2\frac{1}{f^3}\frac{df}{dz}\,dz \qquad (9.3\text{-}18)$$

where

$$R_n = \left(\frac{2K_0 r_0^2}{K_2 A_0^2}\right)^{1/2}, \qquad R_d = \frac{1}{k r_0^2} \qquad (9.3\text{-}19)$$

R_n represents the characteristic length for nonlinear focusing and R_d the characteristic length for diffraction.* On carrying out the integration, we obtain

$$\left(\frac{df}{dz}\right)^2 = (R^{-2} - R_c^{-2}) + \frac{R_c^{-2}}{f^2} \qquad (9.3\text{-}20)$$

where

$$R_c^{-2} = R_n^{-2} - R_d^{-2} \qquad (9.3\text{-}21)$$

and we have used the boundary condition that at $z = 0$,

$$f = 1 \qquad \text{and} \qquad \frac{df}{dz} = \frac{1}{R} \qquad (9.3\text{-}22)$$

Equation (9.3-20) can be rewritten in the form

$$\frac{f\,df}{[(R^{-2} - R_c^{-2})f^2 + R_c^{-2}]^{1/2}} = \pm dz \qquad (9.3\text{-}23)$$

where the upper and lower signs correspond to $R > 0$ and $R < 0$, respectively (this is obvious from the fact that at $z = 0$, df/dz is positive if $R > 0$ and conversely). On integrating Eq. (9.3-23), we obtain

$$[(R^{-2} - R_c^{-2})f^2 + R_c^{-2}]^{1/2} = \pm(R^{-2} - R_c^{-2})z + |R^{-1}| \qquad (9.3\text{-}24)$$

On simplification, we get

$$f^2(z) = 1 + \frac{2}{R}z + (R^{-2} - R_c^{-2})z^2 \qquad (9.3\text{-}25)$$

For an incident plane wave, $1/R = 0$ and one obtains

$$f^2(z) = 1 - R_c^{-2}z^2 = 1 + (R_d^{-2} - R_n^{-2})z^2$$
$$= 1 + \left(\frac{1}{k^2 r_0^4} - \frac{1}{2}\frac{K_2}{K_0}\frac{A_0^2}{r_0^2}\right)z^2 \qquad (9.3\text{-}26)$$

* By the characteristic diffraction length we imply the distance over which the beam should propagate for appreciable diffraction divergence [see, e.g., Eq. (9.3-27)].

We may first consider three cases.

1. In the absence of any nonlinear effects, i.e., for $K_2 = 0$,

$$f^2(z) = 1 + \frac{1}{k^2 r_0^4} z^2 \tag{9.3-27}$$

showing that the beam width monotonically increases with z. The increase is more rapid as the wavelength increases or the initial spot size r_0 decreases. This is nothing but diffraction divergence. Incidentally, Eqs. (9.3-6) and (9.3-27) show that a Gaussian beam propagating through a homogeneous medium remains Gaussian. (There is nothing like an Airy pattern!) (See also Problem 4.8.) From Eq. (9.3-27) it is obvious that the characteristic distance for diffraction divergence is $1/kr_0^2$.

2. In the geometrical optics approximation (i.e., $\lambda \to 0$) Eq. (9.3-26) becomes

$$f^2(z) = 1 - \frac{1}{2} \frac{K_2}{K_0} \frac{A_0^2}{r_0^2} z^2 \tag{9.3-28}$$

The beam gets focused due to nonlinear effects; this is the self-focusing of the beam. Notice that if K_2 is negative the beam will get defocused. Indeed if the nonlinearity is due to thermal effects, one usually obtains a defocusing of the beam (see Section 9.4).

3. Notice that as f decreases, the width of the beam decreases [see Eq. (9.3-6)], and in the limit of $f \to 0$, the intensity is zero at all points except near $r = 0$. Such a point will correspond to the focal point.*

In general, the diffraction effects tend to defocus the beam and the nonlinear effects tend to focus the beam; if the nonlinear effects predominate, self-focusing will occur and conversely.

Problem 9.1. (a) For an incident plane wave, calculate in the geometrical optics approximation the distance at which the beam width will become zero.* This is known as the trapping length.

(b) Garmire *et al.* (1966) reported that the trapping length of a 90 kW ruby laser beam ($\lambda = 0.6943\ \mu$m) was about 12 cm. The initial beamwidth ($=2r_0$) was 0.5 mm. Assuming $n_0 = 1.6276$ and $n_2 = 2 \times 10^{-19}$ mks units, make a theoretical estimate of the trapping length.

* Note that the results reported in Section 9.2 are valid in the paraxial approximation, i.e., for values of r that satisfy the inequality $r/(r_0 f) \ll 1$. Clearly, the theory will not be valid near the focal point where $f \to 0$. Furthermore, near the focal point, the nonlinear medium breaks down and the problem becomes very complicated.

Solution. (a) From Eq. (9.3-26), we see that $f = 0$ when

$$z = z_f \simeq \left(\frac{2K_0 r_0^2}{K_2 A_0^2} \right)^{1/2} = \left(\frac{n_0 r_0^2}{n_2 A_0^2} \right)^{1/2} \tag{9.3-29}$$

where z_f is known as the trapping length.

(b) The power of the beam is given by

$$P = \frac{\pi c}{2} \varepsilon_0 n_0 A_0^2 r_0^2 = 90 \times 10^3 \mathrm{W} \tag{9.3-30}$$

Substituting the value of A_0^2 obtained from Eq. (9.3-30) into Eq. (9.3-29), we can calculate z_f to be about 5 cm.

9.4. *Thermal Self-Focusing/Defocusing of Laser Beams*

When an intense electromagnetic wave propagates through an absorbing medium the beam gets absorbed, and if there is an intensity distribution along its wavefront a temperature gradient is set up. Since the refractive index depends on the temperature, the temperature gradient leads to a refractive-index gradient that will tend either to focus or to defocus the beam.

A beam propagating through an absorbing medium will undergo exponential attenuation of its power. Thus if $P(z)$ denotes the power of the beam after it has propagated through a distance z, then

$$P(z) = P_0 \exp(-\alpha z)$$

where P_0 denotes the power at $z = 0$ and α is the intensity absorption coefficient. We assume that a laser beam having Gaussian intensity distribution along its wavefront is incident at $z = 0$; in the paraxial approximation, the intensity will remain Gaussian (see Section 9.3):

$$I(r, z) = \frac{I_0}{f^2} \exp\left(-\frac{r^2}{r_0^2 f^2} \right) \exp(-\alpha z) \tag{9.4-1}$$

In Eq. (9.4-1), f is the dimensionless beam width parameter and

$$f(z = 0) = 1 \tag{9.4-2}$$

Thus, the fractional power p of the beam incident on a circle of radius r is given by

$$p(r) = \frac{\int_0^r 2\pi r\, dr \exp(-r^2/r_0^2 f^2)}{\int_0^\infty 2\pi r\, dr \exp(-r^2/r_0^2 f^2)} = 1 - \exp\left(-\frac{r^2}{r_0^2 f^2(z)} \right) \tag{9.4-3}$$

In order to obtain the temperature gradient, we write the heat conduction

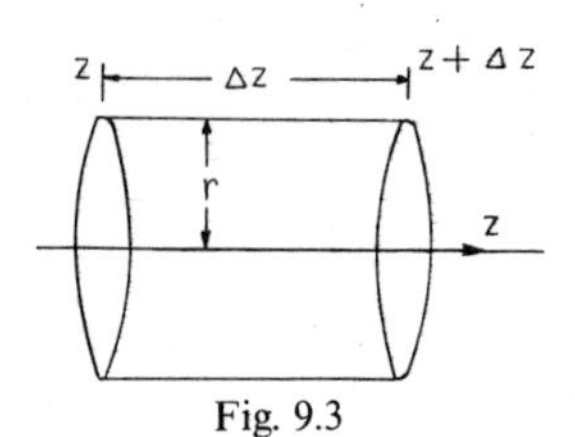

Fig. 9.3

equation and consider a cylinder of radius r and of length Δz (see Fig. 9.3). The power lost in traversing the distance Δz is

$$P_0 e^{-\alpha z} - P_0 e^{-\alpha(z+\Delta z)} \simeq \alpha \Delta z\, e^{-\alpha z} P_0$$

Thus the power lost in the cylinder shown in Fig. 9.3 is $\alpha \Delta z P_0 e^{-\alpha z} p(r)$. Assuming that this energy is conducted away along the curved surface whose area is $2\pi r \Delta z$, we have

$$-2\pi r \Delta z\, \kappa\, dT/dr = \alpha \Delta z P_0 e^{-\alpha z} p(r) \tag{9.4-4}$$

where κ represents the thermal conductivity. In writing Eq. (9.4-4) we have assumed that transmission of heat along the z axis is negligibly small. Thus

$$\begin{aligned} \frac{\partial T}{\partial r} &= -\frac{\alpha P_0}{2\pi\kappa} \frac{[1 - \exp(-r^2/r_0^2 f^2)]}{r} e^{-\alpha z} \\ &= -\frac{\alpha P_0}{2\pi\kappa}\left[\frac{r}{r_0^2 f^2} - \frac{r^3}{2r_0^4 f^4} + \cdots\right] e^{-\alpha z} \end{aligned} \tag{9.4-5}$$

Hence

$$\left.\frac{\partial T}{\partial r}\right|_{r=0} = 0, \qquad \left.\frac{\partial^2 T}{\partial r^2}\right|_{r=0} = -\frac{\alpha P_0}{2\pi\kappa r_0^2 f^2} e^{-\alpha z} \tag{9.4-6}$$

For points close to the axis, we may write

$$T(r) = T(0) + r \left.\frac{\partial T}{\partial r}\right|_{r=0} + \frac{r^2}{2} \left.\frac{\partial^2 T}{\partial r^2}\right|_{r=0} \tag{9.4-7}$$

where $T(0) = T(r = 0)$. Thus

$$T(r) - T(0) = -\left[\frac{1}{4\pi} \frac{\alpha P_0}{\kappa r_0^2 f^2} e^{-\alpha z}\right] r^2 \tag{9.4-8}$$

If K_0 represents the dielectric constant on the axis, then*

* Notice that the temperature variation (and hence the dielectric constant variation) is parabolic, which justifies the assumption that a Gaussian beam will remain Gaussian.

$$K(r, z) = K_0 + [T(r) - T(0)] \frac{dK}{dT}$$

$$= K_0 \left\{ 1 - \left[\frac{1}{K_0} \frac{\alpha P_0 e^{-\alpha z}}{4\pi\kappa r_0^2 f^2} \frac{dK}{dT} \right] r^2 \right\} \tag{9.4-9}$$

(In general, there will be a weak z dependence of K_0.)

If we compare Eq. (9.4-9) with Eq. (9.3-11), we find that the radial variation of the dielectric constant is of the same type, and therefore we can write the differential equation satisfied by $f(z)$ [cf. Eq. (9.3-7)]:

$$\frac{1}{f} \frac{d^2 f}{dz^2} = \frac{1}{k^2 r_0^4 f^4(z)} - \frac{\alpha P_0 e^{-\alpha z}}{4\pi\kappa r_0^2 f^2(z)} \frac{1}{K_0} \frac{dK}{dT} \tag{9.4-10}$$

In most media, $dK/dT < 0$, and we have defocusing of the beam. Furthermore, in experiments reported in the literature (see, e.g., Akhmanov *et al.*, 1967, 1968*a*) thermal defocusing dominates over diffraction divergence. Under such conditions, we may neglect the first term* on the right-hand side of Eq. (9.4-10) to obtain

$$\frac{d^2 f}{dz^2} = -\frac{\alpha P_0 e^{-\alpha z}}{4\pi\kappa r_0^2 f^2(z)} \frac{1}{K_0} \frac{dK}{dT} \tag{9.4-11}$$

In general, the defocusing is weak and $f(z)$ remains close to unity. For such a case very little error will be involved if we replace $f(z)$ by unity on the right-hand side of Eq. (9.4-11). If we do so, then a simple integration of Eq. (9.4-11) will give

$$\frac{df}{dz} = -\frac{(1 - e^{-\alpha z}) P_0}{4\pi K_0 r_0^2 \kappa} \frac{dK}{dT} \tag{9.4-12}$$

where we have assumed the incident wavefront to be plane, i.e., $(df/dz)_{z=0}$ is equal to zero. A further integration will give

$$f(z) = 1 + C\left[z - \frac{1}{\alpha}(1 - e^{-\alpha z}) \right] \tag{9.4-13}$$

where we have used the boundary condition that $f(z = 0) = 1$, and

$$C = -\frac{P_0}{4\pi K_0 r_0^2 \kappa} \frac{dK}{dT} = -\frac{P_0}{2\pi r_0^2 \kappa} \frac{1}{n} \frac{dn}{dT} \tag{9.4-14}$$

Because $dK/dT < 0$ for most media, C is a positive quantity, and Eq. (9.4-14) tells us that for large values of z, $f(z)$ increases linearly with z. Thus the

* This is equivalent to the geometrical optics approximation (see also Problem 9.2).

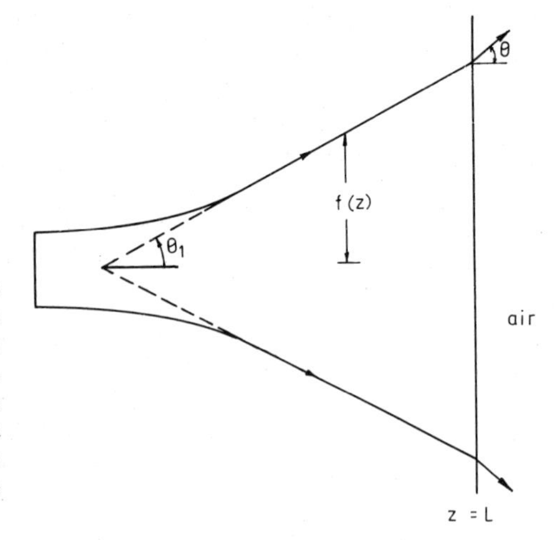

Fig. 9.4. Defocusing of a beam due to thermal effects. The curved line is the locus of points where the intensity is a certain fraction [say $(1/e)$th] of the intensity on the axis ($r = 0$). Notice that at large values of z, the curve tends to a straight line, making an angle θ_1 with the axis. The plane $z = L$ separates the absorbing medium from air.

width of the beam $w(z)$ will be given by

$$w(z) = r_0 f(z) \simeq r_0 C z \tag{9.4-15}$$

(for large values of z). Therefore,

$$\tan \theta_1 = dw(z)/dz \simeq r_0 C \tag{9.4-16}$$

where θ_1 is the angle that the ray makes with the axis (see Fig. 9.4). The angle θ that the emergent ray makes with the axis, is related to θ_1 through Snell's law; for small angles, Snell's law gives us

$$\theta \simeq n\theta_1 \simeq n r_0 C = -\frac{P_0 \, dn/dT}{2\pi r_0 \kappa} \tag{9.4-17}$$

Akhmanov *et al.* (1968*a*) have carried out measurements on thermal

Table 9.1. Data on Thermal Defocusing of an Argon Laser Beam[a,b]

Liquid	dn/dT (deg^{-1})	κ (W/cm-deg)	n_0	P_0 (W)	r_0 (cm)	θ Experimental	θ Theoretical, using Eq. (9.4-17)
Water	-0.8×10^{-4}	6.0×10^{-3}	1.33	0.10	0.09	10.0 ± 0.7	11.2
				0.15	0.09	14.8 ± 0.7	15.8
Acetone	-5.0×10^{-4}	1.6×10^{-3}	1.36	0.10	0.07	28.0 ± 1.5	33.2
				0.15	0.07	41.0 ± 1.5	50.6
Alcohol	-4.0×10^{-4}	1.7×10^{-3}	1.37	0.10	0.07	20.5 ± 1.5	25.4
				0.15	0.07	30.5 ± 1.5	38.0

[a] $\lambda = 0.488\ \mu$m. [b] After Akhmanov *et al.* (1967).

self-defocusing of an argon laser beam ($\lambda = 0.488\ \mu$m) in water, acetone, and alcohol. Small quantities of dyes ($KMnO_4$ in water, fuchsine in alcohol and acetone) that are strongly absorbent at the $0.488\ \mu$m wavelength were added to the pure liquid. The value of α was about $1\ m^{-1}$. The relevant experimental data and the corresponding theoretical prediction [using Eq. (9.4-17)] are displayed in Table 9.1. The agreement between theory and calculations is good. Notice that the defocusing is indeed small, which justifies our assumption that $f(z)$ stays close to unity.

Problem 9.2. In the experimental data reported in Table 9.1, the lengths of the medium (L) were 84 and 44 cm. Assuming $L = 84$ cm and using any one of the experimental data given in Table 9.1, show that the divergence of the beam due to thermal effects is much greater than the diffraction divergence. [*Hint:* Use Eqs. (9.3-25) and (9.4-13).]

9.5. *Solution of the Scalar Wave Equation with Weak Nonlinearity*

In this section we will obtain a solution of the equation

$$\nabla^2 E + \frac{\omega^2}{c^2}\left[K_0 + \Phi\left(\frac{1}{2}E_0^2\right)\right]E = 0 \qquad (9.5\text{-}1)$$

where $\Phi(\frac{1}{2}E_0^2)$ corresponds to the nonlinearity in the medium. The factor $\frac{1}{2}$ in the argument of Φ is introduced for consistency with the existing literature. Equation (9.3-4) is a particular form of Eq. (9.5-1) and is obtained when

$$\Phi(\tfrac{1}{2}E_0^2) = \tfrac{1}{2}K_2 E_0^2 \qquad (9.5\text{-}2)$$

Following Akhmanov *et al.* (1968*b*) we assume an azimuthally symmetric solution of the form

$$E = E_1 \exp[i(\omega t - kz)] \qquad (9.5\text{-}3)$$

where $k = (\omega/c)K_0^{1/2}$. Since azimuthal symmetry implies that the fields are independent of the φ coordinate, we have

$$\nabla^2 = \frac{\partial^2}{\partial z^2} + \frac{\partial^2}{\partial r^2} + \frac{1}{r}\frac{\partial}{\partial r} \qquad (9.5\text{-}4)$$

and we obtain

$$2ik\frac{\partial E_1}{\partial z} = \frac{\partial^2 E_1}{\partial r^2} + \frac{1}{r}\frac{\partial E_1}{\partial r} + \frac{k^2}{K_0}\Phi\left(\frac{1}{2}E_1^2\right)E_1 \qquad (9.5\text{-}5)$$

where we have neglected $\partial^2 E_1/\partial z^2$, which is justified when

$$\partial^2 E_1/\partial z^2 \ll k \cdot \partial E_1/\partial z$$

or

$$\frac{1}{k(\partial E_1/\partial z)} \frac{\partial}{\partial z}\left(\frac{\partial E_1}{\partial z}\right) \ll 1 \tag{9.5-6}$$

i.e., the distance over which $\partial E_1/\partial z$ changes appreciably is large compared to the wavelength, which is indeed true for all practical cases. For example, if we assume $\partial E_1/\partial z$ to vary as e^{-z/z_0}, then Eq. (9.5-6) implies $kz_0 \gg 1$ or $z_0 \gg \lambda$. Clearly z_0 is the characteristic distance over which $\partial E_1/\partial z$ changes appreciably. (See also Problem 9.1 where it has been shown that the trapping length is typically of the order of a few centimeters for optical beams.)

To solve Eq. (9.5-5) we write

$$E_1(r, z) = E_0(r, z) \exp[-ikS(r, z)] \tag{9.5-7}$$

where $E_0(r, z)$ and $S(r, z)$ are real functions of r and z. Thus $E_0(r, z)$ represents the amplitude of the wave and E_0^2 will be proportional to the intensity. If we substitute the expression for E_1 from Eq. (9.5-7) into Eq. (9.5-5) and equate real and imaginary parts, we obtain

$$2\frac{\partial S}{\partial z} + \left(\frac{\partial S}{\partial r}\right)^2 = \frac{1}{k^2 E_0}\left(\frac{\partial^2 E_0}{\partial r^2} + \frac{1}{r}\frac{\partial E_0}{\partial r}\right) + \frac{1}{K_0}\Phi\left(\frac{1}{2}E_0^2\right) \tag{9.5-8}$$

and

$$\frac{\partial E_0^2}{\partial z} + \frac{\partial S}{\partial r}\frac{\partial E_0^2}{\partial r} + E_0^2\left(\frac{\partial^2 S}{\partial r^2} + \frac{1}{r}\frac{\partial S}{\partial r}\right) = 0 \tag{9.5-9}$$

For a slowly converging/diverging beam, we assume the eikonal $S(r, z)$ to be of the form (see Problem 9.3)

$$S = \tfrac{1}{2}r^2\beta(z) + \varphi(z) \tag{9.5-10}$$

where $1/\beta(z)$ represent the radius of curvature of the wavefront (see Problem 9.3). If $S(r, z)$ is given by Eq. (9.5-10) then it can easily be shown (see Problem 9.4) that the general solution of Eq. (9.5-9) is of the form

$$E_0^2(r, z) = \frac{A_0^2}{f^2(z)}\mathscr{F}\left[\frac{r}{r_0 f(z)}\right] \tag{9.5-11}$$

where $\mathscr{F}$ is an arbitrary function of its argument, A_0 and r_0 are constants, and $f(z)$ is related to $\beta(z)$ through the following equation:

$$\frac{1}{f}\frac{df}{dz} \equiv \beta(z) \tag{9.5-12}$$

From Eq. (9.5-11) we can infer that if the intensity distribution at $z = 0$ is Gaussian, then the intensity distribution at an arbitrary point z will also be a Gaussian with a different width [see Eqs. (9.3-5) and (9.3-6)]. In order to determine the z dependence of $f(z)$ we substitute the expressions for $S(r, z)$ and $E_0^2(r, z)$ [from Eqs. (9.5-10) and (9.3-6), respectively] into Eq. (9.5-8):

$$2\left(\frac{r^2}{2}\frac{d\beta}{dr} + \frac{d\varphi}{dz}\right) + [r\beta(z)]^2 = \frac{1}{K_0}\left[\frac{1}{2}K_2\frac{A_0^2}{f^2(z)}\exp\left(-\frac{r^2}{r_0^2 f^2(z)}\right)\right] + \frac{r^2}{k^2 r_0^4 f^4(z)} - \frac{2}{k^2 r_0^2 f^2(z)} \tag{9.5-13}$$

where we have assumed the form of Φ to be given by Eq. (9.5-2) and have used the relation

$$\frac{1}{k^2 E_0}\left(\frac{\partial^2 E_0}{\partial r^2} + \frac{1}{r}\frac{\partial E_0}{\partial r}\right) = \frac{r^2}{k^2 r_0^4 f^4(z)} - \frac{2}{k^2 r_0^2 f^2(z)} \tag{9.5-14}$$

In Eq. (9.5-13) if we expand the exponential and retain terms of $O[r^2/r_0^2 f^2(z)]$ and then equate the coefficient of r^2 on both sides of the equation, we obtain

$$\frac{d\beta}{dz} + \beta^2 = \frac{1}{f}\frac{d^2 f}{dz^2} = \frac{1}{k^2 r_0^4 f^4(z)} - \frac{1}{2}\frac{K_2}{K_0}\frac{A_0^2}{r_0^2 f^4(z)} \tag{9.5-15}$$

where we have used Eq. (9.5-12). The solution of Eq. (9.5-15) is discussed in Section 9.3.

Problem 9.3. The first term on the right-hand side of Eq. (9.5-8) is responsible for the diffraction of the beam and goes to zero as $k \to \infty$. The second term is due to nonlinear effects. Show that in the absence of the diffraction and nonlinear terms, the solution of Eq. (9.5-8) can be written in the form

$$S(r, z) = r^2/2(z + R) \tag{9.5-16}$$

which approximately represents a spherical wave whose radius of curvature is R at $z = 0$. Thus, when the effects due to diffraction and nonlinearities are expected to be small, one may assume a solution of the type given by Eq. (9.5-10), which has a more general form than Eq. (9.5-16). [*Hint:* Use the method of separation of variables.]

Problem 9.4. Using Eqs. (9.5-10) and (9.5-12), Eq. (9.5-9) assumes the form

$$\frac{\partial E_0^2}{\partial z} + r\left(\frac{1}{f}\frac{df}{dz}\right)\frac{\partial E_0^2}{\partial r} + 2\left(\frac{1}{f}\frac{df}{dz}\right)E_0^2 = 0 \tag{9.5-17}$$

Introducing two new variables

$$\eta = \frac{r}{r_0}\frac{1}{f}, \qquad \zeta = \frac{r}{r_0} f \tag{9.5-18}$$

show that Eq. (9.5-17) transforms to

$$\frac{\partial}{\partial \zeta}(\zeta E_0^2) = 0 \tag{9.5-19}$$

the solution of which is given by Eq. (9.5-11).

Problem 9.5. If the nonlinear dielectric constant variation is characterized by the general relation

$$K = K_0 + \Phi(\tfrac{1}{2}E_0^2)$$

then for weak nonlinearities, one can show that a Gaussian beam remains Gaussian; however, the beam width parameter satisfies the equation*

$$\frac{1}{f}\frac{d^2 f}{dz^2} = \frac{1}{k^2 r_0^4 f^4} - \frac{E_0^2}{2K_0 r_0^2 f^4}\Phi'\left(\frac{A_0^2}{2f^2}\right) \tag{9.5-20}$$

where the prime denotes differentiation with respect to the argument. Assuming a saturating dielectric constant profile (see Problem 9.7) of the form†

$$K = K_0 + \frac{\frac{1}{2}K_2 A_0^2}{1 + (K_2/2K_s)A_0^2} \tag{9.5-21}$$

show that for an incident plane wave

$$\left(\frac{df}{dz}\right)^2 = \frac{1 - f^2}{f^2}\frac{[(\alpha - \beta) f^2 - \beta p]}{(f^2 + p)} \tag{9.5-22}$$

where

$$\beta = \frac{1}{k^2 r_0^4}, \qquad \alpha = \frac{b}{r_0^2(1 + p)} \tag{9.5-23}$$

$$b = \frac{K_2}{2K_0}E_0^2, \quad p = \frac{K_2}{2K_s}A_0^2 \tag{9.5-24}$$

Qualitatively discuss the solution of Eq. (9.5-22) and show that for $\alpha/\beta > 1$, the system acts as an oscillatory waveguide, and for $\alpha/\beta < 1$, the diffraction divergence dominates and there is a continuous defocusing of the beam. *Reference:* Ghatak *et al.* (1972).

* Notice that when

$$\Phi(\tfrac{1}{2}E_0^2) = \tfrac{1}{2}K_2 E_0^2$$

then

$$\Phi(E_0^2/2f^2) = K_2 E_0^2/2f^2, \qquad \Phi'(E_0^2/2f^2) = K_2$$

Thus Eq. (9.5-20) reduces to Eq. (9.3-7).

† For weak fields $K \simeq K_0 + \frac{1}{2}K_2 A_0^2$, and for strong fields $K \to K_0 + K_s$.

9.6. General Problems on the Calculation of the Nonlinear Dielectric Constant

In this section we have outlined, through two problems, the technique for calculating the nonlinear variation of the dielectric constant due to the Kerr effect. The nonlinear dielectric constant variation due to electrostriction has been discussed by Sodha *et al.* (1974).*

Problem 9.6. Consider a system of N noninteracting molecules having uniaxial symmetry in a volume V. In general the induced dipole moment $\mathbf{P}$ is related to the electric field $\mathbf{E}$ through the relation

$$P_i = \sum_j \alpha_{ij} E_j, \qquad i, j = x, y, z \tag{9.6-1}$$

where the nine quantities α_{xx}, α_{xy}, etc., constitute the polarizability tensor.† There always exists a coordinate system (fixed in the molecule) for which the polarizability tensor is diagonal, i.e.,

$$(\alpha_{ij})^0 = \begin{pmatrix} \alpha_\perp & 0 & 0 \\ 0 & \alpha_\perp & 0 \\ 0 & 0 & \alpha_\parallel \end{pmatrix} \tag{9.6-2}$$

where the fact that two of the diagonal elements are the same is a manifestation of the uniaxial symmetry of the molecule. Show that the dielectric constant variation is given by

$$K(E) = K_0 + \chi_\pm^{\mathrm{NL}}(E) \tag{9.6-3}$$

where

$$\chi_\pm^{\mathrm{NL}}(E) = \frac{N|\alpha_\parallel - \alpha_\perp|}{2V}\left[\mp\frac{2}{3} - \frac{1}{\zeta^2} + \frac{1}{\zeta G_\pm(\zeta)}\right] \tag{9.6-4}$$

$$G_\pm(\zeta) = \exp(\pm\zeta^2)\int_0^\zeta \exp(\pm x^2)\,dx \tag{9.6-5}$$

$$\zeta = \tfrac{1}{2}E_0\left[|\alpha_\parallel - \alpha_\perp|/k_B T\right]^{1/2} \tag{9.6-6}$$

K_0 is the field-independent dielectric constant, E_0 the strength of the electric field, k_B Boltzmann's constant, and T the absolute temperature. The upper and lower signs in Eq. (9.6-4) correspond to "cigar-shaped" molecules (such as CS_2) for which $\alpha_\parallel > \alpha_\perp$ and "disk-shaped" molecules (such as benzene) for which $\alpha_\parallel < \alpha_\perp$. *Reference:* Wagner *et al.* (1968).

* In recent years there has been a considerable amount of interest in the calculation of the nonlinear dielectric constant variation in plasmas and semiconductors. Interested readers should see Sodha *et al.* (1976).

† Notice that in general $\mathbf{P}$ is not in the direction of the electric field $\mathbf{E}$. This is the anisotropic behavior.

Hint: The energy of one molecule in an electric field is given by

$$W = -\frac{1}{2}\sum \alpha_{ij}\langle E_i E_j\rangle_t \tag{9.6-7}$$

where $\langle\ \rangle$ denotes time average. Denote the principal axis system by x', y', z' and the initial system by x, y, z; transform the x, y, z system to the x', y', z' system by the three Eulerian angles* θ, φ, and ψ (choose, without any loss of generality, $\psi = 0$); assume the electric field to be along the z axis and show

$$W = -\tfrac{1}{4}E_0^2[\alpha_\perp + (\alpha_\parallel - \alpha_\perp)\cos^2\theta] \tag{9.6-8}$$

Then the mean polarizability will be given by

$$\bar{\alpha}_\pm = \frac{\iint[\alpha_\perp + (\alpha_\parallel - \alpha_\perp)\cos^2\theta]\,e^{-W/kT}\sin\theta\,d\theta\,d\varphi}{\iint e^{-W/kT}\sin\theta\,d\theta\,d\varphi}$$

$$= \alpha + \frac{1}{2}|\alpha_\parallel - \alpha_\perp|\left[\mp\frac{2}{3} - \frac{1}{\zeta^2} + \frac{1}{\zeta G_\pm(\zeta)}\right] \tag{9.6-9}$$

$$\alpha = \tfrac{1}{3}(\alpha_\parallel + 2\alpha_\perp) \tag{9.6-10}$$

$$K_0 = \frac{1}{4\pi\varepsilon_0}\alpha\frac{N}{V} \tag{9.6-11}$$

$$\chi_\pm^{\mathrm{NL}} = \frac{1}{4\pi\varepsilon_0}\frac{N}{V}\bar{\alpha}_\pm \tag{9.6-12}$$

Problem 9.7. (a) In the previous problem, show that for small ζ

$$\chi_\pm^{\mathrm{NL}} = \frac{1}{4\pi\varepsilon_0}\frac{N|\alpha_\parallel - \alpha_\perp|}{2V}\left[\frac{8}{45}\zeta^2 \pm \frac{16}{945}\zeta^4 + \cdots\right] \tag{9.6-13}$$

showing that for weak fields $\chi_\pm^{\mathrm{NL}}$ is proportional to the intensity. Also for large ζ

$$\chi_+^{\mathrm{NL}} \to \frac{1}{4\pi\varepsilon_0}\frac{N|\alpha_\parallel - \alpha_\perp|}{2V}\frac{4}{3} \tag{9.6-14}$$

$$\chi_-^{\mathrm{NL}} \to -\frac{1}{4\pi\varepsilon_0}\frac{N|\alpha_\parallel - \alpha_\perp|}{2V}\frac{2}{3} \tag{9.6-15}$$

showing that for strong fields, the susceptibility variation shows a saturation behavior.

* See, for example, Goldstein (1950, Chapter 4).

(b) For CS_2

$$\rho = N/V = 10^{22}\,\text{cm}^{-3} = 10^{28}\,\text{m}^{-3}$$

$$\alpha_{\parallel} \simeq 3.24 \times 10^{-6}\ \text{mks units}\ (\simeq 3.6 \times 10^{-22}\ \text{cgs units}) \tag{9.6-16}$$

$$\alpha_{\perp} \simeq 6.3 \times 10^{-7}\ \text{mks units}\ (\simeq 0.7 \times 10^{-22}\ \text{cgs units})$$

show that for zero field the dielectric constant has a value of about 2.7, which saturates to about 4.6 for high fields.*

(c) Show that the saturating values are reached for about $\zeta \simeq 20$; what will be the corresponding field?

* In order to obtain the corresponding results in Gaussian units, the factor $4\pi\varepsilon_0$ appearing in the expression for K_0 and $\chi_{\pm}^{NL}$ should be removed.

10

Graded-Index Waveguides

10.1. Introduction

One of the most exciting applications of a laser lies in the field of communication. Due to the extremely high frequency of the laser light ($\sim 10^{15}$ Hz), the information-carrying capacity of such a beam is orders of magnitude greater than that of radio waves and microwaves. The concept of communication using light beams is old, but due to the absence of a powerful, coherent and monochromatic beam of light, it could not be put into practice. Unlike radio waves, one cannot transmit information using light beams through the atmosphere, due to attenuation and scattering caused by rain, dust, clouds, etc. Hence one requires a medium through which one can transmit information-carrying light beams without much distortion and attenuation. Fibers made of glass form such a medium and can be used as waveguides for optical beams. The simplest kind of fiber is a dielectric-clad cylindrical fiber in which the refractive-index variation is of the form

$$n = \begin{cases} n_1 & r < a \\ n_2 & r > a \end{cases} \tag{10.1-1}$$

with $n_2 < n_1$ (see Fig. 10.1a); here r represents the radial cylindrical coordinate. The waveguiding action of the dielectric-clad fiber is based on the principle that if a ray is incident at the core–cladding interface ($r = a$) at an angle greater than the critical angle, then the ray will undergo total internal reflection and will be trapped in the core of the fiber (Fig. 10.1a). However, such a fiber has two main disadvantages:

(1) When the fiber is used under multimode operation, different rays take different times in traversing the fiber; this generally leads to a large pulse dispersion. Although single-mode cladded fibers are characterized by small pulse dispersions, it is extremely difficult to couple two such fibers (without introducing considerable losses) because of their small dimensions.

(2) If there is any imperfection at the core–cladding interface, the rays may even come back to the input end of the fiber (see Fig. 10.1b).

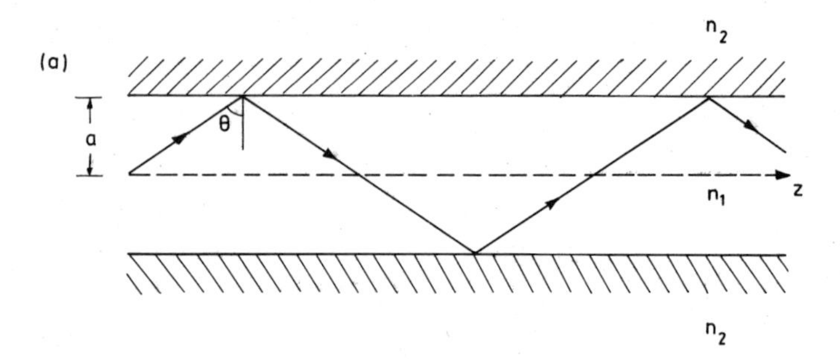

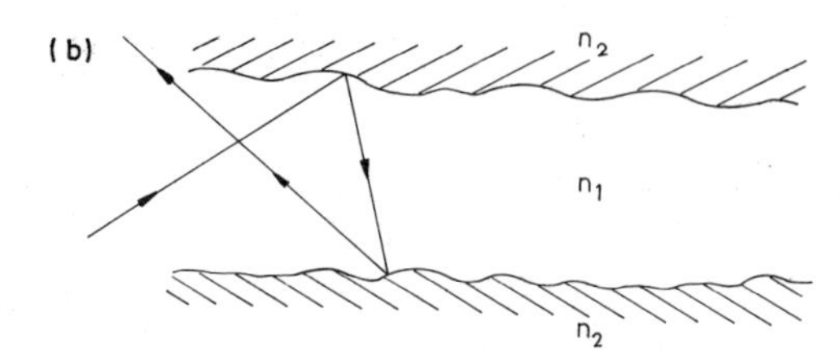

Fig. 10.1 (a) A cladded fiber of radius a; the refractive indices of the core and cladding are n_1 and n_2, respectively. (b) Irregularities in the core–cladding interface may even lead to return of some rays.

In recent years considerable work has been reported on the fabrication of graded-index fibers in which the refractive-index variation is of the form

$$n = \begin{cases} n_0 - n_2 r^2 & r < a \\ n_1 & r > a \end{cases} \tag{10.1-2}$$

Fibers characterized by such a refractive-index variation are known as Selfoc fibers.* The ray paths in such a medium were obtained in Section 1.3, and in the meridional plane the ray paths are sinusoidal (see Fig. 10.2). It may be seen that if the launching angle† is less than a critical value, the rays will not hit the boundary of the core and, as such, will not be affected by the irregularities in the core–cladding interface. Also, in the paraxial approximation, all the ray paths take the same amount of time in traversing the fiber. For example, the sinusoidal ray path shown in Fig. 10.2 takes the same amount of time as the ray that goes along the z axis. This follows from the fact that although the sinusoidal path corresponds to a larger geometrical distance, the optical pathlength is the same and hence the time taken [=(optical pathlength)/(speed of light in free space)] is the same. This follows physically from the fact that as we move away from the axis, the refractive index becomes smaller and the speed becomes larger. Thus

* Selfoc, a contraction of self-focusing, is a registered trade name of the Nippon Electric Company and the Nippon Sheet Glass Company and has been introduced by Uchida *et al.* (1970). Such media are also called GRIN fibers (from graded index).

† This essentially implies that the amplitude of oscillation of the ray, which is equal to $(\sin \gamma_0)/\alpha$ [see Eq. (1.3-24)], is less than the radius of the fiber; γ_0 is the launching angle.

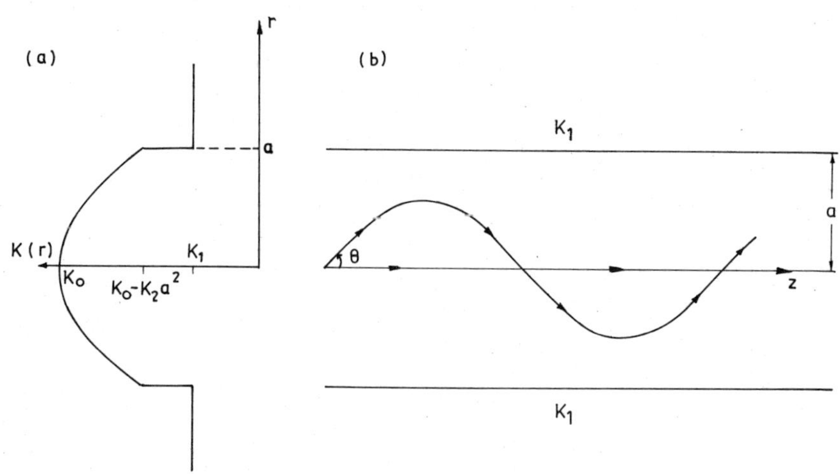

Fig. 10.2. (a) The refractive-index variation of a typical Selfoc fiber; $r = a$ corresponds to the boundary of the core. (b) The path of a typical meridional ray in the fiber.

the larger pathlength is compensated by the greater velocity of the ray. Because of this, the broadening of a temporal pulse is extremely small (see Section 10.4), and this leads to a very high information-carrying capacity.

In this chapter we will discuss the propagation characteristics of Selfoc fibers, which are characterized by the refractive-index variation given by Eq. (10.1-2).* In Sections 10.2 and 10.3 we study the propagation of electromagnetic waves through such fibers. Pulse dispersion is discussed in Section 10.4, and Section 10.5 gives a brief account of the fabrication of such fibers.

10.2. Modal Analysis

In this section, we will solve the scalar wave equation

$$\nabla^2\psi + K\frac{\omega^2}{c^2}\psi = 0 \tag{10.2-1}$$

where ψ is assumed to represent any one of the transverse components of

* Since Selfoc fibers are expected to have much wider applications than dielectric-clad fibers, we are restricting ourselves only to the analysis of Selfoc fibers. The analysis of other optical waveguides can be made in a similar manner. Fabrication techniques and modal analysis of dielectric-clad waveguides [characterized by Eq. (10.1-1)] and of other inhomogeneous optical waveguides can be found in Allan (1973), Kapany (1967), Kapany and Burke (1972), Marcuse (1972), Sodha and Ghatak (1977).

the electromagnetic field and K represents the dielectric constant. The spatial variation of the dielectric constant will be assumed to be given by

$$K = K_0 - K_2 r^2 = K_0 - K_2(x^2 + y^2) \tag{10.2-2}$$

The time dependence of the fields has been assumed to be of the form $\exp(i\omega t)$. It should be pointed out that Eq. (10.2-1) is rigorously valid only for a homogeneous medium. However, if the inhomogeneities are weak (i.e., the variation of the refractive index is small in a distance of the order of the wavelength) then the scalar wave equation can be shown to give fairly accurate results (see Problem 10.2). Further, under such an approximation, the magnitudes of the longitudinal fields are small compared to the transverse fields, and as the beam propagates through the medium the transverse character of the beam is approximately maintained. In an actual Selfoc fiber the dielectric constant variation is of the form

$$K = \begin{cases} K_0 - K_2(x^2 + y^2) & x^2 + y^2 < a^2 \\ K_1 & x^2 + y^2 > a^2 \end{cases} \tag{10.2-3}$$

where a represents the radius of the fiber. However, if the width of the propagating laser beam is small compared to the radius of the fiber (which is indeed the case for practical systems), Eq. (10.2-2) may be assumed to be valid for all values of x and y. This is often referred to as the infinite-medium approximation, which we will employ to carry out the modal analysis.* Thus substituting for the dielectric constant variation as given by Eq. (10.2-2) into Eq. (10.2-1), we obtain

$$\left(\frac{\partial^2}{\partial x^2} + \frac{\partial^2}{\partial y^2} + \frac{\partial^2}{\partial z^2}\right)\psi + [K_0 - K_2(x^2 + y^2)]\frac{\omega^2}{c^2}\psi = 0 \tag{10.2-4}$$

In order to solve this equation, we use the method of separation of variables and write

$$\psi(x, y, z) = X(x)\,Y(y)\,Z(z) \tag{10.2-5}$$

On substitution into Eq. (10.2-4) and subsequent division by XYZ, we obtain

$$\left[\frac{1}{X}\frac{d^2X}{dx^2} - K_2\frac{\omega^2}{c^2}x^2\right] + \left[\frac{1}{Y}\frac{d^2Y}{dy^2} - K_2\frac{\omega^2}{c^2}y^2\right] + \left[\frac{1}{Z}\frac{d^2Z}{dz^2} + K_0\frac{\omega^2}{c^2}\right] = 0 \tag{10.2-6}$$

* In Section 1.3 we showed that the ray paths are sinusoidal. In the language of ray optics, the infinite-medium approximation will be valid when the rays do not hit the boundary of the core (see Fig. 10.2). The effect of cladding on the modes of Selfoc fiber has been discussed by Kumar and Ghatak (1976).

The variables have indeed separated out, and each of the variables inside the brackets must be equal to a constant. If we set

$$\frac{1}{X}\frac{d^2X}{dx^2} - K_2\frac{\omega^2}{c^2}x^2 = -\gamma_1^2 \tag{10.2-7}$$

$$\frac{1}{Y}\frac{d^2Y}{dy^2} - K_2\frac{\omega^2}{c^2}y^2 = -\gamma_2^2 \tag{10.2-8}$$

then $Z(z)$ will satisfy the equation

$$\frac{d^2Z}{dz^2} + \left(K_0\frac{\omega^2}{c^2} - \gamma_1^2 - \gamma_2^2\right)z = 0 \tag{10.2-9}$$

The solution of Eq. (10.2-9) is of the form

$$Z(z) = e^{\mp i\beta z} \tag{10.2-10}$$

where

$$\beta = \left(K_0\frac{\omega^2}{c^2} - \gamma_1^2 - \gamma_2^2\right)^{1/2} \tag{10.2-11}$$

is known as the propagation constant. The minus and plus signs in Eq. (10.2-10) correspond to the waves propagating in the $+z$ and $-z$ directions, respectively; we have omitted the factor $\exp(i\omega t)$. We rewrite Eq. (10.2-7) in the form

$$\frac{d^2X(\xi)}{d\xi^2} + (\mu_1 - \xi^2)X(\xi) = 0 \tag{10.2-12}$$

where

$$\xi = \alpha x, \qquad \alpha = \left(K_2\frac{\omega^2}{c^2}\right)^{1/4}, \qquad \mu_1 = \frac{\gamma_1^2 c}{\omega K_2^{1/2}} \tag{10.2-13}$$

Equation (10.2-12) is of the same form as one encounters in the linear harmonic oscillator problem in quantum mechanics (see Appendix C). For the function $X(\xi)$ to remain bounded as $x \to \pm\infty$, we must have $\mu_1 = 2n + 1$, where* $n = 0, 1, 2, \ldots$. Thus γ_1 takes discrete values that are given by

$$\gamma_1^2 = (2n+1)\frac{\omega}{c}K_2^{1/2}, \qquad n = 0, 1, 2, \ldots \tag{10.2-14}$$

* For $\mu_1 \neq 1, 2, 3, \ldots$, $X(\xi)$ will behave as $\exp(\xi^2/2)$ and will therefore become unbounded as $x \to \pm\infty$ (see Appendix C).

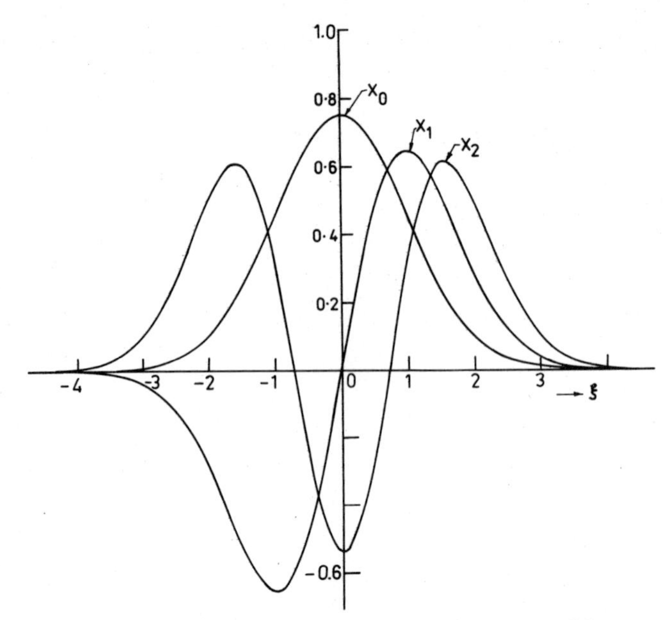

Fig. 10.3. Hermite–Gauss functions of orders 0, 1, and 2.

The corresponding field patterns are given by

$$X_n(x) = N_n H_n(\alpha x) \exp\left(-\frac{\alpha^2 x^2}{2} \right), \qquad N_n = \frac{\alpha^{1/2}}{(2^n n! \pi^{1/2})^{1/2}} \tag{10.2-15}$$

where $H_n(\xi)$ represent the Hermite polynomials. The first few polynomials are

$$H_0(\xi) = 1, \qquad H_1(\xi) = 2\xi, \qquad H_2(\xi) = 4\xi^2 - 2 \tag{10.2-16}$$

Figure 10.3 shows the field patterns given by Eq. (10.2-15) for $n = 0, 1$, and 2. The normalization constant N_n is such that

$$\int_{-\infty}^{\infty} X_n(x)\, X_m(x)\, dx = \delta_{nm} \tag{10.2-17}$$

where

$$\delta_{nm} = \begin{cases} 1 & n = m \\ 0 & n \neq m \end{cases} \tag{10.2-18}$$

In an exactly similar manner, it can be shown that

$$\gamma_2^2 = (2m + 1) \frac{\omega}{c} K_2^{1/2}, \qquad m = 0, 1, 2, \ldots \tag{10.2-19}$$

Using Eqs. (10.2-11), (10.2-14), and (10.2-19) one obtains the following discrete values of the propagation constants:

$$\beta_{nm} = \frac{\omega}{c} K_0^{1/2} \left[1 - 2(n + m + 1) \frac{c}{\omega} \frac{K_2^{1/2}}{K_0} \right]^{1/2} \tag{10.2-20}$$

The field pattern $Y_m(y)$ is of a form similar to $X_n(x)$, and therefore the modal pattern corresponding to the propagation constant β_{nm} will be given by

$$\psi_{nm} = \frac{\alpha}{(2^{n+m} m! n! \pi)^{1/2}} H_n(\alpha x) H_m(\alpha y) \exp\left[-\frac{\alpha^2}{2}(x^2 + y^2) \right] \exp(-i\beta_{nm} z) \tag{10.2-21}$$

For a typical Selfoc fiber (see, e.g., Uchida *et al.*, 1970), $K_0 \simeq 2.25$, $K_2 \simeq 0.5 \times 10^6\, m^{-2}$, and assuming $\lambda_0 \simeq 0.6\ \mu\text{m}$, we obtain

$$\frac{c}{\omega} \frac{K_2^{1/2}}{K_0} = \frac{\lambda_0}{2\pi} \frac{K_2^{1/2}}{K_0} \simeq 10^{-4}$$

Thus for a large number of modes, one is justified in making a binomial expansion in Eq. (10.2-20) to obtain

$$\begin{aligned} \beta_{nm} &= \frac{\omega}{c} K_0^{1/2} \left[1 - (n + m + 1) \frac{c}{\omega} \frac{K_2^{1/2}}{K_0} + \cdots \right] \\ &= \frac{\omega}{c} K_0^{1/2} - \left(\frac{K_2}{K_0} \right)^{1/2} (n + m + 1) + \cdots \end{aligned} \tag{10.2-22}$$

where we have retained terms up to the first order. Thus if we neglect material dispersion, i.e., if we assume K_0 and K_2 to be independent of ω, we obtain

$$\frac{d\beta_{nm}}{d\omega} \simeq \frac{K_0^{1/2}}{c} \tag{10.2-23}$$

To this approximation, we notice the remarkable property that $d\beta_{nm}/d\omega$ (which is the inverse of the group velocity) is simply $K_0^{1/2}/c$ and independent of the mode number. In the language of ray optics, this corresponds to the fact that all ray paths take almost the same amount of time. This fact leads to small pulse dispersions.

Returning to Eq. (10.2-21), we note that any incident field distribution of the form ψ_{nm} remains unchanged as it propagates through the Selfoc fiber. In general, we may write

$$\psi = \sum_{n,m=0,1,2,\ldots}^{\infty} A_{nm} X_n(x) Y_m(y) e^{-i\beta_{nm} z} \tag{10.2-24}$$

where the coefficients A_{nm} are determined from the field distribution at the entrance aperture of the fiber, which we may assume to be $z = 0$. The above expansion follows from the fact that since ψ_{nm} given by Eq. (10.2-21) forms a complete orthonormal set of functions, it must be possible to expand any function in terms of them. Also since ψ_{nm} represents a mode of the fiber, the power associated with the (n, m)th mode is proportional to $|A_{nm}|^2$. In the next section, we will use Eq. (10.2-24) to study the propagation of electromagnetic waves through Selfoc fibers.

10.3. Propagation through a Selfoc Fiber

In order to study the propagation of an electromagnetic wave through a Selfoc fiber, we must calculate the coefficients A_{nm} in Eq. (10.2-24) (such an explicit evaluation has been done in Problem 10.1). However, in this section, we will show that in general, the field distribution at an arbitrary value of z can be related to the field distribution at $z = 0$ through the relation

$$\psi(x, y, z) = \int_{-\infty}^{\infty}\int_{-\infty}^{\infty} \psi(x', y', z = 0)\, K(x, y; x', y'; z)\, dx'\, dy' \quad (10.3\text{-}1)$$

where the function K is often referred to as the kernel of the system. Referring to Eq. (10.2-24), at $z = 0$ we have

$$\psi(x, y, z = 0) = \sum_{n,m} A_{nm} X_n(x)\, Y_m(y) \quad (10.3\text{-}2)$$

Using the orthonormal properties of the Hermite–Gauss functions [Eq. (10.2-17)], we get

$$A_{nm} = \int_{-\infty}^{\infty}\int_{-\infty}^{\infty} \psi(x', y', z = 0)\, X_n(x')\, Y_m(y')\, dx'\, dy' \quad (10.3\text{-}3)$$

Substituting for A_{nm} from Eq. (10.3-3) into Eq. (10.2-24), we get Eq. (10.3-1) with

$$K(x, y; x', y'; z) = \sum_{n,m} X_n(x)\, X_n(x')\, Y_m(y)\, Y_m(y')\, e^{-i\beta_{nm} z} \quad (10.3\text{-}4)$$

The summations appearing in Eq. (10.3-4) can be performed if we substitute the approximate expression for β_{nm} from Eq. (10.2-22). If we do so we obtain

$$K(x, y; x', y'; z) = \exp\left\{ -i\left[\frac{\omega}{c} K_0^{1/2} - \left(\frac{K_2}{K_0}\right)^{1/2}\right] z \right\} G_1(x, x', z)\, G_2(y, y', z) \quad (10.3\text{-}5)$$

where

$$G_1(x, x', z) = \sum_{n=0,1,2,\ldots}^{\infty} X_n(x)\, X_n(x')\, e^{i\delta nz}$$

$$= \frac{\alpha}{\pi^{1/2}} \sum_n \frac{H_n(\alpha x)\, H_n(\alpha x')}{2^n n!} e^{i\delta nz}$$

$$= \frac{\alpha}{\pi^{1/2}} \frac{1}{(1-\gamma^2)^{1/2}} \exp\left[\frac{2\xi\xi'\gamma}{1-\gamma^2} - \frac{(\xi^2+\xi'^2)(1+\gamma^2)}{2(1-\gamma^2)}\right] \tag{10.3-6}$$

where $\delta = (K_2/K_0)^{1/2}$, $\gamma = \exp(i\delta z)$, $\xi' = \alpha x'$, and use has been made of Mehler's formula (see Erdelyi, 1953) to carry out the summation. In a similar manner one can write a closed-form expression for $G_2(y, y', z)$. On substituting these expressions into Eq. (10.3-5) one obtains

$$K(x, y; x', y'; z) = \frac{i\alpha^2}{2\pi \sin \delta z} \exp\left(-i\frac{\omega}{c} K_0^{1/2} z\right)$$

$$\times \exp\left[\frac{i\alpha^2}{\sin \delta z}(xx' + yy') - \frac{i\alpha^2}{2}(x^2 + x'^2 + y^2 + y'^2) \cot \delta z\right] \tag{10.3-7}$$

Using the above kernel, a general treatment for the propagation of a beam has been given by Ghatak and Thyagarajan (1975); we will consider two specific simple cases.

10.3.1. *Propagation of a Gaussian Beam Launched Symmetrically about the Axis*

As a simple application of the above result, we consider the propagation of a Gaussian beam incident (symmetrically about the axis) at the entrance aperture of the fiber (i.e., $z = 0$). The field distribution of the incident beam is given by

$$\psi(x, y, z = 0) = \psi_0 \exp\left(-\frac{x^2 + y^2}{2w_i^2}\right) \tag{10.3-8}$$

where w_i represents a measure of the width of the incident beam. Substituting Eqs. (10.3-7) and (10.3-8) into Eq. (10.3-1), we obtain the field distribution in any transverse plane as

$$\psi(x, y, z) = \frac{i\alpha^2}{2\pi \sin \delta z} \psi_0 \exp\left(-i\frac{\omega}{c} K_0^{1/2} z\right) \exp\left[-\frac{i\alpha^2}{2}(x^2 + y^2)\cot \delta z\right]$$

$$\times \int_{-\infty}^{\infty} dx' \exp\left[-x'^2\left(\frac{1}{2w_i^2} + \frac{i\alpha^2}{2}\cot \delta z\right) + \frac{i\alpha^2}{\sin \delta z} xx'\right]$$

$$\times \int_{-\infty}^{\infty} dy' \exp\left[-y'^2\left(\frac{1}{2w_i^2} + \frac{i\alpha^2}{2}\cot \delta z\right) + \frac{i\alpha^2}{\sin \delta z} yy'\right] \tag{10.3-9}$$

Using the standard integral

$$\int_{-\infty}^{\infty} \exp(-px^2 + qx)\, dx = \left(\frac{\pi}{p}\right)^{1/2} \exp\left(\frac{q^2}{4p}\right) \tag{10.3-10}$$

we obtain

$$\psi(x, y, z) = \frac{\tau^2 \psi_0 e^{-i\varphi(z)}}{(\tau^4 \cos^2 \delta z + \sin^2 \delta z)^{1/2}} \exp\left[-\frac{\tau^2 \alpha^2 (x^2 + y^2)}{2(\sin^2 \delta z + \tau^4 \cos^2 \delta z)}\right] \tag{10.3-11}$$

where

$$\varphi(z) = \frac{\omega}{c} K_0^{1/2} z - \tan^{-1}\left(\frac{\tan \delta z}{\tau^2}\right) - \frac{\alpha^2 (x^2 + y^2)(\tau^4 - 1)\sin 2\delta z}{4(\sin^2 \delta z + \tau^4 \cos^2 \delta z)} \tag{10.3-12}$$

and $\tau = w_i \alpha$. The intensity distribution in any plane z is given by

$$|\psi(x, y, z)|^2 = |\psi_0|^2 \frac{w_i^2}{w^2(z)} \exp\left[-\frac{(x^2 + y^2)}{w^2(z)}\right] \tag{10.3-13}$$

where

$$w^2(z) = \frac{w_i^2}{\tau^4}(\sin^2 \delta z + \tau^4 \cos^2 \delta z) = w_i^2\left(\frac{c^2}{K_2 \omega^2 w_i^4} \sin^2 \delta z + \cos^2 \delta z\right) \tag{10.3-14}$$

is a measure of the beam width of the Gaussian beam in any plane z. We may make the following observations:

(1) The transverse intensity distribution of an incident Gaussian beam remains Gaussian with its width modulating with a period $\delta/2$ (see Fig. 10.4a). Notice that the center of the Gaussian beam lies on the axis for all values of z.

(2) If $\tau = 1$, i.e., $w_i = 1/\alpha$ then $w(z) = w_0$ (independent of z). This shows that the beam propagates as the fundamental mode of the fiber.

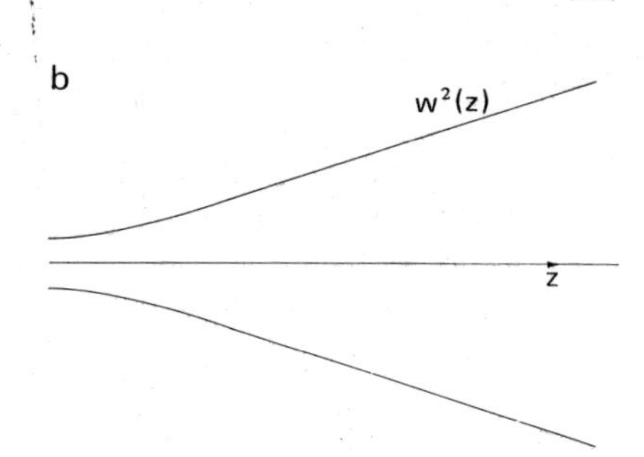

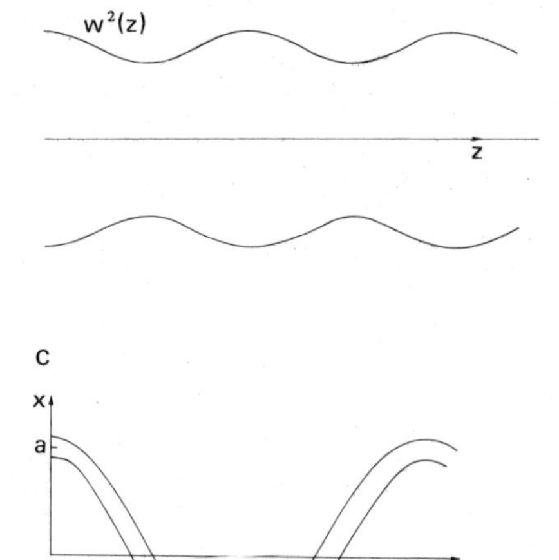

Fig. 10.4. (a) The variation of the width of an incident Gaussian beam (launched along the axis) as it propagates through the fiber. (b) The diffraction divergence of a Gaussian beam in a homogeneous medium. (c) Propagation of a Gaussian beam with half-width $1/\alpha$, launched at an off-axis point.

(3) It is instructive to study the z variation of the beam width when $\delta \to 0$ (the case $\delta = 0$ corresponds to a homogeneous medium). Rewriting Eq. (10.3-14) as

$$w^2(z) = w_i^2 \left\{ \frac{c^2 z^2}{K_0 \omega^2 w_i^4} \left[\frac{\sin\left(\dfrac{K_2}{K_0}\right)^{1/2} z}{\left(\dfrac{K_2}{K_0}\right)^{1/2} z} \right]^2 + \cos^2\left(\frac{K_2}{K_0}\right)^{1/2} z \right\} \tag{10.3-15}$$

we observe that, in the limit $K_2 = 0$,

$$w^2(z) = w_i^2 \left(1 + \frac{c^2 z^2}{K_0 \omega^2 w_i^4} \right) \tag{10.3-16}$$

Since $K_2 = 0$ implies a homogeneous medium, Eq. (10.3-16) gives us the diffraction divergence of a Gaussian beam (see Fig. 10.4b). The physical significance of Eq. (10.3-16) is discussed in Section 9.3 (see also Problem 4.8).

10.3.2. Propagation of a Gaussian Beam Launched at an Off-Axis Point Parallel to the Axis

For a Gaussian beam launched at an off-axis point and incident parallel to the z axis, the field pattern at $z = 0$ is given by

$$\psi(x, y, z = 0) = \psi_0 \exp\left[-\frac{(x-a)^2 + y^2}{2w_i^2} \right] \tag{10.3-17}$$

where the displacement a is assumed to be along the x direction without any loss of generality. We consider the case when $w_i = 1/\alpha$ (see Problem 10.1). On substitution in Eq. (10.3-1), using Eq. (10.3-7), and carrying out the integration, we obtain

$$\psi(x, y, z) = \psi_0 \exp\left[-\frac{(x - a\cos\delta z)^2 + y^2}{2w_i^2}\right]\exp\left[-i\left(\frac{\omega}{c}K_0^{1/2} - \delta\right)z\right]$$
$$\times \exp\left[\frac{i}{2w_i^2}\left(2xa\sin\delta z - \frac{a^2}{2}\sin 2\delta z\right)\right] \qquad (10.3\text{-}18)$$

The intensity pattern will be given by

$$|\psi(x, y, z)|^2 = |\psi_0|^2 \exp\left[-\frac{(x - a\cos\delta z)^2 + y^2}{w_i^2}\right] \qquad (10.3\text{-}19)$$

It follows from Eq. (10.3-19) that the center of the Gaussian beam follows the path given by

$$x(z) = a\cos\delta z, \qquad y(z) = 0 \qquad (10.3\text{-}20)$$

(see Fig. 10.4c); this is essentially the ray path in the paraxial approximation, which can also be obtained by directly solving the ray equation (see Section 1.3).

It should be noted that the kernel given by Eq. (10.3-7) was obtained by making a binomial expansion of β_{nm} and retaining terms up to first order in Eq. (10.2-22). It is known, by solving the ray equation that the path given by Eq. (10.3-20) is valid for paraxial rays alone. Thus the approximation in Eq. (10.2-22) is equivalent to a paraxial approximation.

Problem 10.1. (a) Solve the wave equation in cylindrical coordinates, and show that for azimuthally symmetric fields the modal patterns are the Laguerre–Gauss functions

$$\psi_p(r) = \frac{1}{\pi^{1/2}w_0} L_p\left(\frac{r^2}{w_0^2}\right)\exp\left(-\frac{r^2}{2w_0^2}\right) \qquad (10.3\text{-}21)$$

where $L_p(r)$ are the Laguerre polynomials of degree p satisfying the equation

$$r\frac{d^2L_p(r)}{dr^2} + (1 - r)\frac{dL_p(r)}{dr} + pL_p(r) = 0 \qquad (10.3\text{-}22)$$

and $w_0 = (\omega K_2^{1/2}/c)^{-1/2}$. These functions are normalized by using

$$\int_0^\infty \psi_p(r)\,\psi_q(r)\,2\pi r\,dr = \delta_{pq}$$

The first few Laguerre polynomials are

$$L_0(r) = 1, \qquad L_1(r) = 1 - r, \qquad L_2(r) = 1 - 2r - \tfrac{1}{2}r^2 \qquad (10.3\text{-}23)$$

The corresponding propagation constants are

$$\beta_p = \frac{\omega}{c} K_0^{1/2} \left[1 - 2(2n + 1) \frac{c}{\omega} \frac{K_2^{1/2}}{K_0} \right]^{1/2} \tag{10.3-24}$$

(b) Since the Hermite–Gauss functions [see Eq. (10.2-15)] represent the complete modes of the scalar wave equation, the azimuthally symmetric field distributions given above can be expressed as linear combinations of the Hermite–Gauss modes. Show this explicitly for $p = 0$, 1, and 2.

(c) For any incident field distribution that is symmetric about the z axis, the modes described by Eq. (10.3-21) can be used to study the propagation of the beam. For an incident Gaussian beam of the form

$$\psi(x, y, z = 0) = \frac{1}{\pi^{1/2} w_i} \exp\left(-\frac{r^2}{2w_i^2} \right) \tag{10.3-25}$$

show that

$$\psi(x, y, z) = \sum_p A_p \left[\frac{1}{\pi^{1/2} w_0} L_p\left(\frac{r^2}{w_0^2} \right) \exp\left(-\frac{r^2}{2w_0^2} \right) \right] \exp(-i\beta_p z) \tag{10.3-26}$$

where

$$A_p = \frac{2w}{1 + w^2} \left[\frac{w^2 - 1}{w^2 + 1} \right]^p, \qquad w = w_0 / w_i \tag{10.3-27}$$

Carry out the summation in Eq. (10.3-26) and show that the results are consistent with the results obtained earlier. $|A_p|^2$ is proportional to the excitation efficiency of the pth mode. *Reference*: Gambling and Matsumura (1973).

Problem 10.2. (a) For a general nonabsorbing, isotropic, inhomogeneous medium characterized by the dielectric constant variation $K(x, y, z)$, show that the equations for $\mathscr{E}_x$, $\mathscr{E}_y$, and $\mathscr{E}_z$ are coupled.* However, for K depending on one spatial coordinate (say x) show that the equations for $\mathscr{E}_x$ and $\mathscr{H}_x$ get decoupled and are of the form

$$\nabla^2 \mathscr{E}_x + \frac{\partial}{\partial x}\left(\frac{1}{K} \frac{dK}{dx} \mathscr{E}_x \right) - \mu_0 \varepsilon_0 K(x) \frac{\partial^2 \mathscr{E}_x}{\partial t^2} = 0 \tag{10.3-28}$$

$$\nabla^2 \mathscr{H}_x - \mu_0 \varepsilon_0 K(x) \frac{\partial^2 \mathscr{H}_x}{\partial t^2} = 0 \tag{10.3-29}$$

Thus one can either choose $\mathscr{E}_x = 0$ or $\mathscr{H}_x = 0$ and express all the field components in terms of $\mathscr{E}_x$ and $\mathscr{H}_x$.

(b) Without any loss of generality, show that the y dependence of the fields may be neglected, thereby obtaining TE (transverse electric) or TM (transverse magnetic) modes.

* Script letters $\mathscr{E}$, $\mathscr{H}$, etc., contain the time dependence. For a time variation of the form $\exp(i\omega t)$, we have

$$\mathscr{E}(x, y, z, t) = E(x, y, z) \exp(i\omega t).$$

(c) For determining the propagation characteristics of the TE modes, one has to solve Eq. (10.3-29), whereas for TM modes, one has to solve Eq. (10.3-28). Show that when the variation of the refractive index is small in distances of the order of a wavelength, the term depending on dK/dx in Eq. (10.3-28) can be neglected.

Solution. (a) For an isotropic, linear, nonconducting and nonmagnetic medium, Maxwell's equations take the form

$$\begin{aligned} \nabla \times \mathscr{E} + \mu_0 \frac{\partial \mathscr{H}}{\partial t} &= 0, \qquad \nabla \cdot \mathscr{D} = 0 \\ \nabla \times \mathscr{H} &= \frac{\partial \mathscr{D}}{\partial t}, \qquad \nabla \cdot \mathscr{H} = 0 \end{aligned} \tag{10.3-30}$$

where

$$\mathscr{D} = \varepsilon \mathscr{E} = \varepsilon_0 K \mathscr{E}, \qquad \varepsilon = \varepsilon_0 K \tag{10.3-31}$$

with ε_0 and μ_0 representing the dielectric permittivity and magnetic permeability of free space, and ε $(=\varepsilon_0 K)$ representing the permittivity of the medium. Thus

$$\nabla \times (\nabla \times \mathscr{E}) = -\mu_0 \frac{\partial}{\partial t}(\nabla \times \mathscr{H}) = -\mu_0 \frac{\partial^2 \mathscr{D}}{\partial t^2}$$

or

$$\nabla(\nabla \cdot \mathscr{E}) - \nabla^2 \mathscr{E} = -\varepsilon_0 \mu_0 K \frac{\partial^2 \mathscr{E}}{\partial t^2} \tag{10.3-32}$$

Furthermore,

$$0 = \nabla \cdot \mathscr{D} = \varepsilon_0 \nabla \cdot (K\mathscr{E}) = \varepsilon_0 (\nabla K \cdot \mathscr{E} + K \nabla \cdot \mathscr{E}) \tag{10.3-33}$$

Thus

$$\nabla \cdot \mathscr{E} = -\frac{\nabla K}{K} \cdot \mathscr{E} \tag{10.3-34}$$

Substituting in Eq. (10.3-32), we obtain

$$\nabla^2 \mathscr{E} + \nabla\left(\frac{\nabla K}{K} \cdot \mathscr{E}\right) - \mu_0 \varepsilon_0 K \frac{\partial^2 \mathscr{E}}{\partial t^2} = 0 \tag{10.3-35}$$

which shows that for an inhomogeneous medium, the equations for $\mathscr{E}_x$, $\mathscr{E}_y$, and $\mathscr{E}_z$ are coupled. In the absence of any inhomogeneity, the second term on the left-hand side vanishes and each Cartesian component of the electric vector satisfies the scalar wave equation. In a similar manner, one can obtain

$$\nabla^2 \mathscr{H} + \frac{1}{K}(\nabla K) \times (\nabla \times \mathscr{H}) - \mu_0 \varepsilon_0 K \frac{\partial^2 \mathscr{H}}{\partial t^2} = 0 \tag{10.3-36}$$

When K depends on the x coordinate only, $\mathscr{E}_x$ and $\mathscr{H}_x$ satisfy Eqs. (10.3-28) and (10.3-29), respectively.

(b) Equations (10.3-28) and (10.3-29) can be solved by the method of separation of variables, and it can easily be seen that the y and z dependences of the fields are of the form $\exp[-i(\beta z + \gamma y)]$. We can choose our z axis along the direction of propagation of the wave. Consequently, we may put, without any loss of generality, $\gamma = 0$. Thus we may write

$$\mathscr{E}_j = E_j(x)\, e^{i(\omega t - \beta z)}, \qquad j = x, y, z \tag{10.3-37}$$

$$\mathscr{H}_j = H_j(x)\, e^{i(\omega t - \beta z)}, \qquad j = x, y, z \tag{10.3-38}$$

Thus Eqs. (10.3-28) and (10.3-29) simplify to

$$\frac{d^2 E_x}{dx^2} + \frac{d}{dx}\left(\frac{1}{K}\frac{dK}{dx} E_x\right) + \left(\frac{\omega^2}{c^2} K(x) - \beta^2\right) E_x = 0 \tag{10.3-39}$$

$$\frac{d^2 H_x}{dx^2} + \left(\frac{\omega^2}{c^2} K(x) - \beta^2\right) H_x = 0 \tag{10.3-40}$$

Furthermore, using Maxwell's equations, we obtain

$$\begin{aligned}
i\beta E_y &= -i\omega\mu_0 H_x, & i\beta H_y &= i\omega\varepsilon_0 K(x)\, E_x \\
-i\beta E_x - \frac{\partial E_z}{\partial x} &= -i\omega\mu_0 H_y, & -i\beta H_x - \frac{\partial H_z}{\partial x} &= i\omega\varepsilon_0 K(x)\, E_y \\
\frac{\partial E_y}{\partial x} &= -i\omega\mu_0 H_z, & \frac{\partial H_y}{\partial x} &= i\omega\varepsilon_0 K(x)\, E_z \\
\frac{1}{K}\left[\frac{\partial}{\partial x}(K E_x)\right] &= i\beta E_z, & \frac{\partial H_x}{\partial x} &= i\beta H_z
\end{aligned} \tag{10.3-41}$$

(i) For $E_x = 0$, from the above equations one obtains

$$\begin{aligned}
E_z &= 0, & E_y &= -\frac{\omega\mu_0}{\beta} H_x \\
H_y &= 0, & H_z &= -\frac{i}{\beta}\frac{\partial H_x}{\partial x}
\end{aligned} \tag{10.3-42}$$

Since $E_z = 0$, such modes are known as TE modes and the propagation characteristics of such modes are determined by solving Eq. (10.3-40). A similar equation is also satisfied by E_y.

(ii) For $H_x = 0$, one obtains

$$\begin{aligned}
H_z &= 0, & H_y &= \frac{\omega\varepsilon_0 K(x)}{\beta} E_x \\
E_y &= 0, & E_z &= -\frac{i}{\beta}\frac{1}{K}\frac{\partial}{\partial x}[K(x)\, E_x]
\end{aligned} \tag{10.3-43}$$

Since $H_z = 0$, such modes are known as TM modes and the propagation characteristics are determined by solving Eq. (10.3-39).

(c) If we make the substitution $\varphi(x) = [K(x)]^{1/2} E_x$ in Eq. (10.3-39), we obtain

$$\frac{d^2\varphi}{dx^2} + \left[\frac{\omega^2}{c^2} K(x) - \frac{3}{4}\left(\frac{1}{K}\frac{dK}{dx}\right)^2 + \frac{1}{2K}\frac{d^2 K}{dx^2} - \beta^2\right)\Bigg]\varphi = 0 \tag{10.3-44}$$

Clearly, if

$$\left|\frac{1}{K(x)}\left[\frac{1}{2K}\frac{d^2K}{dx^2}-\frac{3}{4}\left(\frac{1}{K}\frac{dK}{dx}\right)^2\right]\right| \ll \frac{4\pi^2}{\lambda_0^2} \tag{10.3-45}$$

then $\varphi(x)$ satisfies the scalar wave equation. In Eq. (10.3-45) λ_0 is the free-space wavelength. Physically, the inequality expressed by Eq. (10.3-45) shows that if the variation of the dielectric constant is small in distances of the order of a wavelength, then the transverse component satisfies a scalar wave equation.

Problem 10.3. Consider a planar waveguide in which the dielectric constant variation is given by

$$K(x) = K_0 - K_2x^2 \tag{10.3-46}$$

Show that if terms of $O(K_2^3x^4/K_0^3)$ are neglected, then Eq. (10.3-44) can be rigorously solved and the propagation constants are given by

$$\beta_n^2 = \frac{\omega^2}{c^2}K_0 - \frac{K_2}{K_0} - (2n+1)\left(\frac{\omega^2}{c^2}K_2 + 4\frac{K_2^2}{K_0^2}\right)^{1/2}, \qquad n = 0, 1, 2, \ldots \tag{10.3-47}$$

Find the corresponding field patterns. Compare these expressions with the one obtained by neglecting the term depending on ∇K in the wave equation. *Reference*: Ghatak and Kraus (1974); Marcuse (1973).

Problem 10.4. Assuming a time dependence of the form $\exp(i\omega t)$, solve Eq. (10.3-35) for a dielectric constant variation given by Eq. (10.2-2) by using first-order perturbation theory, assuming $\nabla[(\nabla K/K)\cdot\mathbf{E}]$ to be the perturbation term. [*Hint:* One has to use degenerate perturbation theory.] *Reference*: Thyagarajan and Ghatak (1974).

Problem 10.5. Consider a dielectric constant variation of the form

$$K(x) = \begin{cases} K_1 & x < 0 \\ K_0 + \Delta K \exp(-x/d) & x > 0 \end{cases} \tag{10.3-48}$$

(see Fig. 10.5). Waveguides with such a dielectric constant variation have been obtained by Kaminow and Carruthers (1973). Show that for TE modes, the field pattern is given by

$$\psi(x) = \begin{cases} AJ_\alpha\left[4\pi(\Delta K)^{1/2}\dfrac{d}{\lambda}\exp\left(-\dfrac{x}{2d}\right)\right]\exp[i(\omega t - \beta z)] & x > 0 \\ B\exp(-p_1x)\exp[i(\omega t - \beta z)] & x < 0 \end{cases} \tag{10.3-49}$$

where

$$\alpha = 2dp_0, \qquad p_0 = \left(\beta^2 - K_0\frac{\omega^2}{c^2}\right)^{1/2}, \qquad p_1 = \left(\beta^2 - K_1\frac{\omega^2}{c^2}\right)^{1/2}$$

The corresponding propagation constants are to be determined from the transcen-

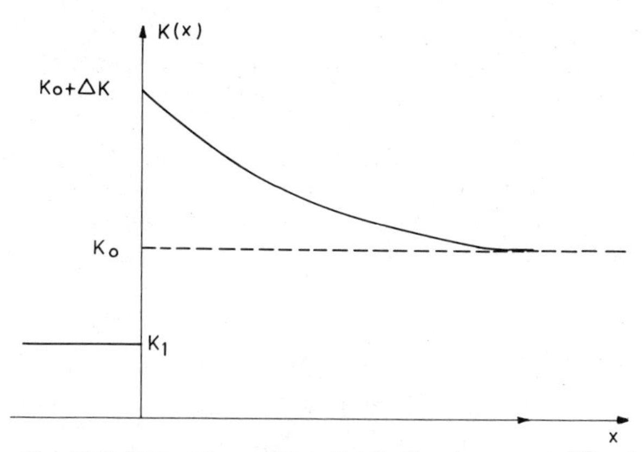

Fig. 10.5. Dielectric constant distribution in an out-diffused waveguide described by Eq. (10.3-48).

dental equation

$$\frac{J_{\alpha+1}(\xi) - J_{\alpha-1}(\xi)}{J_\alpha(\xi)} = \frac{p_1 \lambda}{\pi(\Delta K)^{1/2}} \tag{10.3-50}$$

where

$$\xi = 2kd(\Delta K)^{1/2}$$

For small values of ΔK, the above transcendental equation assumes the form $J_\alpha(\xi) \simeq 0$. Show that for $\Delta K = 0.3$, $\lambda = 0.633$ μm, $d = 2.5$ μm, the values of α corresponding to the modes are approximately given by 21.5, 17.4, 14.1, 11.2, 8.5, 6.1, 3.8, and 1.7. *Reference*: Conwell (1973); see also Khular and Ghatak (1976).

Problem 10.6. Show that in a medium characterized by a dielectric constant variation of the form given by Eq. (10.2-2), for an incident field distribution of the form

$$\psi = \begin{cases} \psi_0 & r \leq a \\ 0 & r > a \end{cases} \tag{10.3-51}$$

the intensity pattern in the focal plane (i.e., $z = \pi/2\delta$) is the Airy pattern:

$$I(r) = I_0 \left| 2J_1(a\alpha^2 r)/a\alpha^2 r \right|^2 \tag{10.3-52}$$

where $I_0 = |\psi_0 a^2 \alpha^2/2|^2$. The above expression may be compared with the focal pattern produced by a lens. *Reference*: Ghatak and Thyagarajan (1975).

Hint: Make the following substitution in Eq. (10.3-7)

$$\xi = r\alpha \cos\theta, \qquad \eta = r\alpha \sin\theta \tag{10.3-53}$$

and show that Eq. (10.3-1) reduces to

$$\psi(r, \theta, z) = i\frac{\exp[-i(\omega/c)K_0^{1/2}z]}{2\pi \sin \delta z}\exp\left(-\frac{i}{2}r^2\alpha^2 \cot \delta z\right)\psi_0$$

$$\times \int_0^a \int_0^{2\pi} dr'\, d\theta' \alpha^2 r' \exp\left[i\frac{rr'\alpha^2}{\sin \delta z}\cos(\theta - \theta') - i\frac{r'^2\alpha^2}{2}\cot \delta z\right] \tag{10.3-54}$$

Problem 10.7. Using the parabolic equation approach (discussed in Section 9.5) solve the scalar wave equation for an incident Gaussian field (launched centrally along the axis) in a medium characterized by the dielectric constant distribution given by Eq. (10.2-2). Show that the final result is identical to the one given by Eq. (10.3-13). *Reference*: Sodha *et al.* (1971*a,b*).

Hint: Adopting a procedure similar to the one followed in Section 9.5 we obtain

$$2\left(\frac{\partial S}{\partial z}\right) + \left(\frac{\partial S}{\partial r}\right)^2 = -\frac{K_2}{K_0}r^2 + \frac{1}{k^2 A_0}\left(\frac{\partial^2 A_0}{\partial r^2} + \frac{1}{r}\frac{\partial A_0}{\partial r}\right) \tag{10.3-55}$$

$$\frac{\partial A_0^2}{\partial z} + \frac{\partial S}{\partial r}\frac{\partial A_0^2}{\partial r} + A_0^2\left(\frac{\partial^2 S}{\partial r^2} + \frac{1}{r}\frac{\partial S}{\partial r}\right) = 0 \tag{10.3-56}$$

These may be compared with Eqs. (9.5-8) and (9.5-9). Equations (10.3-55) and (10.3-56) can be solved in an exactly similar manner; the intensity distribution is

$$A_0^2(r, z) = \frac{E_0^2}{f^2(z)}\exp\left[-\frac{r^2}{w_0^2 f^2(z)}\right] \tag{10.3-57}$$

where $f(z)$ satisfies

$$\frac{1}{f}\frac{d^2 f}{dz^2} = -\frac{K_2(z)}{K_0} + \frac{1}{k^2 w_0^4 f^4} \tag{10.3-58}$$

The main advantage of the direct solution lies in the fact that the z dependence of K_2 can also be taken into account. At the same time it must be noted that the treatment is restricted to azimuthally symmetric fields. When K_2 is independent of z, Eq. (10.3-58) can be solved by multiplying by $2f(df/dz)\,dz$ and integrating:

$$\left(\frac{df}{dz}\right)^2 = -\frac{K_2}{K_0}f^2 - \frac{1}{k^2 w_0^4 f^2} + \left[\frac{K_2}{K_0} + \frac{1}{k^2 w_0^4}\right] \tag{10.3-59}$$

where we have used the boundary condition that at $z = 0$, $f = 1$ and $df/dz = 0$. Equation (10.3-59) can easily be solved and one obtains an expression for $w^2(z)$ $[\equiv w_i^2 f^2(z)]$ identical to the one given by Eq. (10.3-14).

10.4. Pulse Propagation

One of the most important characteristics in the use of a fiber in the field of optical communication is its dispersion characteristic. By dispersion, we imply the broadening of a temporal pulse as it propagates through a certain length of the fiber. Clearly, if we have a large number of pulses incident at the entrance aperture of the fiber, then in general the pulses will broaden and as they reach the output end of the fiber, they may start overlapping. Consequently, the smaller the broadening, the higher will be the information-carrying capacity of the fiber. As such, a study of the broadening of a pulse as it propagates through the fiber is of great relevance to its potential use in the field of optical communications. In this section, we will calculate the broadening of a temporal pulse as it propagates through a Selfoc fiber. As discussed in Section 10.3, for a pure harmonic wave incident on the fiber at $z = 0$,

$$\Psi^{\mathrm{P}}(x, y, z = 0, t) = \varphi(x, y)\exp(i\omega_c t) = \sum_p A_p \psi_p(x, y)\exp(i\omega_c t) \tag{10.4-1}$$

where the superscript P refers to the fact that we are considering pure harmonic waves and the subscript p in ψ is a composite index in place of (n, m). At an arbitrary value of z, the field is given by

$$\Psi^{\mathrm{P}}(x, y, z, t) = \sum_p A_p \psi_p(x, y)\exp[i(\omega_c t - \beta_p z)] \tag{10.4-2}$$

The function $\exp(-i\beta_p z)$ is usually termed the amplitude transfer function (ATF). For a temporal pulse, we assume that at $z = 0$, which is the entrance aperture of the fiber, the field pattern can be written in the form

$$\Psi(x, y, z = 0, t) = \varphi(x, y) f(t) \tag{10.4-3}$$

We Fourier analyze $f(t)$ and obtain the Fourier components (see Appendix B)

$$F(\omega) = \frac{1}{(2\pi)^{1/2}} \int_{-\infty}^{\infty} f(t)\, e^{-i\omega t}\, dt \tag{10.4-4}$$

Furthermore,

$$f(t) = \frac{1}{(2\pi)^{1/2}} \int_{-\infty}^{\infty} F(\omega)\, e^{i\omega t}\, d\omega$$

Hence

$$\Psi(x, y, z = 0, t) = \int_{-\infty}^{\infty} \frac{1}{(2\pi)^{1/2}} \varphi(x, y)\, F(\omega)\, e^{i\omega t}\, d\omega \tag{10.4-5}$$

which implies that the incident field is a superposition of pure harmonic waves of amplitude $(2\pi)^{-1/2}\,\varphi(x, y)\,F(\omega)$. Expanding $\varphi(x, y)$ in terms of the modes, we obtain

$$\varphi(x, y) = \sum_p A_p(\omega)\,\psi_p(x, y, \omega) \tag{10.4-6}$$

Using Eq. (10.4-6) in Eq. (10.4-5), we obtain

$$\Psi(x, y, z = 0, t) = \frac{1}{(2\pi)^{1/2}} \sum_p \int_{-\infty}^{\infty} A_p(\omega)\,\psi_p(x, y, \omega)\,F(\omega)\,e^{i\omega t}\,d\omega \tag{10.4-7}$$

When the field propagates through a distance z, then each mode gets multiplied by a factor $\exp[-i\beta_p(\omega)\,z]$. Thus

$$\Psi(x, y, z, t) = \frac{1}{(2\pi)^{1/2}} \sum_p \int_{-\infty}^{\infty} A_p(\omega)\,\psi_p(x, y, \omega)\,F(\omega)\exp\{i[\omega t - \beta_p(\omega)\,z]\}\,d\omega \tag{10.4-8}$$

In order to study the propagation of the pulse we assume $f(t)$ to be of the form (see Fig. 10.6a)

$$f(t) = \exp\left(-\frac{t^2}{2\tau^2}\right)\exp(i\omega_c t) \tag{10.4-9}$$

where ω_c is the carrier frequency. The time distribution is Gaussian, which is indeed the case for most practical systems. The corresponding Fourier components are given by

$$\begin{aligned} F(\omega) &= \frac{1}{(2\pi)^{1/2}} \int_{-\infty}^{\infty} \exp\left[-\frac{t^2}{2\tau^2} + i\omega_c t - i\omega t\right] dt \\ &= \tau \exp\left[-\frac{\tau^2}{2}(\omega - \omega_c)^2\right] \end{aligned} \tag{10.4-10}$$

For all practical cases* $\tau \gg 1/\omega_c$ and the Fourier spectrum is very sharply peaked around $\omega = \omega_c$ (see Fig. 10.6b). Now, in general $A_p(\omega)$ and $\psi_p(x, y, \omega)$ are slowly varying functions of ω, and since $F(\omega)$ is very sharply peaked around $\omega = \omega_c$, we may take them outside the integral at $\omega = \omega_c$. Thus, we obtain

$$\Psi(x, y, z, t) = \sum_p A_p(\omega_c)\,\psi_p(x, y, \omega_c)\,\theta_p(t) \tag{10.4-11}$$

* For a typical nanosecond optical pulse $\omega_c\tau \simeq 10^6$.

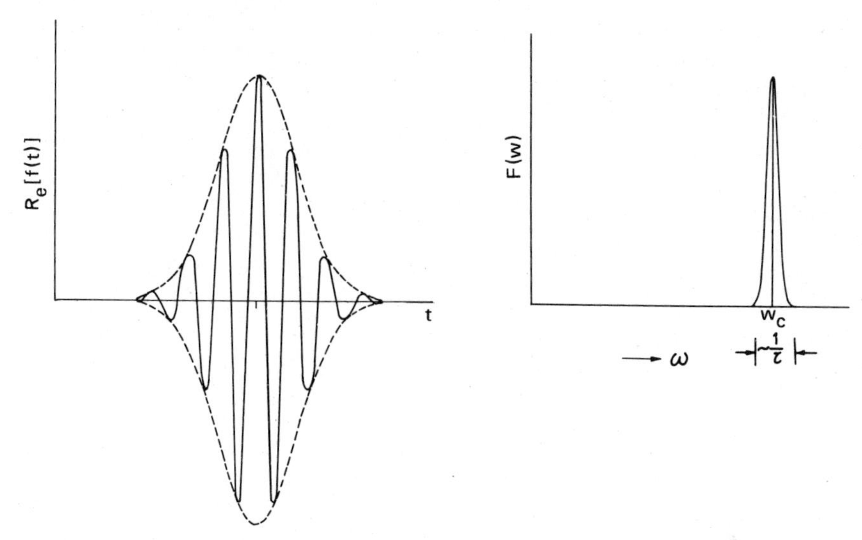

Fig. 10.6. A typical variation of the field for a Gaussian temporal pulse [what we have plotted is essentially $\exp(-t^2/\tau^2)\cos\omega_c t$, which is the real part of Eq. (10.4-9)]. The dashed curve shows the envelope of the pulse whose width is $\sim\tau$. The corresponding $F(\omega)$ is shown in (b). Figures are not to scale.

where

$$\theta_p(t) = \frac{1}{(2\pi)^{1/2}} \int_{-\infty}^{\infty} F(\omega)\exp\{i[\omega t - \beta_p(\omega)\,z]\}\,d\omega \qquad (10.4\text{-}12)$$

represents the time dependence of the output pulse corresponding to the pth mode. For a Gaussian temporal dependence, we therefore have

$$\theta_p(t) = \frac{\tau}{(2\pi)^{1/2}} \int_{-\infty}^{\infty} \exp\left[-\frac{\tau^2}{2}(\omega-\omega_c)^2 + i(\omega t - \beta_p z)\right] d\omega \qquad (10.4\text{-}13)$$

Since $F(\omega)$ is a sharply peaked function around $\omega = \omega_c$, we may expand β_p in a Taylor series in ascending powers of $(\omega - \omega_c)$ about $\omega = \omega_c$ and retain terms up to second order:

$$\beta_p = \beta_p(\omega_c) + \gamma_{1,p}(\omega - \omega_c) + \gamma_{2,p}\frac{(\omega-\omega_c)^2}{2!} \qquad (10.4\text{-}14)$$

where

$$\gamma_{1,p} = \left.\frac{\partial\beta_p}{\partial\omega}\right|_{\omega=\omega_c}, \qquad \gamma_{2,p} = \left.\frac{\partial^2\beta_p}{\partial\omega^2}\right|_{\omega=\omega_c} \qquad (10.4\text{-}15)$$

Thus

$$\theta_p(t) = \frac{\tau}{(2\pi)^{1/2}} \int_{-\infty}^{\infty} \exp\left[-\frac{(\omega-\omega_c)^2}{2}(\tau^2 + i\gamma_{2,p}z) + i(t-\gamma_{1,p}z)(\omega-\omega_c)\right] \times \exp\{i[\omega_c t - \beta_p(\omega_c)\, z]\}\, d\omega \tag{10.4-16}$$

which on integration yields

$$\theta_p(t) = \left[1 + i\frac{\gamma_{2,p}z}{\tau^2}\right]^{-1/2} \exp\{i[\omega_c t - \beta_p(\omega_c)\, z]\} \exp\left[-\frac{(t-\gamma_{1,p}z)^2}{2(\tau^2 + i\gamma_{2,p}z)}\right] \tag{10.4-17}$$

The intensity will be proportional to $|\theta_p(t)|^2$, which is given by

$$|\theta_p(t)|^2 = \left[1 + \left(\frac{L\gamma_{2,p}}{\tau^2}\right)^2\right]^{-1/2} \exp\left\{-\frac{(t-\gamma_{1,p}L)^2}{\tau^2\left[1 + \left(\frac{\gamma_{2,p}L}{\tau^2}\right)^2\right]}\right\} \tag{10.4-18}$$

where $z = L$ is the output plane. The temporal dependence of $|\theta_p(t)|^2$ at $z = 0$ and at $z = L$ is shown in Fig. 10.7. We may note the following points:

(1) The maximum of the intensity occurs at

$$t = L\gamma_{1,p} \tag{10.4-19}$$

which is also to be expected from the fact that $\gamma_{1,p}$ is essentially the group velocity of the pth mode ($=\partial\omega/\partial\beta$) and hence $L\gamma_{1,p}$ must be the time that the pulse should take in traversing a distance L.

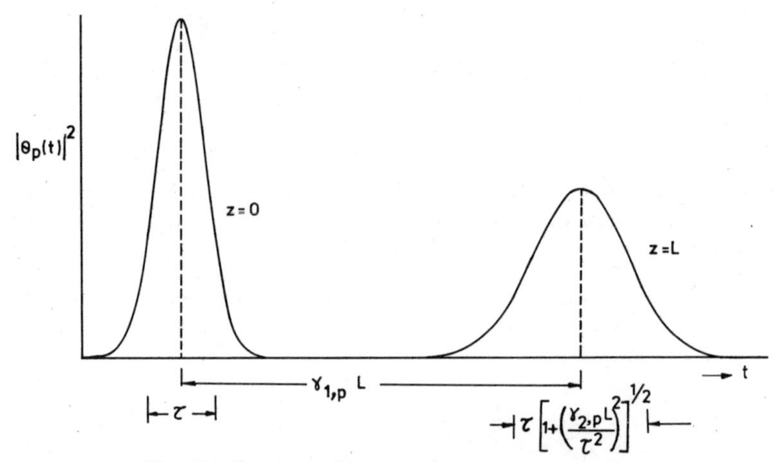

Fig. 10.7. Envelope of the pulse at $z = 0$ and at $z = L$.

(2) The total power contained in any mode is given by

$$P = \int_{-\infty}^{\infty} |\theta_p(t)|^2 \, dt = \pi^{1/2}\tau \tag{10.4-20}$$

which is independent of z. This is a manifestation of the fact that we have assumed the medium to be lossless, and therefore the energy contained in a particular mode remains constant as it propagates through the fiber.*

(3) The second derivative $\gamma_{2,p} \equiv \partial^2\beta/\partial\omega^2|_{\omega=\omega_c}$ is responsible for the distortion of the pulse. The width† of the pulse that was proportional to τ at $z = 0$ becomes proportional to $\tau[1 + (\gamma_{2,p}L/\tau^2)^2]^{1/2}$ after traversing a distance L.

Up to now, our analysis has been valid for any transmitting channel. In particular, for a Selfoc fiber, we have

$$\beta_{nm} \simeq \frac{\omega}{c} K_0^{1/2} \left[1 - (n+m+1)\frac{c}{\omega}\frac{K_2^{1/2}}{K_0} - \frac{1}{2}(n+m+1)^2 \frac{c^2}{\omega^2}\frac{K_2}{K_0^2} + \cdots \right] \tag{10.4-21}$$

and thus

$$\frac{d^2\beta_{nm}}{d\omega^2} \simeq -(n+m+1)^2 \frac{c}{\omega^3}\frac{K_2}{K_0^{3/2}} \tag{10.4-22}$$

where we are neglecting material dispersion, i.e., we have assumed K_0 and K_2 to be independent of ω. The contribution from material dispersion is discussed in Problem 10.8. Using typical values of various parameters,

$$K_0 = 1.5, \quad K_2 = 4.7 \times 10^6 \text{ m}^{-2}, \quad \omega_c = 3 \times 10^{15} \text{ sec}^{-1}, \quad L = 1 \text{ km}, \quad \tau = 100 \text{ psec} \tag{10.4-23}$$

we obtain

$$(\gamma_{2,p}L/\tau^2)^2 = (n+m+1)^4 \times 10^{-17} \tag{10.4-24}$$

* Even in a lossless medium, mode conversion can occur due to imperfections in the waveguide (a typical imperfection may be the z dependence of K_2). Mode conversion is a phenomenon by which the power in one particular mode gets transferred to other modes. Since we are assuming a perfect waveguide, no mode conversion can occur. For an account of mode conversion see Marcuse (1973, 1974), Sodha and Ghatak (1977), and Ghatak *et al.* (1976).

† The width of the pulse is usually defined as the full width at half-maximum and for a pulse characterized by a temporal dependence of intensity of the form $\exp(-t^2/\tau^2)$ the width will be $2\tau(\ln 2)^{1/2}$. The effect of input pulse shape on pulse dispersion has been studied by Goyal *et al.* (1976).

Thus for almost all modes, the fractional increase in the width of the pulse is negligible. This is the chief characteristic of a Selfoc fiber. The dispersion Δ is defined as the difference in the widths between the output and the input pulses. Thus

$$\Delta = 2\tau(\ln 2)^{1/2}\left\{\left[1+\left(\frac{L\gamma_{2,p}}{\tau^2}\right)^2\right]^{1/2}-1\right\} \tag{10.4-25}$$

For the values of various parameters given by Eq. (10.4-23), one obtains (for $n = 0, m = 0$)

$$\Delta \simeq 10^{-18}\,\text{nsec/km} \tag{10.4-26}$$

Equation (10.4-18) corresponds to the temporal dependence of the pth mode. The total intensity distribution as recorded by the detector is

$$|\theta(t)|^2 = \sum_p |\theta_p(t)|^2\,|A_p|^2 \tag{10.4-27}$$

For an incident Gaussian beam, considering the Laguerre–Gauss functions, we get (see Problem 10.1)

$$|A_p|^2 = \frac{4w^2}{(1+w^2)^2}\left[\frac{w^2-1}{w^2+1}\right]^{2p} \tag{10.4-28}$$

Thus the output pulse is given by

$$\begin{aligned}|\theta(t)|^2 = {} & \frac{4w^2}{(1+w^2)^2}\sum_p\left[\frac{w^2-1}{w^2+1}\right]^{2p}\left[1+\left(\frac{L\gamma_{2,p}}{\tau^2}\right)^2\right]^{-1/2} \\ & \times \exp\left\{-\frac{(t-\gamma_{1,p}L)^2}{\tau^2\left[1+\left(\frac{L\gamma_{2,p}}{\tau^2}\right)^2\right]}\right\}\end{aligned} \tag{10.4-29}$$

The above summation must be carried out numerically. Such an evaluation has been done by Gambling and Matsumura (1973).

Problem 10.8. Assuming

$$\beta \simeq \frac{\omega}{c}n_0 \tag{10.4-30}$$

where $n_0 = K_0^{1/2}$ [see Eq. (10.2-22)], show that

$$\frac{d^2\beta}{d\omega^2} = \frac{\lambda_0}{2\pi c^2}\left(\lambda_0^2\frac{d^2n_0}{d\lambda_0^2}\right) \tag{10.4-31}$$

where $\lambda_0 = 2\pi c/\omega$ is the free-space wavelength. Equation (10.4-31) leads to what is known as material dispersion. Assuming

$$n_0 = 1.5329 + 2.12 \times 10^{-7}\omega, \quad \text{for} \quad \omega \approx \omega_c = 2.979 \times 10^{15}\,\text{Hz} \tag{10.4-32}$$

$$K_2/K_0 = 3.13 \times 10^6\,\text{m}^{-2}, \qquad \tau \simeq 0.325\,\text{ns} \tag{10.4-33}$$

show that the pulse dispersion for a 1 km Selfoc fiber is about 10^{-15} sec. The above parameters correspond to the Selfoc fiber used by Gambling and Matsumura (1973).

Problem 10.9. Using Eq. (10.4-25) show that for large z the material dispersion is given by

$$\Delta \simeq \frac{L}{c}\frac{\Delta\lambda_0}{\lambda_0}\left(\lambda_0^2 \frac{d^2 n}{d\lambda_0^2}\right) \tag{10.4-34}$$

where $\Delta\lambda_0$ is the spread in wavelength of the carrier components. Assuming a spread of $\Delta\lambda_0 = 0.036\ \mu$m, show that in a silica fiber the pulse spread is about 4 nsec/km. Assume $\lambda_0 = 0.8\ \mu$m, $\lambda_0^2 d^2 n_0/d\lambda_0^2 = 0.024$. *Reference*: Gloge (1974).

Problem 10.10. Consider a Selfoc fiber of radius a and assume that only those modes whose propagation constants satisfy the inequality

$$\beta < n(a)\,k_0 \tag{10.4-35}$$

are propagated through the fiber ($n = K^{1/2}$). Assuming that all these modes are equally excited, calculate the dispersion of an impulse when it propagates through a distance L. [An impulse has a time variation given by $\delta(t)$, where $\delta(t)$ represents the Dirac delta function.] *Reference*: Marcuse (1973).

Solution: From Eq. (10.2-20) we have

$$\beta_{nm} \simeq \frac{\omega}{c} K_0^{1/2}\left[1 - (n + m + 1)\frac{c}{\omega}\frac{K_2^{1/2}}{K_0} - \frac{1}{2}(n + m + 1)^2 \frac{c^2}{\omega^2}\frac{K_2}{K_0^2}\right] \tag{10.4-36}$$

where we have made a binomial expansion and have retained terms up to second order. The time taken by the modes to travel a distance L is

$$\tau = L\frac{d\beta}{d\omega} = L\frac{K_0^{1/2}}{c}\left[1 + \frac{1}{2}(n + m + 1)^2 \frac{K_2}{K_0^2}\frac{c^2}{\omega^2}\right] \tag{10.4-37}$$

where $d\beta/d\omega$ represents the group velocity of the mode. To calculate the largest value of $(m + n)$ allowed by the cutoff condition $\beta = n(a)\,\omega/c$, we use the complete expression for β and find

$$n^2(a)\,k_0^2 = (K_0 - K_2 a^2)\frac{\omega^2}{c^2} = K_0\frac{\omega^2}{c^2} - 2(m + n + 1)\,K_2^{1/2}\frac{\omega}{c} \tag{10.4-38}$$

i.e.,

$$(m + n)_{\max} = \frac{K_2^{1/2}}{2} k_0 a^2 \tag{10.4-39}$$

where we have neglected unity in comparison to $(m + n)$, because $(m + n) \gg 1$. It can also be seen from Eq. (10.4-37) that all modes with equal values of $(m + n)$

arrive at the end of the fiber at the same time. In a time

$$d\tau = L\frac{K_0^{1/2}}{2}\frac{K_2}{K_0^2}\frac{c}{\omega^2}d(m+n)^2 \tag{10.4-40}$$

the number of modes arriving is $(m+n)\,d(m+n)\,[=\frac{1}{2}d(m+n)^2]$. Assuming equal power in all modes and neglecting mode conversion, we obtain the power arriving at the output end in a unit time

$$\frac{dP}{d\tau} = \frac{K_0^2\omega^2}{LK_0^{1/2}K_2c}P_0 \tag{10.4-41}$$

where P_0 is the power in each mode. Thus the total power P is given by

$$P = NP_0 = \tfrac{1}{8}K_2k_0^2a^4P_0 \tag{10.4-42}$$

where we have used the fact that the total number of modes N is given by

$$N = \int_0^{(m+n)_{\max}} \frac{1}{2}d(m+n)^2 = \frac{1}{8}K_2k_0^2a^4$$

Thus the impulse response $F(\tau)$, which is equal to the relative amount of power arriving in unit time at the output end of the fiber ($z = L$), is given by

$$F(\tau) = \frac{1}{P}\frac{dP}{d\tau} = \frac{8cK_0^{3/2}}{LK_2^2a^4} \tag{10.4-43}$$

The lowest modes arrive at $\tau = K_0^{1/2}L/c$ and the last modes arrive at

$$\tau = L\frac{K_0^{1/2}}{c}\left(1 + \frac{K_2^2a^4}{8K_0^2}\right)$$

Thus the impulse response is

$$F(\tau) = \begin{cases} 0 & \tau < \dfrac{LK_0^{1/2}}{c} \\[2ex] \dfrac{8cK_0^{3/2}}{LK_2^2a^4} & LK_0^{1/2}/c < \tau < \dfrac{LK_0^{1/2}}{c}\left(1 + \dfrac{K_2^2a^4}{8K_0^2}\right) \\[2ex] 0 & \tau > \dfrac{LK_0^{1/2}}{c}\left(1 + \dfrac{K_2^2a^4}{8K_0^2}\right) \end{cases} \tag{10.4-44}$$

i.e., the impulse response is rectangular.

Problem 10.11. In an actual fiber one should also take into account the presence of the r^4 term in the dielectric constant variation:

$$K(r) = K_0 - K_2r^2 + K_4r^4 \tag{10.4-45}$$

Using first-order perturbation theory, calculate the propagation constants for the azimuthally symmetric modes and study the dispersion as a function of the parameter K_4. [*Hint:* The scalar wave equation is

$$\frac{1}{r}\frac{\partial}{\partial r}\left(r\frac{\partial\psi}{\partial r}\right)+\frac{\partial^2\psi}{\partial z^2}+\frac{\omega^2}{c^2}(K_0-K_2r^2+K_4r^4)\,\psi=0 \tag{10.4-46}$$

The mode pattern and the propagation constants, when $K_4 = 0$, have already been obtained in Problem 10.1.] *Reference*: Gambling and Matsumura (1973).

10.5. Fabrication

The fabrication of Selfoc fibers is based on the ion exchange technique, which makes use of the principle that a change in composition of glass leads to a change in refractive index. If a small ion like Li^+ replaces a larger ion like Na^+ or K^+, then the resulting glass will have a higher refractive index because the network of atoms will relax around the smaller ion to produce a more compact structure. Conversely, if a large ion (Na^+ or K^+) replaces a smaller ion (Li^+) then one obtains a decrease in the refractive index. For example, a silicate glass (having 70 mole % of SiO_2) doped with 30 mole % of Li_2O has a refractive index of 1.53. On the other hand, if the doping material is Na_2O, K_2O, or Tl_2O, then the corresponding refractive index of the composite material is 1.50, 1.51, or 1.83, respectively. Thus if Li^+ ions replace Na^+ ions, we have an increase in the refractive index, and if Li^+ ions replace Tl^+ ions, then we have a decrease in the refractive index. Similarly, if a silicate glass (having 60 mole % of SiO_2) is doped with 40 mole % of PbO, the refractive index is 1.81; on the other hand, if it is doped with BaO, the refractive index is 1.68.

In a typical experiment, carried out by Kitano *et al.* (1970) a glass fiber (having a composition of 16% Tl_2O, 12% Na_2O, 25% PbO, and 45% SiO_2 by weight) was steeped in a KNO_3 bath at 460°C for 432 hours. The concentrations of K^+, Na^+, and Tl^+ ions at $t = 24$ hours and $t =$

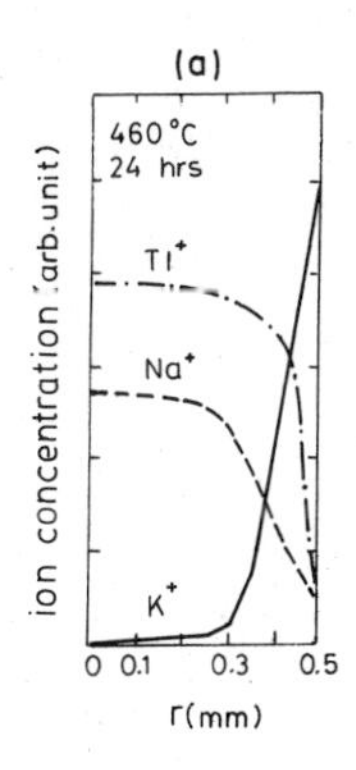

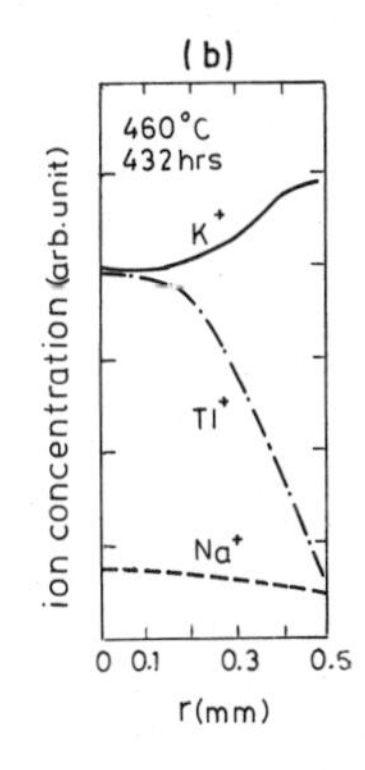

Fig. 10.8. Ion concentration distribution in the rod after (a) 24 hours and (b) 432 hours of ion exchange. Ion exchange temperature was 460°C; r represents the transverse distance from the axis of the rod (after Kitano *et al.*, 1970).

432 hours are shown in Fig. 10.8. The variation of ion concentration leads to a refractive-index distribution. In this process the K^+ ion replaces Tl^+ and Na^+ ions, and since the K^+ ion concentration is a maximum near the periphery of the glass rod, the refractive index is a minimum there and increases toward the axis. The corresponding refractive-index variation was indeed found to be parabolic in nature. In a typical experiment, the fiber diameter was about 160 μm and $a_2 = 0.16 \times 10^6$ m^{-2}; the coefficient a_2 is defined by $n = n_0(1 - \frac{1}{2}a_2r^2)$.

In order to show that the refractive-index variation is indeed parabolic in nature, we have plotted in Fig. 10.9 a typical refractive-index profile obtained by Pearson *et al.* (1969) using the ion exchange technique. A glass rod of diameter 1.9×10^{-3} m containing 30 mole % Li_2O, 15 mole % Al_2O_3, and 55 mole % SiO_2 was heated in a fused salt bath containing 50 mole % $NaNO_3$ and 50 mole % $LiNO_3$ at 470°C for 50 hours. One can immediately see that the refractive-index variation is very close to parabolic.

Recently Koizumi *et al.* (1974) have been able to produce Selfoc fibers by a continuous process that produces kilometer lengths of fibers.

An approximate parabolic refractive-index variation is also obtained in a gas lens that consists of a hollow warm metal tube through which a gas at room temperature is passed (see Fig. 10.10). Typically the diameter

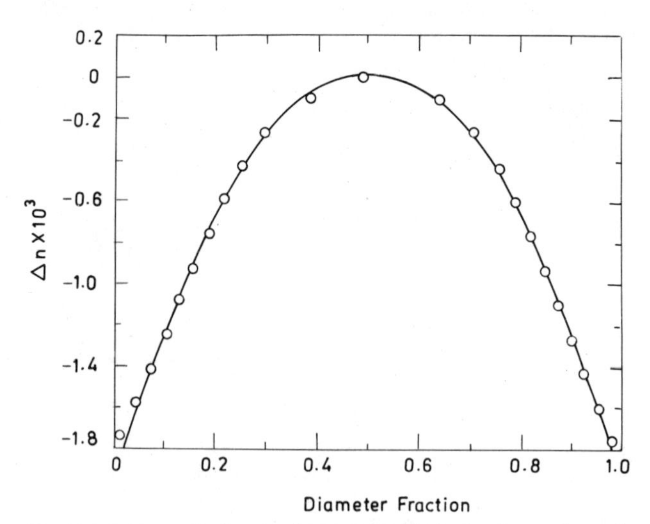

Fig. 10.9. Measured refractive-index profile (○) of an ion exchanged rod, normalized to a maximum of zero. The solid line is a parabola fitted to the experimental points by the least-squares method. This shows the accuracy of the parabolic approximation to the actual profile (after Pearson *et al.*, 1969).

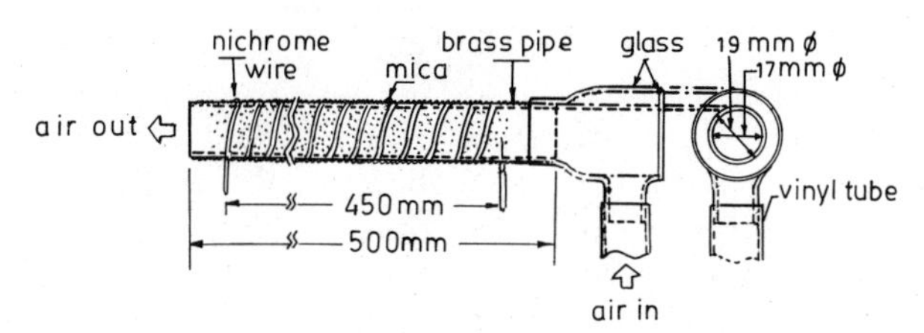

Fig. 10.10. A gas lens (after Aoki and Suzuki, 1967).

of the tube is ~1 cm, the length of the tube may be ≃ 50 cm, the temperature of the outer surface of the tube may be 70°C, and the rate of flow of gas through the tube is a few liters per minute. The gas near the surface will have a higher temperature than the gas at the axis, and hence a temperature gradient is produced. Since the refractive index of a gas decreases with increase of temperature, the refractive index is a maximum on the axis and is found to decrease parabolically as one moves away from the axis (Marcatili, 1964; Ghatak *et al.*, 1973). Such a system is known as a gas lens and has been used for aberrationless imaging of distant objects (Aoki and Suzuki, 1967).

11

Evanescent Waves and the Goos–Hänchen Effect

11.1. Introduction

The reflection of electromagnetic waves from a dielectric interface is usually studied by assuming the incidence of plane waves having infinite spatial extent; in addition, the radiation is also assumed to be purely monochromatic (see, e.g., Reitz and Milford, 1962). Both these concepts are introduced for simplicity in theoretical considerations. In practice, one encounters a finite-width beam that extends over a finite time and hence does not remain monochromatic. In this chapter we shall study the total internal reflection of a bounded beam by a plane dielectric interface and we shall show that the beam appears to undergo a lateral shift as shown in Fig. 11.1a. This displacement of the beam was experimentally studied by Goos and Hänchen. Hence this effect is called the Goos–Hänchen effect.* This shift cannot be explained from ray-optic considerations and is a consequence of the evanescent field set up in the rarer medium. Since evanescent fields are of great importance in electromagnetic theory,† we thought it worthwhile devoting a chapter to a phenomenon that brings out many salient features associated with the evanescent field.

The Goos–Hänchen shift suffered by a beam is very small (about one wavelength) and depends on the state of polarization of the beam, the shift for the electric vector lying in the plane of incidence being about twice the shift when the electric vector lies normal to it. Since the shift is of the order

* A comprehensive review of the Goos–Hänchen effect has been given by Lotsch (1970, 1971).

† The study of evanescent waves is of great importance in many areas, e.g., in order to couple light energy into a thin-film waveguide (of thickness $\lesssim 1\ \mu m$), one uses the prism-film coupler arrangement (see Section 11.6). Other applications include modulators, continuously variable beam splitters, and internal reflection spectroscopy. (For further details, see Harrick, 1967.) The study of evanescent waves is also of great importance in image formation, image processing, and image transfer (see, e.g., the review by Bryngdahl, 1973).

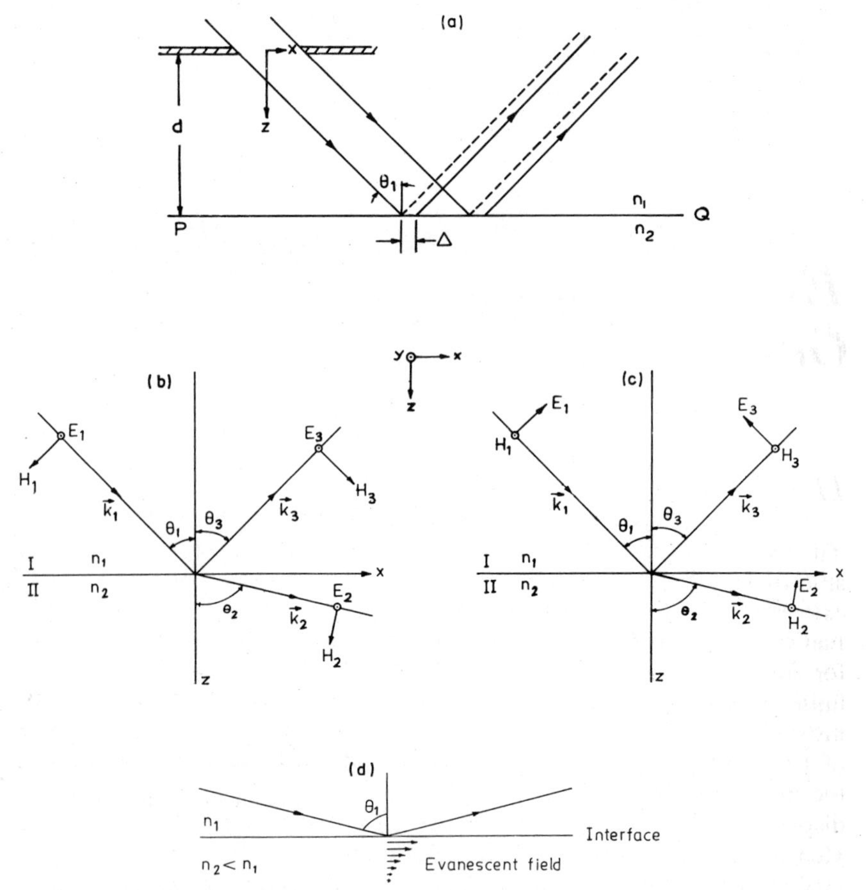

Fig. 11.1. (a) Total internal reflection of a beam of finite width by a dielectric interface separating media of refractive indices n_1 and n_2 ($n_1 > n_2$). The plane of the paper represents the x-z plane and the y axis is perpendicular to the plane of the paper. The beam extends from $-\infty$ to $+\infty$ in the y direction. The dashed lines show the reflected beam obtained by geometrical considerations and Δ is the Goos–Hänchen shift. Reflection of an infinite plane wave at a dielectric interface for perpendicular (b) and parallel (c) polarizations, respectively. (d) For $\theta_1 > \theta_c$, the field in the second medium is an evanescent wave (shown by arrows), which propagates along the x direction and decays exponentially along the z direction.

of a wavelength, if one performs an experiment using optical beams, the beam has to undergo a large number of reflections to obtain a measurable shift. The dependence of the shift on wavelength suggests the use of longer-wavelength radiation to obtain a larger shift. Indeed, the Goos–Hänchen shift has been measured by using microwaves ($\lambda \sim 1$ cm) and by letting the

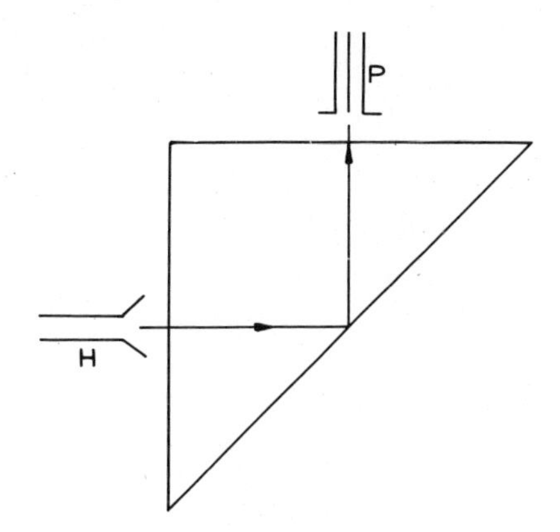

Fig. 11.2. Experimental arrangement for measuring the Goos–Hänchen shift suffered by a microwave beam ($\lambda \sim 1$ cm). *H* is the microwave horn and *P* denotes the probe for detecting the reflected radiation. The prism is made of a low-loss plastic.

beam undergo a single reflection. Figure 11.2 shows a typical experimental arrangement of Read *et al.* (1972) in which polarized microwaves are fed into a 45–45–90° prism made of a low-loss plastic having an index of refraction 1.685 and a critical angle of 36.4° at the microwave frequency. The polarized microwaves were produced by an electromagnetic horn and the radiation was detected by a dipole probe. An aluminum sheet is first kept touching the base and the reflected waves are received by a microwave detector. Because a metallic reflection is taking place, in the present case no Goos–Hänchen shift is observed. Then the aluminum sheet is removed and the shift is measured.

In Section 11.2, we show the existence of evanescent waves when an electromagnetic wave undergoes total internal reflection. In Section 11.3, we give a derivation of the Goos–Hänchen shift using the plane-wave expansion method. In Section 11.4 we deal with the physical understanding of this phenomenon and show that the radiation actually enters the rarer medium and travels through it. Such a shift is observed in the interaction of any kind of waves with boundaries; in particular, Problem 11.4 deals with the quantum-mechanical reflection of a particle by a potential barrier. In Section 11.5, we discuss the Goos–Hänchen effect in a planar waveguide. Finally, in Section 11.6, we briefly discuss how evanescent waves can be used to couple light energy into a thin-film waveguide.

11.2. Existence of Evanescent Waves

Let us first consider the reflection of plane waves at an interface between two dielectrics with refractive indices n_1 and n_2 (see Figs. 11.1b and

11.1c). We assume both media to be nonabsorbing and nonmagnetic. Let

$$\mathbf{E}_1 = \mathbf{E}_{10} \exp[i(\omega t - \mathbf{k}_1 \cdot \mathbf{r})] \tag{11.2-1}$$

$$\mathbf{E}_2 = \mathbf{E}_{20} \exp[i(\omega t - \mathbf{k}_2 \cdot \mathbf{r})] \tag{11.2-2}$$

$$\mathbf{E}_3 = \mathbf{E}_{30} \exp[i(\omega t - \mathbf{k}_3 \cdot \mathbf{r})] \tag{11.2-3}$$

represent the electric fields associated with the incident, the transmitted, and the reflected waves, respectively. Without any loss of generality, we assume that $\mathbf{k}_1$ (and hence $\mathbf{k}_2$ and $\mathbf{k}_3$) lies in the x-z plane. Now, the fields have to satisfy certain boundary conditions at the interface*; for example, if we consider the case when $\mathbf{E}$ is perpendicular to the plane of incidence (see Fig. 11.1b) then the continuity of E_y (which represents the tangential component of $\mathbf{E}$) across the interface gives us

$$E_{1y} + E_{3y} = E_{2y}$$

or

$$E_{10} \exp[-i(k_{1x}x + k_{1z}d)] + E_{30} \exp[-i(k_{3x}x + k_{3z}d)] = E_{20} \exp[-i(k_{2x}x + k_{2z}d)] \tag{11.2-4}$$

Equation (11.2-4) has to be valid for all values of x. Thus we must have

$$k_{1x} = k_{2x} = k_{3x}$$

or

$$k_1 \sin\theta_1 = k_2 \sin\theta_2 = k_3 \sin\theta_3 \tag{11.2-5}$$

Now, in a homogeneous medium, each Cartesian component of $\mathbf{E}$ (or $\mathbf{H}$) must satisfy the wave equation (see Section 10.3)

$$\nabla^2 \mathbf{E} = \varepsilon\mu\, \partial^2 \mathbf{E}/\partial t^2 \tag{11.2-6}$$

where ϵ and μ, respectively, represent the dielectric permittivity and the magnetic permeability of the medium. On substituting the fields expressed by Eqs. (11.2-1)–(11.2-3) into Eq. (11.2-6), we immediately obtain

$$k_1^2 = k_3^2 = \frac{\omega^2}{c^2} n_1^2, \qquad k_2^2 = \frac{\omega^2}{c^2} n_2^2 \tag{11.2-7}$$

where

$$n_1 = c(\varepsilon_1 \mu_0)^{1/2}, \qquad n_2 = c(\varepsilon_2 \mu_0)^{1/2} \tag{11.2-8}$$

represent the refractive indices of the two media (ε_1 and ε_2 are the corresponding dielectric permittivities). The media are assumed to be non-

* The boundary conditions are the continuity of the tangential components of $\mathbf{E}$ and $\mathbf{H}$ and normal components of $\mathbf{D}$ $(=\varepsilon\mathbf{E})$ and $\mathbf{B}$.

magnetic, so that $\mu_1 \simeq \mu_2 \simeq \mu_0$. Thus, Eq. (11.2-5) gives us $\sin\theta_1 = \sin\theta_3$ (i.e., the angle of incidence is equal to the angle of reflection) and

$$n_1 \sin\theta_1 = n_2 \sin\theta_2 \tag{11.2-9}$$

which is Snell's law. It can immediately be seen that if $n_2 < n_1$, then $\sin\theta_2$ will be greater than unity for $\theta_1 > \theta_c$, where

$$\theta_c = \sin^{-1}(n_2/n_1) \tag{11.2-10}$$

is known as the critical angle. This is the well-known phenomenon of total internal reflection. It should be mentioned that for $\theta_1 > \theta_c$, Eq. (11.2-9) (which still remains valid) gives us a complex value of θ_2 with $\sin\theta_2 > 1$ and $\cos\theta_2$ a pure imaginary quantity. Thus,

$$\mathbf{k}_2 \cdot \mathbf{r} = \frac{\omega}{c} n_2(x \sin\theta_2 + z\cos\theta_2) = \frac{\omega}{c}\left[(n_1 \sin\theta_1)x - iz(n_1^2 \sin^2\theta_1 - n_2^2)^{1/2}\right] \tag{11.2-11}$$

and Eq. (11.2-2) becomes

$$\mathbf{E}_2 = \mathbf{E}_{20} \exp\left\{ i\left[\omega t - \left(\frac{\omega}{c} n_1 \sin\theta_1\right)x\right]\right\} \times \exp\left[-\frac{\omega}{c} z(n_1^2 \sin^2\theta_1 - n_2^2)^{1/2}\right] \tag{11.2-12}$$

which represents a wave propagating in the $+x$ direction with its amplitude decaying exponentially along the z direction. Such a wave is known as an evanescent wave (see Fig. 11.1d). In Section 11.3, we shall discuss the total internal reflection of a bounded beam and show that the presence of the evanescent field results in the Goos–Hänchen shift; however, before we discuss that, it is necessary to know the amplitude reflection coefficient of plane waves.

The amplitude reflection and transmission coefficients can be readily obtained by equating the tangential components of **H** and **E** across the interface. We first consider the case in which the electric vector is perpendicular to the plane of incidence. The continuity of the tangential component of **E** gave us Eq. (11.2-4), which can be written in the form

$$E_{10} \exp(-ik_{1z}d) + E_{30} \exp(ik_{1z}d) = E_{20} \exp(-ik_{2z}d) \tag{11.2-13}$$

where we have used Eq. (11.2-5). In order to calculate **H**, we use one of Maxwell's equations [see Eq. (10.3-30)]:

$$\nabla \times \mathbf{E} + \mu_0 \frac{\partial \mathbf{H}}{\partial t} = 0 \tag{11.2-14}$$

to obtain

$$\mathbf{H} = \frac{1}{\omega\mu_0}(\mathbf{k} \times \mathbf{E}) \tag{11.2-15}$$

where we have assumed the space and time dependence of the electric and magnetic fields to be of the form $\exp[i(\omega t - \mathbf{k}\cdot\mathbf{r})]$ [see Eq. (11.2-1)]. Thus

$$H_x = -\frac{1}{\omega\mu_0}k_z E_y \tag{11.2-16}$$

Since **E** is assumed to be perpendicular to the plane of incidence, **H** will lie in the plane of incidence (see Fig. 11.1b). Thus the continuity of H_x (which represents the tangential component of the magnetic field) will yield

$$H_{1x} + H_{3x} = H_{2x} \tag{11.2-17}$$

or

$$k_{1z}E_{10}\exp[-i(k_{1x}x + k_{1z}d)] + k_{3z}E_{30}\exp[-i(k_{3x}x + k_{3z}d)] = k_{2z}E_{20}\exp[-i(k_{2x}x + k_{2z}d)]$$

or

$$E_{10}\exp(-ik_{1z}d) - E_{30}\exp(ik_{1z}d) = \frac{k_{2z}}{k_{1z}}E_{20}\exp(-ik_{2z}d) \tag{11.2-18}$$

Dividing Eq. (11.2-13) by Eq. (11.2-18), we obtain

$$\frac{E_{10}\exp(-ik_{1z}d) + E_{30}\exp(ik_{1z}d)}{E_{10}\exp(-ik_{1z}d) - E_{30}\exp(ik_{1z}d)} = \frac{1 + (E_{30}/E_{10})\exp(2ik_{1z}d)}{1 - (E_{30}/E_{10})\exp(2ik_{1z}d)} = \frac{k_{1z}}{k_{2z}}$$

or

$$\frac{E_{30}}{E_{10}}\exp(2ik_{1z}d) = \frac{k_{1z} - k_{2z}}{k_{1z} + k_{2z}} \tag{11.2-19}$$

The incident and the reflected fields on the surface ($z = d$) are

$$E_1(z = d) = E_{10}\exp[i(\omega t - k_{1x}x - k_{1z}d)] \tag{11.2-20}$$

$$E_3(z = d) = E_{30}\exp[i(\omega t - k_{3x}x - k_{3z}d)]$$
$$= E_{30}\exp[i(\omega t - k_{1x}x + k_{1z}d)] \tag{11.2-21}$$

where we have used Eq. (11.2-5). Thus, the amplitude reflection coefficient $\Gamma(k_x)$ is

$$\Gamma(k_x) = \frac{E_3(z = d)}{E_1(z = d)} = \frac{E_{30}}{E_{10}}\exp(2ik_{1z}d) = \frac{k_{1z} - k_{2z}}{k_{1z} + k_{2z}} = \frac{n_1\cos\theta_1 - n_2\cos\theta_2}{n_1\cos\theta_1 + n_2\cos\theta_2} \tag{11.2-22}$$

Similarly, when the electric vector lies in the plane of incidence (see Fig. 11.1c) we obtain, by equating the tangential components of **E** and **H**, the following equations:

$$E_{10}\exp(-ik_{1z}d) - E_{30}\exp(ik_{1z}d) = \frac{\cos\theta_2}{\cos\theta_1}E_{20}\exp(-ik_{2z}d) \qquad (11.2\text{-}23)$$

$$E_{10}\exp(-ik_{1z}d) + E_{30}\exp(ik_{1z}d) = \frac{k_2}{k_1}E_{20}\exp(-ik_{2z}d) \qquad (11.2\text{-}24)$$

Thus the amplitude reflection coefficient is given by

$$\Gamma(k_x) = \frac{E_{30}\exp(ik_{1z}d)}{E_{10}\exp(-ik_{1z}d)} = \frac{n_2\cos\theta_1 - n_1\cos\theta_2}{n_2\cos\theta_1 + n_1\cos\theta_2} \qquad (11.2\text{-}25)$$

Equations (11.2-22) and (11.2-25) can be put in the form

$$\Gamma(k_x) = \frac{n_1\cos\theta_1 - mn_2\cos\theta_2}{n_1\cos\theta_1 + mn_2\cos\theta_2} = \frac{(k_1^2 - k_x^2)^{1/2} - m(k_2^2 - k_x^2)^{1/2}}{(k_1^2 - k_x^2)^{1/2} + m(k_2^2 - k_x^2)^{1/2}} \qquad (11.2\text{-}26)$$

where $m = n^2$ $(= n_1^2/n_2^2)$ for **E** lying in the plane of incidence and $m = 1$ for **E** perpendicular to the plane of incidence. Notice that there is no suffix on k_x because $k_{1x} = k_{2x} = k_{3x}$ [see Eq. (11.2-5)].

For $\theta_1 > \theta_c$, $\cos\theta_2$ is pure imaginary and hence $|\Gamma(k_x)| = 1$; thus, we may write

$$\Gamma(k_x) = \exp[-i\chi(k_x)] \qquad (11.2\text{-}27)$$

where

$$\chi(k_x) = -2\tan^{-1}\left[m\frac{(k_x^2 - k_2^2)^{1/2}}{(k_1^2 - k_x^2)^{1/2}}\right] \qquad (11.2\text{-}28)$$

is the phase shift introduced by reflection.

Problem 11.1. Show that for the evanescent wave $\langle S_z \rangle = 0$ (however, $\langle S_x \rangle \neq 0$) where angular brackets denote time averaging and $\mathbf{S} = \mathbf{E} \times \mathbf{H}$ represents the Poynting vector. Thus, associated with the evanescent wave there is no energy flow in the z direction.

11.3. Total Internal Reflection of a Bounded Beam

In a homogeneous dielectric, a plane wave can be written

$$\psi = A\exp[-ik_1(x\sin\theta + z\cos\theta)] = A\exp\{-i[k_x x + z(k_1^2 - k_x^2)^{1/2}]\} \qquad (11.3\text{-}1)$$

where $k_1 = \omega n_1/c$, ψ represents a transverse component of the field, and we have assumed $\mathbf{k}_1$ to lie in the x-z plane; θ is the angle between $\mathbf{k}_1$ and the z axis. An arbitrary wave in such a medium can be expressed as a linear superposition of the plane waves described by Eq. (11.3-1),

$$\psi(x, z) = \frac{1}{(2\pi)^{1/2}} \int_{-\infty}^{+\infty} \Phi(k_x) \exp\{-i[k_x x + z(k_1^2 - k_x^2)^{1/2}]\}\, dk_x \qquad (11.3\text{-}2)$$

where for the sake of convenience, we are assuming no y dependence of the fields. For each value of k_x, the integrand represents a plane wave of amplitude $(2\pi)^{-1/2}\,\Phi(k_x)$ traveling along a direction determined by k_x. For $|k_x| < k_1$, these waves are homogeneous plane waves, while for $|k_x| > k_1$, these waves are inhomogeneous plane waves or evanescent waves. As can easily be seen, the evanescent waves propagate along the x direction and decay exponentially along the z direction. Hence their contribution to the integral as z increases becomes smaller and smaller. Let $\psi(x, z = 0)$ represent the amplitude distribution at $z = 0$. Thus

$$\psi(x, z = 0) = \frac{1}{(2\pi)^{1/2}} \int_{-\infty}^{\infty} \Phi(k_x) \exp(-ik_x x)\, dk_x \qquad (11.3\text{-}3)$$

Clearly $\Phi(k_x)$ is the Fourier transform of $\psi(x, z = 0)$ and hence

$$\Phi(k_x) = \frac{1}{(2\pi)^{1/2}} \int_{-\infty}^{+\infty} \psi(x, z = 0) \exp(ik_x x)\, dx \qquad (11.3\text{-}4)$$

Consequently,

$$\psi(x, z) = \frac{1}{2\pi} \iint_{-\infty}^{+\infty} \psi(x', z = 0) \exp\{-i[k_x(x - x') + (k_1^2 - k_x^2)^{1/2} z]\}\, dx'\, dk_x \qquad (11.3\text{-}5)$$

Equation (11.3-5) tells us that if we know the amplitude distribution at an arbitrary plane $z = 0$, then we can determine the amplitude distribution at any other plane. For example, for a slit of width $2a$ placed along the x axis with the origin lying at the midpoint of the slit, we may have

$$\psi(x, z = 0) = \begin{cases} A \exp(-ik_1 x \sin\theta) & |x| < a \\ 0 & |x| > a \end{cases} \qquad (11.3\text{-}6)$$

On substituting in Eq. (11.3-4) and carrying out the integration, we obtain

$$\Phi(k_x) = A\left(\frac{2}{\pi}\right)^{1/2} \frac{\sin[(k_1 \sin\theta - k_x)\, a]}{(k_1 \sin\theta - k_x)} \qquad (11.3\text{-}7)$$

It can easily be seen that $\Phi(k_x)$ is sharply peaked around $k_x = k_1 \sin\theta$ and the width of the peak is $\sim 1/a$ (see Fig. 11.3). Thus as a (the width of

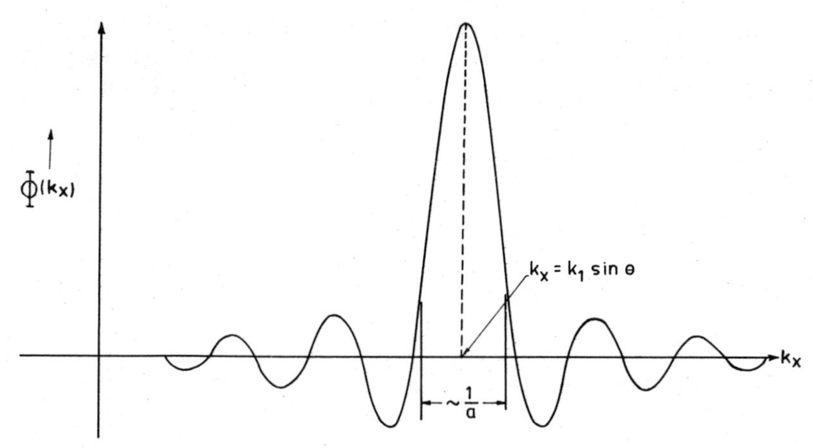

Fig. 11.3. $\Phi(k_x)$ as given by Eq. (11.3-7). With an increase in the value of a, the curve will become more sharply peaked at $k_x = k_1 \sin \theta$.

the beam) increases, $\Phi(k_x)$ becomes more sharply peaked. On substitution in Eq. (11.3-2) we finally get

$$\psi(x, z) = \frac{A}{\pi} \int_{-\infty}^{+\infty} \frac{\sin[(k_1 \sin \theta - k_x) a]}{k_1 \sin \theta - k_x} \exp\{-i[k_x x + z(k_1^2 - k_x^2)^{1/2}]\}\, dk_x \tag{11.3-8}$$

Evaluation of the above integral will give the diffraction pattern due to the slit.

We next consider the reflection of such a bounded beam by a medium of refractive index n_2. Let $z = d$ represent the interface shown as PQ in Fig. 11.1a. Then, using Eq. (11.3-2), the amplitude distribution at the surface is given by

$$\psi(x, z = d) = \frac{1}{(2\pi)^{1/2}} \int_{-\infty}^{+\infty} \Phi(k_x) \exp[-i(k_x x + k_z d)]\, dk_x \tag{11.3-9}$$

Now, each component plane wave gets reflected from the surface with a reflection coefficient $\Gamma(k_x)$. Also since the reflected beam is propagating in the negative z direction we have, for the complete reflected field,

$$\psi_r(x, z = d) = \frac{1}{(2\pi)^{1/2}} \int_{-\infty}^{+\infty} \Gamma(k_x)\, \Phi(k_x) \exp[-i(k_x x - k_z d)]\, dk_x \tag{11.3-10}$$

The incident beam is made up of plane waves, some of which are incident at an angle less and some of which are incident at an angle greater than the critical angle. Now, if the angle θ_i (as shown in Fig. 11.1a) is greater than the critical angle and the beam width is large in comparison to the wavelength, then only those plane waves will predominantly contribute that propagate at angles approximately equal to θ_i. This essentially implies that $\Phi(k_x)$ is sharply peaked around $k_x = k_1 \sin\theta_i$ [see Eq. (11.3-7)]. We may then assume $\Gamma(k_x)$ to be given by Eq. (11.2-27) in the entire region of integration in Eq. (11.3-10), and we may also make a Taylor expansion of $\chi(k_x)$ about $k_x = k_1 \sin\theta_i$ and retain terms up to first order (cf. Problem 11.2) to obtain

$$\chi(k_x) \simeq \chi_0 + \chi_0' \bar{k}_x \tag{11.3-11}$$

where

$$\chi_0 = \chi(k_x = k_1 \sin\theta_i), \qquad \bar{k}_x = k_x - k_1 \sin\theta_i, \qquad \chi_0' = d\chi_0/dk_x|_{\bar{k}_x = 0}$$

Thus Eq. (11.3-10) becomes

$$\begin{aligned}\psi_r(x) = \frac{1}{(2\pi)^{1/2}} \exp[-i(\chi_0 + k_1 x \sin\theta_i)] \int_{-\infty}^{+\infty} \Phi(k_x) \exp[-i(x + \chi_0')\bar{k}_x] \\ \times \exp\{id[k_1^2 - (\bar{k}_x + k_1 \sin\theta_i)^2]^{1/2}\}\, d\bar{k}_x\end{aligned} \tag{11.3-12}$$

On the other hand, under geometrical considerations, we should have expected the reflected beam to be of the form (dashed lines of Fig. 11.1a)

$$\begin{aligned}\psi_{r_0}(x) = \frac{1}{(2\pi)^{1/2}} \exp[-i(\chi_0 + k_1 x \sin\theta_i)] \int_{-\infty}^{\infty} \Phi(k_x) \\ \times \exp(-i\{\bar{k}_x x - d[k_1^2 - (\bar{k}_x + k_1 \sin\theta_i)^2]^{1/2}\})\, d\bar{k}_x\end{aligned} \tag{11.3-13}$$

which shows that the beam suffers a phase shift of $\exp(-i\chi_0)$ and emerges from the point of incidence. A comparison of Eqs. (11.3-12) and (11.3-13) shows that the beam undergoes a lateral shift in the x direction by an amount

$$\Delta = -\left.\frac{\partial\chi}{\partial k_x}\right|_{k_x = k_1 \sin\theta_i} \tag{11.3-14}$$

This is the Goos–Hänchen shift suffered by a beam when it undergoes total internal reflection. The beam displacement is shown in Fig. 11.1a. Notice that this shift is in the plane of incidence. To obtain an explicit expression for the beam shift, we use Eq. (11.2-28) and find

$$\Delta = \frac{\lambda_1 m(n^2 - 1)}{\pi n^2} \frac{\tan\theta_i}{(\sin^2\theta_i - \sin^2\theta_c)^{1/2}[\cos^2\theta_i + m^2(\sin^2\theta_i - \sin^2\theta_c)]} \tag{11.3-15}$$

where $\lambda_1 = \lambda_0/n_1$; specifically, for polarization parallel to or normal to the plane of incidence, we get

$$\Delta_{\parallel} = \frac{\lambda_1}{\pi}(n^2 - 1)\frac{\tan\theta_i}{(\sin^2\theta_i - \sin^2\theta_c)^{1/2}\left[\cos^2\theta_i + n^4(\sin^2\theta_i - \sin^2\theta_c)\right]} \tag{11.3-16}$$

$$\Delta_{\perp} = \frac{\lambda_1}{\pi}\frac{\tan\theta_i}{(\sin^2\theta_i - \sin^2\theta_c)^{1/2}} \tag{11.3-17}$$

For angles of incidence near the critical angle, $\sin\theta_i \simeq \sin\theta_c$, and Eq. (11.3-16) reduces to

$$\Delta_{\parallel} = \frac{\lambda_1 n^2}{\pi}\frac{\tan\theta_i}{(\sin^2\theta_i - \sin^2\theta_c)^{1/2}} \tag{11.3-18}$$

It is interesting to note from the above equations that the shift is independent of the width of the beam. From Eqs. (11.3-16) and (11.3-17) it is clear that, as the angle of incidence nears the critical angle, the shift becomes larger and larger. It should be pointed out that our analysis (and hence the above expressions for $\Delta_{\parallel}$ and $\Delta_{\perp}$) are not valid very close to the critical angle.* It is also evident from Eqs. (11.3-17) and (11.3-18) that $\Delta_{\parallel} > \Delta_{\perp}$. Since any arbitrary polarization can be resolved into two linear perpendicular polarizations, one in the plane of incidence and the other perpendicular to it, when an unpolarized beam is incident on a surface, the components parallel to the plane of incidence shift by an amount $\Delta_{\parallel}$ and the components normal to it shift by an amount $\Delta_{\perp}$. If the same beam is allowed to get totally reflected many times, the difference $(\Delta_{\parallel} - \Delta_{\perp})$ gets magnified and one expects two beams in place of one. Such an experiment was performed by Mazet *et al.* (1971).

Problem 11.2. In the Taylor expansion of the phase shift $\chi(k_x)$ we retained terms up to the first order [see Eq. (11.3-11)]. Show that if we retain the second-order terms in the expansion of $\chi(k_x)$, i.e., if we assume

$$\chi(k_x) = \chi_0 + \chi_0'\bar{k}_x + \frac{1}{2}\frac{\partial^2\chi}{\partial k_x^2}\bigg|_{\bar{k}_x = 0}\bar{k}_x^2$$

then the beam undergoes a distortion in addition to the shift. In particular, consider a Gaussian beam of the form

$$\psi(x, z = d) = A\exp(-x^2/\alpha^2)\exp(-ik_1 x\sin\theta_i) \tag{11.3-19}$$

* This is because of the assumption that only those plane waves that are incident at $\theta_i > \theta_c$ contribute and hence $|\Gamma(k_x)| = 1$. But when θ_i is very near to θ_c, one has contributions even from plane waves that are incident at $\theta_i < \theta_c$; for these plane waves $|\Gamma(k_x)| \neq 1$.

and show that the reflected beam is given by

$$\psi_r(x, z = d) = \frac{A \exp[-i(\chi_0 + k_1 x \sin\theta_i)]}{(1 + 2i\chi_0''/\alpha^2)^{1/2}} \exp\left[-\frac{(x + \chi_0')^2}{\alpha^2(1 + 4\chi_0''^2/\alpha^4)}\right]$$

$$\times \exp\left[\frac{2i\chi_0''(x + \chi_0')^2}{(\alpha^4 + 4\chi_0''^2)}\right] \qquad (11.3\text{-}20)$$

where

$$\chi_0'' = \left.\frac{\partial^2 \chi}{\partial k_x^2}\right|_{k_x = k_1 \sin\theta_i}$$

Hint: From Eq. (11.3-9) we get

$$\Phi(k_x)\exp(-ik_z d) = \frac{1}{(2\pi)^{1/2}} \int_{-\infty}^{+\infty} \psi(x, z = d)\exp(ik_x x)\, dx$$

which using Eq. (11.3-19) gives us

$$\Phi(k_x)\exp(-ik_z d) = \frac{\alpha A}{2^{1/2}} \exp\left[-\frac{\alpha^2(k_x - k_1 \sin\theta_i)^2}{4}\right] \qquad (11.3\text{-}21)$$

Using the second-order approximation, we may write for the reflected field

$$\psi_r(x, z = d) = \frac{1}{(2\pi)^{1/2}} \int_{-\infty}^{+\infty} \Gamma(k_x)\Phi(k_x)\exp[-i(k_x x + k_z d)\, dk_x$$

$$= \frac{A\alpha}{2\pi^{1/2}} \exp(-ik_1 x \sin\theta_i) \int_{-\infty}^{+\infty} \exp\left[-\frac{\alpha^2}{4}\bar{k}_x^2 - i\chi_0 - i\chi_0'\bar{k}_x\right.$$

$$\left. - i\chi_0''\frac{\bar{k}_x^2}{2} - i\bar{k}_x x\right] dk_x \qquad (11.3\text{-}22)$$

which on integration gives Eq. (11.3-20). From Eq. (11.3-20) the following observations can be made:

(1) The center of the totally reflected beam is shifted by an amount $-\chi_0'$; this is the Goos–Hänchen shift.

(2) The beam width that was α for the incident beam becomes

$$\alpha_r = \alpha(1 + 4\chi_0''^2/\alpha^4)^{1/2} \qquad (11.3\text{-}23)$$

Thus χ_0'' determines the distortion in the beam. This distortion has been recently measured in an experiment conducted with ultrasonic waves (see Breazeale *et al.*, 1974). Notice that since sound waves also satisfy a scalar wave equation, one also obtains exactly similar shifts in acoustics.

(3) It is also evident that the phase fronts that were plane before reflection have been modified. Under first-order approximation, when χ_0'' can be neglected, the totally reflected phase fronts are also plane. Thus the effect of the term χ_0'' is to distort both the beam and the phase fronts.

Problem 11.3. Consider the total internal reflection of a pulse whose time dependence is of the form

$$f(t) = \exp(-t^2/\tau^2)\, e^{i\omega_c t} \qquad (11.3\text{-}24)$$

where ω_c represents the carrier frequency of the wave and τ represents the width of the pulse. The function $\exp(-t^2/\tau^2)$ determines the envelope of the pulse. The beam may be assumed to have infinite spatial extent.

(a) Following an exactly similar analysis as done for a bounded beam, show that to a first order of approximation the pulse suffers a shift in time given by

$$\Delta_t = \left.\frac{\partial \chi}{\partial \omega}\right|_{\omega = \omega_c} \tag{11.3-25}$$

(b) To a second order of approximation, show that the pulse suffers a distortion also.

Solution. If we Fourier analyze the function $f(t)$,

$$f(t) = \frac{1}{(2\pi)^{1/2}} \int_{-\infty}^{+\infty} F(\omega)\, e^{i\omega t}\, d\omega \tag{11.3-26}$$

then for $f(t)$ given by Eq. (11.2-24), we obtain

$$F(\omega) = \frac{\tau}{2^{1/2}} \exp\left[-\frac{\tau^2(\omega - \omega_c)^2}{4}\right] \tag{11.3-27}$$

Since $\tau \gg 1/\omega_c$, the function $F(\omega)$ is very sharply peaked about $\omega = \omega_c$. For example, for a 1 nsec optical pulse ($\omega_c \simeq 10^{15}$ sec^{-1}), $\omega_c \tau \simeq 10^6$. The reflected pulse is of the form

$$f_r(t) = \frac{1}{(2\pi)^{1/2}} \int_{-\infty}^{+\infty} \Gamma(\omega)\, F(\omega)\, e^{i\omega t}\, d\omega \tag{11.3-28}$$

where $\Gamma(\omega)$ is the reflection coefficient; its dependence on ω has been explicitly indicated. Since $\Gamma(\omega) = \exp[-i\chi(\omega)]$, we can, in light of the above discussions, Taylor expand $\chi(\omega)$ about $\omega = \omega_c$

$$\chi(\omega) = \chi^0 + \chi'\bar{\omega} + \tfrac{1}{2}\chi''\bar{\omega}^2 + \cdots \tag{11.3-29}$$

where

$$\bar{\omega} = \omega - \omega_c, \qquad \chi^0 = \chi(\omega_c), \qquad \chi' = \left.\frac{\partial \chi}{\partial \omega}\right|_{\bar{\omega}=0}, \qquad \chi'' = \left.\frac{\partial^2 \chi}{\partial \omega^2}\right|_{\bar{\omega}=0}$$

On substituting $\Gamma(\omega)$ under this approximation into Eq. (11.3-28) and using Eq. (11.3-27), we obtain

$$f_r(t) = \frac{\tau}{(\tau^2 + 2i\chi'')^{1/2}} \exp\left\{-i\left[\chi^0 - \omega_c t + \frac{2(t - \chi')^2 \chi''}{(\tau^4 + 4\chi''^2)}\right]\right\}$$

$$\times \exp\left[-\frac{(t - \chi')^2 \tau^2}{(\tau^4 + 4\chi''^2)}\right] \tag{11.3-30}$$

The reflected pulse is thus shifted in time by an amount $\Delta_t = \chi' = \partial\chi/\partial\omega|_{\omega=\omega_c}$. If we neglect the term depending on χ'', then there is no distortion of the pulse.

11.4. Physical Understanding of the Goos–Hänchen Shift

Since the reflection coefficient of a plane wave is known, the incident beam was expressed as a superposition of plane waves of varying amplitudes and each of the component plane waves was made to reflect from the boundary with the proper reflection coefficient. Since on reflection, each plane wave suffers different amounts of phase shifts, when the reflected beam is formed by adding the reflected plane-wave components, a lateral shift occurs in the reflected beam. However, from the above analysis, it is not evident how the energy redistribution takes place to produce a shifted beam. In order to understand the energy redistribution, Lotsch (1968) made an explicit calculation of the Poynting vector, which gives the direction of energy flow, and showed that a part of the beam indeed enters the rarer medium and reemerges at another point to produce the observed shift. He used the pictorial representation of Schaefer and Pich (1937) to understand the physics involved. The representation is shown in Fig. 11.4. *ABCD* represents the incident beam, which is homogeneous in the region *BC* and dies off to zero from *B* to *A* and *C* to *D*. Lotsch (1968) showed using explicit calculations that for values of $z \simeq \lambda$ in the rarer medium, the Poynting vector in the regions *PQ*, *QR*, and *RS* is directed toward the rarer medium, parallel to the surface, and away from the rarer medium, respectively (see Fig. 11.4). Thus energy seems to be entering the rarer medium from the leading portion of the beam and reemerging from the rarer medium in the trailing portion after reflection.

This phenomenon of the energy entering and leaving the rarer medium can be understood as follows. The boundary conditions imposed on the fields require that there be an evanescent field in the rarer medium, which decays exponentially away from the boundary (see Section 11.2). When the leading portion of the beam strikes the interface, since there is no

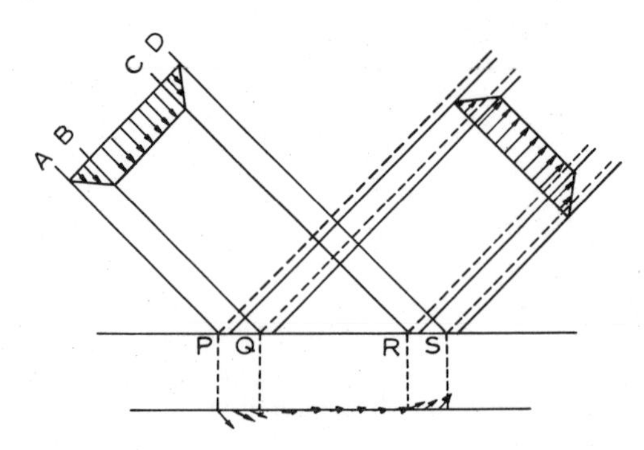

Fig. 11.4. The Poynting vector direction in different parts of a beam *ABCD*. In the region *PQ*, energy enters the rarer medium; in the region *QR*, energy travels parallel to the surface; and in the region *RS*, energy travels out of the rarer medium (shown by arrows in the lower part of the figure).

evanescent field, it enters the rarer medium to set up one. Because of this, the beam undergoes a displacement. An expression for the beam shift from such an analysis was obtained by Renard (1964), who calculated the distance over which the energy incident equals the energy in the evanescent wave and showed that this distance is indeed the required Goos–Hänchen shift.

Problem 11.4. Consider the quantum-mechanical reflection of a particle of mass m from a potential barrier of the following type:

$$V = \begin{cases} V_1, & z < 0 \\ V_2, & z > 0 \end{cases} \tag{11.4-1}$$

with $V_2 > V_1$. Solve the Schrödinger equation and show that there is a Goos–Hänchen shift in the reflected wave. *Reference*: Renard (1964).

Solution: The solution of the Schrödinger equation

$$\nabla^2 \psi + \frac{2m}{\hbar^2}[E - V(z)]\psi = 0 \tag{11.4-2}$$

which corresponds to the reflection and transmission of the particle, is given by*

$$\psi = \begin{cases} A \exp[i(k_{1x}x + k_{1z}z)] + B \exp[i(k_{2x}x - k_{2z}z)] & \text{for} \quad z < 0 \\ C \exp[i(k_{3x}x + k_{3z}z)] & \text{for} \quad z > 0 \end{cases} \tag{11.4-3}$$

where without any loss of generality we have assumed the x axis to lie in the plane of incidence and therefore set $k_y = 0$. Furthermore,

$$k_{1x}^2 + k_{1z}^2 = k_{2x}^2 + k_{2z}^2 = \frac{2m}{\hbar^2}(E - V_1) \tag{11.4-4}$$

$$k_{3x}^2 + k_{3z}^2 = \frac{2m}{\hbar^2}(E - V_2) \tag{11.4-5}$$

The wave function has to be continuous at $z = 0$, for all values of x. As such, $k_{1x} = k_{2x} = k_{3x}$ [cf. Eq. (11.2-5)]. If i_1 and i_2 represent the angles of incidence and refraction, respectively, then one immediately obtains

$$\frac{\sin i_1}{\sin i_2} = \left(\frac{E - V_2}{E - V_1}\right)^{1/2} \tag{11.4-6}$$

The critical angle i_c will therefore be given by

$$\sin i_c = \left(\frac{E - V_2}{E - V_1}\right)^{1/2} \tag{11.4-7}$$

* It should be pointed out that in quantum mechanics the time dependence is chosen to be of the form $\exp(-i\omega t)$.

Continuity of ψ and $\partial\psi/\partial z$ at $z = 0$ yields

$$\frac{B}{A} = \frac{k_{1z} - k_{3z}}{k_{1z} - k_{3z}} = \frac{\cos i_1 - (\sin^2 i_c - \sin^2 i_1)^{1/2}}{\cos i_1 + (\sin^2 i_c - \sin^2 i_1)^{1/2}} \tag{11.4-8}$$

$$\frac{C}{A} = \frac{2\cos i_1}{\cos i_1 + (\sin^2 i_c - \sin^2 i_1)^{1/2}} \tag{11.4-9}$$

For $i_1 > i_c$, $|B/A| = 1$, implying the total reflection of the particle. Now, the particle current density is given by

$$\mathbf{J} = \frac{\hbar}{2mi}[\psi^*\nabla\psi - \psi\nabla\psi^*] \tag{11.4-10}$$

For $i_1 > i_c$, the field in the region $z > 0$ is given by

$$\psi = C\exp(i\kappa_2 x \sin i_2)\exp\left[-\kappa_2 z\frac{(\sin^2 i_1 - \sin^2 i_c)^{1/2}}{\sin i_c}\right] \tag{11.4-11}$$

where $\kappa_2 = [2m(E - V_2)]^{1/2}/\hbar$, and since $i_1 > i_c$, the wave decays exponentially in the z direction. It can easily be seen that in the region $z > 0$

$$J_x = \frac{\hbar\kappa_2}{m}\frac{\sin i_1}{\sin i_c}C^*C\exp\left[-2\kappa_2\frac{(\sin^2 i_1 - \sin^2 i_c)^{1/2}}{\sin i_c}z\right] \tag{11.4-12}$$

and $J_y = J_z = 0$. Thus the flux of particles contained in this wave is (considering a unit length in the y direction)

$$F_{\text{ev}} = \int_0^\infty J_x\,dz = \frac{\hbar}{2m}\frac{\sin i_1}{(\sin^2 i_1 - \sin^2 i_c)^{1/2}}C^*C \tag{11.4-13}$$

This flux must have come from the incident beam. The incident particle current density along the z direction is given by

$$J_z = \frac{\hbar\kappa_1}{m}\cos i_1\; A^*A \tag{11.4-14}$$

where $\kappa_1 = [2m(E - V_1)]^{1/2}/\hbar$. Thus the incident flux in a length d must equal F_{ev}; d then represents the shift. Thus

$$d = \frac{J_z}{F_{\text{ev}}} = \frac{\lambda_1}{\pi}\frac{\tan i_1\cos^2 i_1}{1 - \sin^2 i_c}\frac{1}{(\sin^2 i_1 - \sin^2 i_c)^{1/2}} \tag{11.4-15}$$

where λ_1 is the de Broglie wavelength of the particle, defined by

$$\lambda_1 = \frac{h}{[2m(E - V_1)]^{1/2}} = \frac{2\pi}{\kappa_1} \tag{11.4-16}$$

For $i_1 \simeq i_c$, we obtain [cf. Eq. (11.3-17)]

$$d \simeq \frac{\lambda_1}{\pi}\frac{\tan i_1}{(\sin^2 i_1 - \sin^2 i_c)^{1/2}} \tag{11.4-17}$$

11.5. The Goos–Hänchen Effect in a Planar Waveguide

A planar waveguide consists of a film of refractive index n_1 surrounded by a medium of refractive index n_2 such that $n_1 > n_2$ (see Fig. 11.5a). If rays are incident inside the film at an angle greater than the critical angle $\theta_c[= \sin^{-1}(n_2/n_1)]$ they would undergo total internal reflection at both the boundaries and would thus be guided by the film. Such waveguides are of great importance in integrated optics. The modes in such waveguides can be thought of as a pair of plane waves getting reflected from the two boundaries.* We will show by an explicit evaluation that the wave analysis and the geometrical optics analysis yield the same group velocity only when the Goos–Hänchen shift and the associated time delays are taken into account. We follow the treatment of Kogelnik and Weber (1974).

Since a mode has a stationary transverse pattern of intensity, standing waves must be formed in the transverse direction. If the plane-wave component is traveling at an angle θ with the normal to the boundaries, then the total phase shift suffered by the wave must be $2\nu\pi$, where ν is an integer, i.e.,

$$2n_1kd\cos\theta + 2\chi_{12} = 2\nu\pi \tag{11.5-1}$$

where d is the thickness of the film and χ_{12} is the phase shift suffered by the wave at the boundary between n_1 and n_2 (see also Problem 11.5). The propagation constant β along the guide axis is given by

$$\beta = n_1 k \sin\theta \tag{11.5-2}$$

We will now determine the group velocity $d\omega/d\beta$. Differentiating Eq. (11.5-1) with respect to β, we get

$$\frac{d\omega}{d\beta}\left(2\frac{\partial\chi_{12}}{\partial\omega} + 2kd\cos\theta\frac{dn_1}{d\omega} + 2\frac{n_1 d}{c}\cos\theta\right) + 2\frac{\partial\chi_{12}}{\partial\beta} - 2n_1kd\sin\theta\frac{d\theta}{d\beta} = 0 \tag{11.5-3}$$

where we have used

$$\frac{d\chi_{12}}{d\beta} = \frac{\partial\chi_{12}}{\partial\beta} + \frac{\partial\chi_{12}}{\partial\omega}\frac{d\omega}{d\beta} \tag{11.5-4}$$

* For a modal analysis of such waveguides see, e.g., Kapany and Burke (1972), Marcuse (1974), and Sodha and Ghatak (1977).

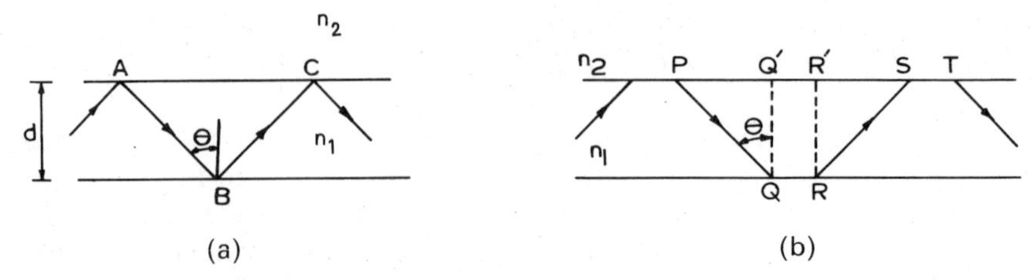

Fig. 11.5. (a) A ray undergoing total internal reflection at the two boundaries of an optical waveguide when the Goos–Hänchen shifts are not considered. (b) A ray in total internal reflection when the Goos–Hänchen shifts (QR and ST) are also taken into account.

Differentiating Eq. (11.5-2) with respect to β we get

$$1 = n_1 k \cos\theta \frac{d\theta}{d\beta} + \frac{d\omega}{d\beta}\frac{m_1 \sin\theta}{c} \tag{11.5-5}$$

where

$$m_1 = n_1 + \omega \frac{dn_1}{d\omega} \tag{11.5-6}$$

represents the effective refractive index of the medium in presence of dispersion. Eliminating $d\theta/d\beta$ from Eqs. (11.5-3) and (11.5-5) we finally obtain

$$\frac{d\omega}{d\beta} = \frac{2d \tan\theta - 2\,\partial\chi_{12}/\partial\beta}{2m_1 d/c\cos\theta + 2\,\partial\chi_{12}/\partial\omega} \tag{11.5-7}$$

Let us first employ the geometrical consideration without the Goos–Hänchen shifts and determine whether the same group velocity would be obtained. The time taken to travel a distance $AC = 2d\tan\theta$ along the guide axis is $2m_1 d/c\cos\theta$, since m_1 represents the effective index (see Fig. 11.5a). Thus the group velocity along the axis is

$$v_g = \frac{2d\tan\theta}{2m_1 d/c\cos\theta} = \frac{c}{m_1}\sin\theta \tag{11.5-8}$$

which does not agree with Eq. (11.5-7). On the other hand if the Goos–Hänchen shifts QR and ST (see Fig. 11.5b) and the associated time delays are taken into account, then the time taken to travel a distance PT will be given by

$$\tau = \frac{2m_1 d}{c\cos\theta} + 2\Delta_T = \frac{2m_1 d}{c\cos\theta} + 2\frac{\partial\chi_{12}}{\partial\omega} \tag{11.5-9}$$

where we have used Eq. (11.3-25) for the time shift. Thus, the group velocity along the guide axis is given by

$$v_g = \frac{PQ' + Q'R' + R'S + ST}{\tau} = \frac{2d \tan\theta + 2\Delta_L}{\tau}$$

$$= \frac{2d \tan\theta - 2\,\partial\chi_{12}/\partial\beta}{2m_1 d/c \cos\theta + 2\,\partial\chi_{12}/\partial\omega} \tag{11.5-10}$$

which agrees exactly with Eq. (11.5-7). Thus it is important to consider the Goos–Hänchen effect in waveguide propagation.

Problem 11.5. Consider a simple planar waveguide in which the refractive-index variation is of the form (see Fig. 11.5a)

$$n = \begin{cases} n_1 & \text{for } |z| < d/2 \\ n_2 & \text{for } |z| > d/2 \end{cases} \tag{11.5-11}$$

Carry out a rigorous modal analysis and show that the propagation constants are indeed given by Eq. (11.5-2), where θ is determined by Eq. (11.5-1).

Solution. Let us consider the TE modes* in which $E_x = E_z = 0$. Furthermore, as discussed in Problem 10.2, we may neglect the y dependence of the fields. Thus, for modes propagating along the x direction we may write [see Eq. (10.3-37)]

$$\mathbf{E} = \hat{\mathbf{y}} E_y(x, z) = \hat{\mathbf{y}} E_y(z) \exp[i(\omega t - \beta x)] \tag{11.5-12}$$

Since $\nabla^2 E_y = \varepsilon\mu\,\partial^2 E_y/\partial t^2$, we have

$$\frac{d^2 E_y}{dz^2} - \beta^2 E_y(z) = \varepsilon\mu(-\omega^2) E_y = -n^2 \frac{\omega^2}{c^2} E_y \tag{11.5-13}$$

where $n = c(\varepsilon\mu_0)^{1/2}$ represents the refractive index. Thus, in the two regions E_y satisfies the equations

$$\frac{d^2 E_y}{dz^2} + \left(\frac{\omega^2}{c^2} n_1^2 - \beta^2\right) E_y = 0 \qquad \text{for } |z| < d/2$$

$$\frac{d^2 E_y}{dz^2} + \left(\frac{\omega^2}{c^2} n_2^2 - \beta^2\right) E_y = 0 \qquad \text{for } |z| > d/2 \tag{11.5-14}$$

For the guided modes, the fields must vanish as $|z| \to \infty$. This will happen if

$$\beta^2 < \frac{\omega^2}{c^2} n_2^2 \tag{11.5-15}$$

* TE and TM modes for planar waveguides have been discussed in Problem 10.2. It may be mentioned that in Section 10.3 the propagation was along the z axis and the refractive-index variation was along the x axis; here the propagation is along the x axis and the refractive-index variation is along the z axis.

For $\beta^2 > (\omega^2/c^2)\, n_2^2$ the fields will be oscillatory even in the region $|z| > d/2$; these modes are known as radiation modes. Thus we introduce the following two parameters,

$$\sigma^2 = \frac{\omega^2}{c^2} n_1^2 - \beta^2, \qquad \kappa^2 = \beta^2 - \frac{\omega^2}{c^2} n_2^2 \tag{11.5-16}$$

where σ and κ are real quantities.* Thus Eq. (11.5-14) can be written in the form

$$\frac{d^2 E_y}{dz^2} + \sigma^2 E_y = 0 \qquad \text{for } |z| < d/2 \tag{11.5-17}$$

$$\frac{d^2 E_y}{dz^2} - \kappa^2 E_y = 0 \qquad \text{for } |z| > d/2 \tag{11.5-18}$$

The solutions of the above equations can be classified under two categories: one in which the fields are symmetric [i.e., $E_y(z) = E_y(-z)$] and one in which the fields are antisymmetric [i.e., $E_y(-z) = -E_y(z)$]; these are known as symmetric and antisymmetric TE modes, respectively. We consider only symmetric TE modes in detail. For symmetric TE modes the solution of Eqs. (11.5-17) and (11.5-18) are of the form

$$E_y = \begin{cases} A\cos(\sigma z) & \text{for } |z| < d/2 \\ A\cos\left(\dfrac{\sigma d}{2}\right)\exp[-\kappa(|z| - d/2)] & \text{for } |z| > d/2 \end{cases} \tag{11.5-19}$$

where we have taken into account the continuity of E_y (which represents a tangential component) at $z = \pm d/2$. From Eq. (11.2-14) we have

$$\mu_0 \frac{\partial H_z}{\partial t} = -(\nabla \times \mathbf{E})_z = -\left(\frac{\partial E_y}{\partial x} - \frac{\partial E_x}{\partial y}\right) = -\frac{\partial E_y}{\partial x}$$

Thus

$$H_z = \frac{i}{\omega\mu_0}\frac{\partial E_y}{\partial x} \tag{11.5-20}$$

Hence for nonmagnetic media, continuity of H_z implies continuity of $\partial E_y/\partial x$; consequently,

$$-A\sigma \sin\left(\frac{\sigma d}{2}\right) = -\kappa A \cos\left(\frac{\sigma d}{2}\right)$$

* It can be shown that β^2 can never be greater than $(\omega^2/c^2)\, n_1^2$, for otherwise the boundary conditions cannot be satisfied.

or

$$\tan\left(\frac{\sigma d}{2}\right) = \frac{\kappa}{\sigma}$$

or

$$\tan\left[\frac{d}{2}\left(n_1^2\frac{\omega^2}{c^2} - \beta^2\right)^{1/2}\right] = \frac{(\beta^2 - n_2^2\omega^2/c^2)^{1/2}}{(n_1^2\omega^2/c^2 - \beta^2)^{1/2}} \tag{11.5-21}$$

Similarly, for antisymmetric TE modes, we obtain

$$-\cot\left[\frac{d}{2}\left(n_1^2\frac{\omega^2}{c^2} - \beta^2\right)^{1/2}\right] = \frac{(\beta^2 - n_2^2\omega^2/c^2)^{1/2}}{(n_1^2\omega^2/c^2 - \beta^2)^{1/2}} \tag{11.5-22}$$

Equations (11.5-21) and (11.5-22) determine the allowed values of β.

Now if we substitute for θ from Eq. (11.5-2) into Eq. (11.5-1) we obtain

$$2d(n_1^2k^2 - \beta^2)^{1/2} - 4\tan^{-1}\left[m\frac{(k_x^2 - k_2^2)^{1/2}}{(k_1^2 - k_x^2)^{1/2}}\right] = 2\nu\pi$$

where we have substituted for χ_{12} from Eq. (11.2-28). Thus

$$-\tan\left[\frac{\nu\pi}{2} - \frac{d}{2}(n_1^2k^2 - \beta^2)^{1/2}\right] = m\frac{(k_x^2 - k_2^2)^{1/2}}{(k_1^2 - k_x^2)^{1/2}} = m\frac{(\beta^2 - n_2^2\omega^2/c^2)^{1/2}}{(n_1^2\omega^2/c^2 - \beta^2)^{1/2}} \tag{11.5-23}$$

where we have used Eq. (11.2-7) and the fact that $\beta = k_x$. For TE modes we must choose $m = 1$; then Eq. (11.5-23) becomes identical to Eqs. (11.5-21) and (11.5-22) for even and odd values of ν. In a similar manner, one can carry out the analysis for TM modes and show that the propagation constants are determined by Eq. (11.5-23) for $m = n^2 = n_1^2/n_2^2$.

11.6. Prism–Film Coupler

Planar waveguides form one of the basic components of integrated optical devices. A typical problem that one encounters in dealing with planar waveguides is the coupling of the energy from a laser beam into the waveguide. The method in which one shines the laser beam into one end of the planar waveguide has associated problems—like critical alignment and scattering due to irregularities at the end of the film—that reduce the overall efficiency of the coupling process. The prism–film coupler, which makes use of the evanescent waves that exist near the surface from which an incident radiation undergoes total internal reflection, overcomes some of the difficulties enumerated above and provides an efficient means of coupling laser beams into planar waveguides. Figure 11.6 shows the basic

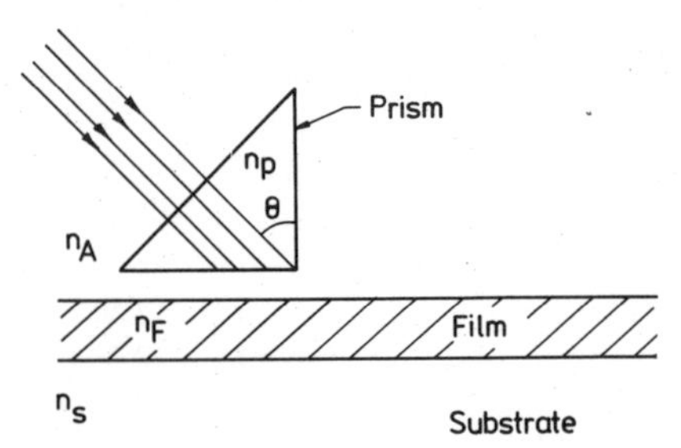

Fig. 11.6. The prism–film coupler arrangement.

arrangement of a prism–film coupler. The incident radiation undergoes total internal reflection at the prism–air space boundary, thus producing an evanescent wave that decays exponentially from the prism–air surface into the planar waveguide. The air space between the prism base and the film is $\sim\lambda/4$. If θ represents the angle of incidence of the laser beam inside the prism (of refractive index n_p), then the propagation constant along the direction parallel to the film will be $kn_p \sin\theta$. Now, a wave analysis of the planar waveguide shows that the waveguide possesses guided modes with discrete propagation constants β. Detailed theory shows that there is a maximum power transfer exclusively to the waveguide mode (characterized by the propagation constant β_1) if the following condition is satisfied (see, e.g., Barnoski, 1974; Midwinter, 1970):

$$\beta_1 = n_p k \sin\theta \tag{11.6-1}$$

Thus by choosing different values of θ, different modes can be excited. Efficiencies of about 80% are attainable with such a configuration. Notice that since the ray paths are reversible, one can use the same configuration for coupling out of energy from the film.

Appendix

A. The Dirac Delta Function

The Dirac delta function is defined through the equations

$$\delta(x - a) = 0 \qquad \text{for } x \neq a \tag{A-1}$$

$$\int_{a-\alpha}^{a+\alpha} \delta(x - a)\, dx = 1, \qquad \alpha > 0 \tag{A-2}$$

Thus at $x = a$, the delta function has an infinite value. For an arbitrary function $f(x)$ that is continuous at $x = a$, one obtains

$$\int_{a-\alpha}^{a+\alpha} f(x)\, \delta(x - a)\, dx = f(a) \tag{A-3}$$

There are many representations of the delta function, e.g., the Gaussian function

$$G_\sigma(x) = \frac{1}{\sigma(2\pi)^{1/2}} \exp\left[-\frac{(x - a)^2}{2\sigma^2} \right] \tag{A-4}$$

has unit area, i.e.,

$$\int_{-\infty}^{\infty} G_\sigma(x)\, dx = 1 \tag{A-5}$$

irrespective of the value of σ. Further, the function $G_\sigma(x)$ has a value of $1/\sigma(2\pi)^{1/2}$ at $x = 0$ and has a width $\sim\sigma$. Thus, in the limit of $\sigma \to 0$ it has all the properties of the delta function (see Fig. A.1) and we may write

$$\delta(x - a) = \lim_{\sigma \to 0} \frac{1}{\sigma(2\pi)^{1/2}} \exp\left[-\frac{(x - a)^2}{2\sigma^2} \right] \tag{A-6}$$

Similarly, one can show that for a large value of g the function $\sin g(x - a)/\pi(x - a)$ is sharply peaked around $x = a$ (see Fig. 4.6a) and has a unit

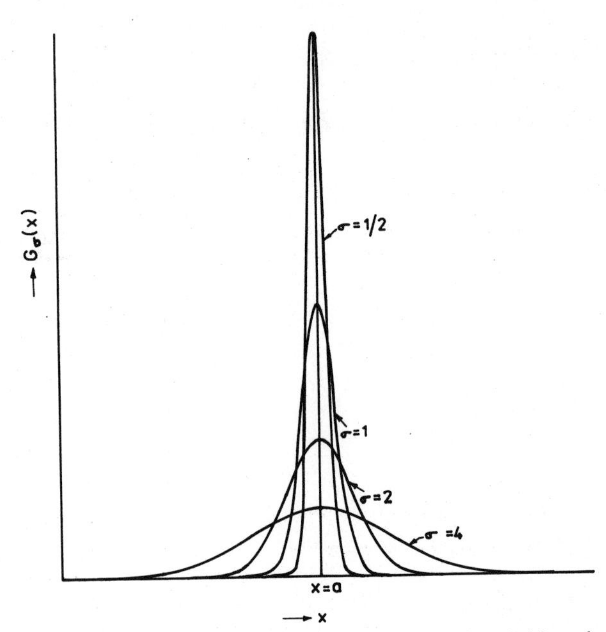

Fig. A.1. In the limit of $\sigma \to 0$, the Gaussian function $G_\sigma(x)$ tends to a delta function.

area irrespective of the value of g. Thus, we may write

$$\delta(x - a) = \lim_{g \to \infty} \frac{\sin g(x - a)}{\pi(x - a)} \tag{A-7}$$

Since

$$\frac{1}{2\pi} \int_{-g}^{g} e^{\pm ik(x-a)}\, dk = \frac{\sin g(x - a)}{\pi(x - a)} \tag{A-8}$$

we obtain another representation for the delta function,

$$\delta(x - a) = \frac{1}{2\pi} \int_{-\infty}^{\infty} e^{\pm ik(x-a)}\, dk = \int_{-\infty}^{\infty} e^{\pm 2\pi iu(x-a)}\, du \tag{A-9}$$

where $u \equiv k/2\pi$. It can easily be seen from Eqs. (A-1)–(A-3) that

$$\delta(x - a) = \delta(a - x) \tag{A-10}$$

and

$$\delta[p(x - a)] = \frac{1}{|p|} \delta(x - a) \tag{A-11}$$

B. The Fourier Transform

From the definition of the delta function, one has (see Appendix A)

$$f(x') = \int_{-\infty}^{\infty} f(x)\,\delta(x - x')\,dx = \int_{-\infty}^{\infty}\int_{-\infty}^{\infty} f(x)\,e^{2\pi iu(x-x')}\,du\,dx \qquad \text{(B-1)}$$

which is known as Fourier's integral theorem. Thus, if we write

$$F(u) = \int_{-\infty}^{\infty} f(x)\,e^{2\pi iux}\,dx = \mathscr{F}[f(x)] \qquad \text{(B-2)}$$

Then

$$f(x) = \int_{-\infty}^{\infty} F(u)\,e^{-2\pi iux}\,du \qquad \text{(B-3)}$$

The function $F(u)$ is known as the Fourier transform of the function $f(x)$ and Eq. (B-3) enables us to calculate the original function from the Fourier transform. In addition,

$$\begin{aligned}\mathscr{F}[f(x-a)] &= \int_{-\infty}^{\infty} f(x-a)\,e^{2\pi iux}\,dx = e^{2\pi iau}\int_{-\infty}^{\infty} f(y)\,e^{2\pi iuy}\,dy \\ &= F(u)\,e^{2\pi iau} \qquad \text{(B-4)}\end{aligned}$$

$$\begin{aligned}\mathscr{F}\{\mathscr{F}[f(x)]\} &= \int_{-\infty}^{\infty} du\,e^{2\pi iux}\int_{-\infty}^{\infty} dx' f(x')\,e^{2\pi iux'} \\ &= \int_{-\infty}^{\infty} f(x')\,dx'\;\delta(x + x') = f(-x) \qquad \text{(B-5)}\end{aligned}$$

As an example, we calculate the Fourier transform of the rectangle function (see Fig. B.1a):

$$\operatorname{rect}\left(\frac{x}{a}\right) = \begin{cases} 1, & |x| < \tfrac{1}{2}a \\ 0, & |x| > \tfrac{1}{2}a \end{cases} \qquad \text{(B-6)}$$

Its Fourier transform will be

$$\begin{aligned}\mathscr{F}\left[\operatorname{rect}\left(\frac{x}{a}\right)\right] &= \int_{-\infty}^{\infty} \operatorname{rect}\left(\frac{x}{a}\right) e^{2\pi iux}\,dx = \int_{-a/2}^{a/2} e^{2\pi iux}\,dx = \frac{\sin(\pi ua)}{\pi u} \\ &= a\,\operatorname{sinc}(au) \qquad \text{(B-7)}\end{aligned}$$

which has been plotted in Fig. B.1b. We next consider a Gaussian function

$$f(x) = \frac{1}{\sigma(2\pi)^{1/2}}\exp\left(-\frac{x^2}{2\sigma^2}\right) \qquad \text{(B-8)}$$

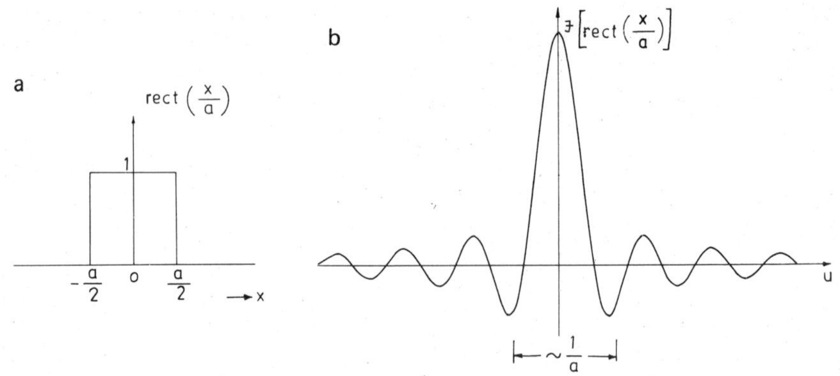

Fig. B.1. (a) The rectangle function. (b) The Fourier transform of the rectangle function.

Its Fourier transform will be

$$F(u) = \mathscr{F}[f(x)] = \frac{1}{\sigma(2\pi)^{1/2}} \int_{-\infty}^{\infty} \exp\left(-\frac{x^2}{2\sigma^2}\right) e^{2\pi i u x}\, dx$$
$$= \exp(-2\pi^2 u^2 \sigma^2) \tag{B-9}$$

Thus, the Fourier transform of a Gaussian is again a Gaussian function and if the original function has a width $\sim\sigma$, then its Fourier transform will have a width $\sim 1/\sigma$.

The Fourier transform of a delta function will be

$$\mathscr{F}[\delta(x-a)] = \int_{-\infty}^{\infty} \delta(x-a)\, e^{2\pi i u x}\, dx = e^{2\pi i a u} \tag{B-10}$$

Thus, the Fourier transform of the function $\tau \sum_{m=-N}^{N} \delta(x - m\tau)$ will be

$$\mathscr{F}\left[\tau \sum_{m=-N}^{N} \delta(x - m\tau)\right] = \tau \sum_{m=-N}^{N} e^{2\pi i u m \tau} = \tau e^{-2\pi i u N \tau}\left[\frac{1 - e^{2\pi i u \tau(2N+1)}}{1 - e^{2\pi i u \tau}}\right]$$
$$= \tau \frac{\sin[\pi u \tau(2N+1)]}{\sin \pi u \tau} \tag{B-11}$$

which is a periodic function with a period $1/\tau$. If we restrict ourselves to the region $-1/2\tau < u < 1/2\tau$, then for a large value of N the above function is sharply peaked around $u = 0$ and we may write [see Eq. (A-7)]

$$\lim_{N\to\infty} \frac{\sin[\pi u \tau(2N+1)]}{\sin \pi u \tau} = \lim_{g\to\infty} \frac{\sin g u \tau}{\pi u \tau} = \delta(u\tau) \tag{B-12}$$

Since $\text{comb}(x/\tau) = \tau \sum_{m=-\infty}^{\infty} \delta(x - m\tau)$, we obtain

$$\mathscr{F}\left[\text{comb}\left(\frac{x}{\tau}\right)\right] = \tau \sum_{m=-\infty}^{\infty} \delta(u\tau - m) = \tau\, \text{comb}(u\tau) \qquad \text{(B-13)}$$

Convolution Theorem

The convolution of two functions $f(x)$ and $g(x)$ is defined by the relation

$$f(x) * g(x) = \int_{-\infty}^{\infty} f(x')\, g(x - x')\, dx' = g(x) * f(x) \qquad \text{(B-14)}$$

Thus

$$\mathscr{F}[f(x) * g(x)] = \int_{-\infty}^{\infty} \int_{-\infty}^{\infty} f(x')\, g(x - x')\, e^{2\pi i u x}\, dx\, dx'$$

$$= \int_{-\infty}^{\infty} dx' f(x')\, e^{2\pi i u x'} \int_{-\infty}^{\infty} g(x - x')\, e^{2\pi i u(x - x')}\, dx$$

$$= F(u)\, G(u) \qquad \text{(B-15)}$$

i.e., the Fourier transform of the convolution of two functions is the product of their Fourier transforms. Similarly,

$$\mathscr{F}[f(x)\, g(x)] = \int f(x)\, g(x)\, e^{2\pi i u x}\, dx = \int f(x) \int G(u')\, e^{2\pi i (u - u')x}\, du'\, dx$$

$$= \int G(u')\, F(u - u')\, du' = F(u) * G(u) \qquad \text{(B-16)}$$

i.e., the Fourier transform of the product of two functions is the convolution of their Fourier transforms. We now consider some examples. The convolution of a function $f(x)$ with a delta function $\delta(x - a)$ is

$$f(x) * \delta(x - a) = \int_{-\infty}^{\infty} f(x')\, \delta(x - x' - a)\, dx' = f(x - a) \qquad \text{(B-17)}$$

As another example, we consider the convolution of two rectangle functions:

$$\mathrm{rect}\left(\frac{x}{a}\right) * \mathrm{rect}\left(\frac{x}{a}\right) = \int_{-\infty}^{\infty} \mathrm{rect}\left(\frac{x'}{a}\right)\mathrm{rect}\left(\frac{x-x'}{a}\right)dx'$$

$$= \int_{-a/2}^{a/2} \mathrm{rect}\left(\frac{x-x'}{a}\right)dx'$$

$$= \begin{cases} 0 & \text{for } |x| > a \\ \int_{x-a/2}^{a/2} dx = a - x & \text{for } 0 < x < a \\ \int_{-a/2}^{x+a/2} dx = a + x & \text{for } -a < x < 0 \end{cases} \tag{B-18}$$

which represents the triangle function. Similarly the convolution between two Gaussian functions will again be another Gaussian with a different width:

$$\exp\left(-\frac{x^2}{\alpha^2}\right) * \exp\left(-\frac{x^2}{\beta^2}\right) = \frac{\alpha\beta\pi^{1/2}}{(\alpha^2+\beta^2)^{1/2}} \exp\left(-\frac{x^2}{\alpha^2+\beta^2}\right) \tag{B-19}$$

Parseval's Theorem

$$\int_{-\infty}^{\infty} |f(x)|^2\,dx = \int_{-\infty}^{\infty} dx \int_{-\infty}^{\infty} du F(u)\, e^{-2\pi iux} \int_{-\infty}^{\infty} du' F^*(u')\, e^{2\pi iu'x}$$

$$= \int_{-\infty}^{\infty} du \int_{-\infty}^{\infty} du' F(u)\, F^*(u')\, \delta(u-u')$$

$$= \int_{-\infty}^{\infty} |F(u)|^2\, du \tag{B-20}$$

C. Solution of Equation (10.2-12)

In this appendix we will obtain bounded solutions of the equation

$$\frac{d^2X}{d\xi^2} + (\mu - \xi^2)\,X = 0 \tag{C-1}$$

If we make the substitution

$$X(\xi) = \exp(-\xi^2/2)\,u(\xi) \tag{C-2}$$

then $u(\xi)$ can be shown to satisfy the equation

$$\frac{d^2u}{d\xi^2} - 2\xi\frac{du}{d\xi} + (\mu - 1)\,u(\xi) = 0 \tag{C-3}$$

Equation (C-3) is known as the Hermite equation. We seek a solution of this equation in the form of a power series,

$$u = \xi^s(a_0 + a_1\xi + a_2\xi^2 + \cdots) = \sum_{r=0}^{\infty} a_r\xi^{r+s} \tag{C-4}$$

with $a_0 \neq 0$ and $s \geq 0$. If we substitute this solution into Eq. (C-3) we obtain

$$\sum[(r+s)(r+s-1)\,a_r\xi^{r+s-2} - 2(r+s)\,a_r\xi^{r+s} + (\mu-1)\,a_r\xi^{r+s}] = 0 \tag{C-5}$$

Since Eq. (C-5) has to be valid for all values of ξ, the coefficients of powers of ξ must be equated to zero; this gives

$$s(s-1)\,a_0 = 0 \tag{C-6}$$

$$s(s+1)\,a_1 = 0 \tag{C-7}$$

and

$$(r+s+2)(r+s+1)\,a_{r+2} = (2r+2s+1-\mu)\,a_r, \qquad r = 0, 1, 2, \ldots \tag{C-8}$$

Since $a_0 \neq 0$, s must be either 0 or 1. Furthermore,

$$\frac{a_{r+2}}{a_r} = \frac{2r+2s+1-\mu}{(r+s+2)(r+s+1)} \to \frac{2}{r} \quad \text{for large } r \tag{C-9}$$

This ratio is the same as that of the coefficients of ξ^{r+2} and ξ^r in the expression of e^{ξ^2}. Hence, if the infinite series in Eq. (C-4) is not terminated it will behave as e^{ξ^2} for large ξ, which will imply that $X(\xi)$ will behave as $e^{\xi^2/2}$ for $\xi \to \pm\infty$. Thus the fields will become unbounded, which will not correspond to a guided mode. Consequently, for the fields to be bounded for $\xi \to \pm\infty$ the series must be terminated, which can only happen if μ is an odd integer:

$$\mu = 2n + 1, \qquad n = 0, 1, 2, \ldots \tag{C-10}$$

Clearly since $a_0 \neq 0$, we must have $a_1 = 0$ (otherwise the series involving a_1, a_3, a_5, etc., will not terminate). Furthermore, for $s = 0$, $u(\xi)$ will be a polynomial in even powers of ξ, and for $s = 1$, $u(\xi)$ will be a polynomial in odd powers of ξ; the degree of the polynomial will be determined by the value of μ. These polynomials are called Hermite polynomials and can easily be shown to be of the form

$$H_0(\xi) = 1, \qquad H_1(\xi) = 2\xi, \qquad H_2(\xi) = 4\xi^2 - 2, \ldots \tag{C-11}$$

Let $X_n(\xi)$ and $X_m(\xi)$ be the fields corresponding to $\mu = 2n + 1$ and $(2m + 1)$, respectively. Then

$$\frac{d^2 X_n}{d\xi^2} + [2n + 1 - \xi^2]\, X_n = 0 \tag{C-12}$$

$$\frac{d^2 X_m}{d\xi^2} + [2m + 1 - \xi^2]\, X_m = 0 \tag{C-13}$$

If we multiply Eq. (C-12) by $X_m(\xi)$ and Eq. (C-13) by $X_n(\xi)$ and subtract, we get

$$\left(X_m \frac{d^2 X_n}{d\xi^2} - X_n \frac{d^2 X_m}{d\xi^2} \right) = 2(m - n)\, X_n X_m$$

or

$$2(m - n) \int_{-\infty}^{\infty} X_n(\xi)\, X_m(\xi)\, d\xi = \int_{-\infty}^{\infty} \frac{d}{d\xi} \left(X_m \frac{dX_n}{d\xi} - X_n \frac{dX_m}{d\xi} \right) d\xi$$

$$= \left[X_m \frac{dX_n}{d\xi} - X_n \frac{dX_m}{d\xi} \right]_{-\infty}^{\infty} = 0 \tag{C-14}$$

Thus if $m \neq n$, then

$$\int_{-\infty}^{\infty} = X_m(\xi)\, X_n(\xi)\, d\xi = 0 \tag{C-15}$$

which is the orthonormality condition. The normalized Hermite–Gauss functions (for $n = 0$, 1, and 2) are plotted in Fig. 10.3.

References

Akhmanov, S. A., Krindach, D. P., Sukhorukov, A. P., and Khokhlov, R. V. (1967), Nonlinear defocusing of laser beams, *JETP Lett.* **6**, 38.

Akhmanov, S. A., Krindach, D. P., Migulin, A. V., Sukhorukov, A. P., and Khokhlov, R. V. (1968*a*), Thermal self-action of laser beams, *IEEE J. Quant. Electr.* **QE-4**, 568.

Akhmanov, S. A., Sukhorukov, A. P., and Khokhlov, R. V. (1968*b*), Self-focusing and diffraction of light in a nonlinear medium, *Sov. Fiz. Usp.* **10**, 609.

Allan, W. B. (1973), *Fiber Optics*, Plenum Press, New York.

Aoki, Y., and Suzuki, M. (1967), Imaging properties of a gas-lens, *IEEE Trans. Micro. Th. Tech.* **MTT-15**, 2.

Barakat, R. (1962), Computation of the transfer function of an optical system from the design data for rotationally symmetric aberrations. Part I: Theory, *J. Opt. Soc. Am.* **52**, 985.

Barakat, R., and Morello, M. V. (1962), Computation of the transfer function of an optical system from the design data for rotationally symmetric aberrations. Part II: Programming and numerical results, *J. Opt. Soc. Am.* **52**, 992.

Barnoski, M. (ed.) (1974), *Introduction to Integrated Optics*, Plenum Press, New York.

Beran, J., and Parrent, G. B., Jr. (1964), *Theory of Partial Coherence*, Prentice–Hall, Englewood Cliffs, New Jersey.

Bergstein, L., and Schachter, H. (1967), On modes of optical resonators with non-planar end reflectors, *in Proc. Symp. Modern Optics* **XVII**, Polytechnic Press, Brooklyn, New York.

Black, P. W. (1976), Fabrication of optical waveguides, *Electrical Commun.* **51**(1), 4.

Born, M., and Wolf, E. (1975), *Principles of Optics*, Pergamon Press, Oxford.

Bouillie, R., Cozannet, A., Steiner, K. H., and Treheux, M. (1974), Ray delay in gradient waveguides with arbitrary symmetric refractive profile, *Appl. Opt.* **13**, 1045.

Boyd, G. D., and Gordon, J. P. (1961), Confocal multimode resonator for millimeter through optical wavelength masers, *Bell Syst. Tech. J.* **40**, 489.

Breazeale, M. A., Adler, L., and Flax, L. (1974), Reflection of a Gaussian ultrasonic beam from a liquid–solid interface, *J. Acoust. Soc. Am.* **56**, 866.

Brooks, R. E., Heflinger, L. O., and Weurker, R. F. (1966). 9A9-Pulsed laser holograms, *IEEE J. Quant. Electr.* **QE-2**, 275.

Bryngdahl, O. (1973), Evanescent waves in optical imaging, in: *Progress in Optics* (E. Wolf, ed.), Vol. XI, p. 169, North Holland Publ. Co., Amsterdam.

Buchdahl, H. A. (1970), *An Introduction to Hamiltonian Optics*, Cambridge Univ. Press, London and New York.

Cathey, W. T. (1974), *Optical Information Processing and Holography*, Wiley, New York.

Chynoweth, A. G. (1976), Lightwave communications; the fiber lightguide, *Phys. Today* **29**(5), 28.

Collier, R. J., Burckhardt, C. B., and Lin, L. H. (1971), *Optical Holography*, Academic Press, New York.

Conwell, E. (1973), Modes in optical waveguides formed by diffusion, *Appl. Phys. Lett.* **23**, 328.

Cutrona, L. J. (1965), Recent developments in coherent optical technology, in: *Optical and Electro-Optical Information Processing*, Tippet, J. T., Berkovitz, D. A., Clapp, L. C., Koester, C. J., and Vanderbugh, A. Jr. (eds.), MIT Press, Cambridge.

De, M. (1955), The influence of astigmatism on the response function of an optical system, *Proc. Roy. Soc. (London)* **A233**, 91.

De, M., and Nath, B. K. (1958), Response of optical systems suffering from primary coma, *Optik* **15**, 739.

Eichmann, G. (1971), Quasi-geometric optics of media with inhomogeneous index of refraction, *J. Opt. Soc. Am.* **61**, 161.

Erdelyi, A. (1953), *Higher Transcendental Functions*, McGraw–Hill, New York.

Felsen, L. B., and Marcuvitz, N. (1973), *Radiation and Scattering of Waves*, Prentice–Hall, Englewood Cliffs, New Jersey.

Feynman, R. P., Leighton, R. B., and Sands, M. (1965), *The Feynman Lectures on Physics*, Addison–Wesley, Reading, Massachusetts.

Francon, M., and Mallick, S. (1967), Measurement of the second order degree of coherence, in: *Progress in Optics* (E. Wolf, ed.), Vol. VI, p. 73, North-Holland Publ. Co., Amsterdam.

Gabor, D. (1948), A new microscopic principle, *Nature* **161**, 777. [A reprint also appears in *An Introduction to Coherent Optics and Holography*, G. W. Stroke, p. 263, Academic Press, New York.]

Gambling, W. A., and Matsumura, H. (1973), Pulse dispersion in a lenslike medium, *Opt. Electr.* **5**, 429.

Gambling, W. A., Payne, D. N., Hammond, C. R., and Norman, S. R. (1976), Optical fibers based on phosphosilicate glass, *Proc. IEE* **123**, 570.

Garmire, E., Chiao, R. V., and Townes, C. H. (1966), Dynamics of characteristics of the self-trapping of intense light beams, *Phys. Rev. Lett.* **16**, 347.

Gerrard, A., and Burch, J. M. (1975), *An Introduction to Matrix Methods in Optics*, Wiley, New York.

Ghatak, A. K., and Kraus, L. A. (1974), Propagation of waves in a medium varying transverse to the direction of propagation, *IEEE J. Quant. Electr.* **QE-10**, 465.

Ghatak, A. K., and Thyagarajan, K. (1975), Ray and energy propagation in graded index media, *J. Opt. Soc. Am.* **65**, 169.

Ghatak, A. K., Goyal, I. C., and Sodha, M. S. (1972), Series solution for steady state self-focusing with saturating nonlinearity, *Opt. Acta* **19**, 693.

Ghatak, A. K., Malik, D. P. S., and Goyal, I. C. (1973), Electromagnetic wave propagation through a gas lens, *Opt. Acta* **20**, 303.

Ghatak, A. K., Goyal, I. C., and Sharma, A. (1976), Mode conversion in dielectric waveguides, *Opt. Quant. Electr.* **8**, 399.

Gloge, D. (1974), Optical fibers for communication, *Appl. Opt.* **13**, 249.

Gloge, D., and Marcuse, D. (1969), Formal quantum theory of light rays, *J. Opt. Soc. Am.* **59**, 1629.

Goldstein, H. (1950), *Classical Mechanics*, Addison–Wesley, Reading, Massachusetts.

Goodbody, A. M. (1958), The influence of spherical aberration on the response function of an optical system, *Proc. Phys. Soc.* **72**, 411.

Goodman, J. W. (1968), *Introduction to Fourier Optics*, McGraw–Hill, New York.

Goodman, J. W., Huntley, W. H., Jackson, D. W., and Lehmann, M. (1966), Wavefront reconstruction imaging through random media, *Appl. Phys. Lett.* **8**, 311.

Goyal, I. C., Kumar, A., and Ghatak, A. K. (1976), Calculation of bandwidth of optical fibers from experiments on dispersion measurement, *Opt. Quant. Electr.* **8**, 79.

Gradshtein, I. S., and Ryzhik, I. M. (1965), *Tables of Integrals, Series, and Products*, Academic Press, New York.

Gupta, A., Thyagarajan, K., Goyal, I. C., and Ghatak, A. K. (1976), Theory of fifth-order aberration of graded-index media, *J. Opt. Soc. Am.* **66**, 1320.

Harrick, N. J. (1967), *Internal Reflection Spectroscopy*, Wiley (Interscience), New York.
Hecht, N. L., Minardi, J. E., Lewis, D., and Fusek, R. L. (1973), Quantitative theory for predicting fringe pattern formation in holographic interferometry, *Appl. Opt.* **12**, 2665.
Heflinger, L. O., Wuerker, R. F., and Brooks, R. E. (1966), Holographic interferometry, *J. Appl. Phys.* **37**, 642.
Hopkins, H. H. (1955), The frequency response of a defocused optical system, *Proc. Roy. Soc. (London)* **A231**, 91.
Hopkins, H. H. (1957), Geometrical optical treatment of frequency response, *Proc. Phys. Soc.* **B70**, 1162.
Hopkins, H. H. (1962), The application of frequency response techniques in optics (21st Thomas Young Oration), *Proc. Phys. Soc.* **79**, 889.
Jenkins, F. A., and White, H. E. (1957), *Fundamentals of Optics*, McGraw–Hill, New York.
Kaminow, I. P., and Carruthers, J. R. (1973), Optical waveguiding layers in $LiNbO_3$ and $LiTaO_3$, *Appl. Phys. Lett.* **22**, 326.
Kapany, N. S. (1967), *Fiber Optics*, Academic Press, New York.
Kapany, N. S., and Burke, J. J. (1972), *Optical Waveguides*, Academic Press, New York.
Katti, P. K., Singh. K., and Kavathekar, A. K. (1969), Modulation of a general periodic object with triangular transmission profile by an annular aperture using incident incoherent light, *Opt. Acta* **16**, 629.
Khular, E., and Ghatak, A. K. (1976), Pulse dispersion in planar waveguides having exponential variation of refractive index, *Optica Acta* **23**, 957.
Kitano, I., Koizumi, K., Matsumura, H., Uchida, T., and Furukawa, M. (1970). A light focusing fiber guide prepared by ion-exchange techniques, *Japan Soc. Appl. Phys. (Suppl.)* **39**, 63.
Kogelnik, H. (1969), Coupled wave theory for thick hologram gratings, *Bell Syst. Tech. J.* **48**, 2909.
Kogelnik, H., and Li, T. (1966), Laser beams and resonators, *Appl. Opt.* **5**, 1550.
Kogelnik, H., and Weber, H. P. (1974), Rays, stored energy and power flow in dielectric waveguides, *J. Opt. Soc. Am.* **64**, 174.
Koizumi, K., Ikeda, Y., Kitano, I., Furukawa, M., and Sumimoto, T. (1974), New light focusing fibers made by a continuous process, *Appl. Opt.* **13**, 255.
Kottler, F., and Perrin, F. H. (1966), Imagery of one-dimensional patterns, *J. Opt. Soc. Am.* **56**, 377.
Kumar, A., and Ghatak, A. K. (1976), Perturbation theory to study the guided propagation through cladded parabolic index fibers, *Optica Acta* **23**, 413.
Leith, E. N., and Upatneiks, J. (1962), Reconstructed wavefronts and communication theory, *J. Opt. Soc. Am.* **52**, 1123.
Lindberg, P. (1954), Measurement of contrast transmission characteristics in optical image formation, *Opt. Acta* **1**, 80.
Linfoot, E. H. (1964), *Fourier Methods in Optical Image Evaluation*, Focal Press, London.
Linfoot, E. H., and Wolf, E. (1956), Phase-distribution near focus in an aberration-free diffraction image, *Proc. Phys. Soc.* **B69**, 823.
Lipsett, M. S., and Mandel, L. (1963), Coherence time measurements of light from ruby optical masers, *Nature* **199**, 553.
Lotsch, H. K. V. (1968), Reflection and refraction of a beam of light at a plane interface, *J. Opt. Soc. Am.* **58**, 551.
Lotsch, H. K. V. (1970), Beam displacement at total reflection: The Goos–Hänchen effect, Parts I and II, *Optik* **32**, 116, 189.
Lotsch, H. K. V. (1971), Beam displacement at total reflection: The Goos–Hänchen effect, Parts III and IV, *Optik* **32**, 299, 553.
Luneburg, R. K. (1964), *Mathematical Theory of Optics*, California Univ. Press, Berkeley, California.

Marcatili, E. A. J. (1964), Modes in a sequence of thick astigmatic lens-like focusers, *Bell. Syst. Tech. J.* **43**, 2887.

Marchand, E. W. (1973), Gradient index lenses, in: *Progress in Optics* (E. Wolf, ed.), Vol. IX, p. 307. North-Holland Publ. Co., Amsterdam.

Marcuse, D. (1972), *Light Transmission Optics*, Van Nostrand Reinhold, New York.

Marcuse, D. (1973*a*), The effect of the ∇n^2 term on the modes of an optical square law medium, *IEEE J. Quant. Electr.* **QE-9**, 958.

Marcuse, D. (1973*b*), The impulse response of an optical fiber with parabolic index profile, *Bell Syst. Tech. J.* **52**, 1169.

Marcuse, D. (1974), *Theory of Dielectric Optical Waveguides*, Academic Press, New York.

Mazet, A., Huard, S., and Imbert, C. (1971), Effet Goos–Hänchen en lumière non-polarisée: La réflexion totale séparée des états de polarisation rectiligné, *C. R. Acad. Sci. Ser. B.* **B273,** 592.

Meier, R. W. (1965), Magnification and third-order aberrations in holography, *J. Opt. Soc. Am.* **55**, 987.

Midwinter, J. E. (1970), Evanescent field coupling into a thin-film waveguide, *IEEE J. Quant. Electr.* **QE-6,** 583.

Miyamoto, K. (1961), Wave optics and geometrical optics in optical design, in: *Progress in Optics* (E. Wolf, ed.), Vol. I, North-Holland Publ. Co., Amsterdam.

Murata, K. (1966), Instruments for the measuring of optical transfer functions, in: *Progress in Optics* (E. Wolf, ed.), Vol. V, North-Holland Publ. Co., Amsterdam.

Nalimov, I. P. (1969), Applications of holography, in: *An Introduction to Coherent Optics and Holography* (G. W. Stroke, ed.), p. 181. Academic Press, New York.

Panofsky, W. K. H., and Phillips, M. (1962), *Classical Electricity and Magnetism*, Addison–Wesley, Reading, Massachusetts.

Parrent, G. B., and Roman, P. (1960), On the matrix formulation of the theory of partial polarization in terms of observables, *Nuovo Cimento* **15**, 370. (A reprint of this paper appears in *Polarized Light* (W. Swindell, ed.), Dowden, Hutchinson, and Ross, Stroudsburg, Pennsylvania.

Pearson, A. D., French, W. G., and Rawson, E. G. (1969), Preparation of a light focusing glass rod by ion-exchange technique, *Appl. Phys. Lett.* **15**, 76.

Pegis, R. J. (1961), The modern development of Hamiltonian optics, in: *Progress in Optics* (E. Wolf, ed.), Vol. I, North-Holland Publ. Co., Amsterdam.

Phillips, R. A. (1969), Spatial filtering experiments for undergraduate laboratories, *Am. J. Phys.* **37**, 537.

Powell, J. L., and Craseman, B. (1961), *Quantum Mechanics*, Addison–Wesley, Reading, Massachusetts.

Powell, R. L., and Stetson, K. A. (1965), Interferometric vibration analysis by wavefront reconstruction, *J. Opt. Soc. Am.* **55**, 1593.

Preston, K. Jr. (1972), *Coherent Optical Computers,* McGraw–Hill, New York.

Read, A. A., Dagg, L. I. R., and Reesor, G. E. (1972), Microwave phase measurements associated with the Goos–Hänchen shift, *Can. J. Phys.* **50**, 52.

Reitz, J. R., and Milford, F. J. (1962), *Foundations of Electromagnetic Theory*, Addison–Wesley, Reading, Massachusetts.

Renard, R. H. (1964), Total reflection: A new evaluation of the Goos–Hänchen shift, *J. Opt. Soc. Am.* **54**, 1190.

Rosenhauer, K. (1967), Measurement of aberrations and optical transfer functions of optical systems, in: *Advanced Optical Techniques* (A. C. S. Van Heel, ed.), North-Holland Publ. Co., Amsterdam.

Saito, H., Yamaguchi, I., and Nakajima, T. (1971), Applications of holographic interferometry to mechanical experiments, in: *Applications of Holography* (E. S. Barrekette, W. E. Kock, T. Ose, J. Tsujiuchi, and G. W. Stroke, eds.), p. 105, Plenum Press, New York.

Schaefer, C., and Pich, R. (1937), *Ann. Physik.* **30**, 245.

Seigman, A. E. (1971), *An Introduction to Lasers and Masers*, McGraw–Hill, New York.

Shulman, A. R. (1970), *Optical Data Processing*, Wiley, New York.

Singh, K. (1973), Vibration and motion analysis using classical interferometry, holography and laser speckling, in: *Optical Behaviour of Materials* (K. L. Chopra, ed.), Thomson Press (India), Delhi.

Slepian, D., and Pollack, H. O. (1961), Prolate spheroidal wave functions—Fourier analysis and uncertainty—I, *Bell Syst. Tech. J.* **40**, 43.

Smith, H. M. (1975), *Principles of Holography*, Wiley, New York.

Snyder, A. W., and Mitchell, D. J. (1974), Generalised Fresnel's laws for determining radiation loss from optical waveguides and curved dielectric structures, *Optik* **40**, 438.

Sodha, M. S. (1973), Theory of nonlinear refraction: self-focusing of laser beams, *J. Phys. Educ. (India)* **1**, 13.

Sodha, M. S., and Ghatak, A. K. (1977), *Inhomogeneous Optical Waveguides*, Plenum Press, New York.

Sodha, M. S., Ghatak, A. K., and Malik, D. P. S. (1971*a*), Electromagnetic wave propagation in radially and axially non-uniform media: Geometrical optics approximation, *J. Opt. Soc. Am.* **61**, 1492.

Sodha, M. S., Ghatak, A. K., and Malik, D. P. S. (1971*b*), Electromagnetic wave propagation in a radially and axially non-uniform medium: Optics of SELFOC fibers and rods; wave optics considerations, *J. Phys. D: Appl. Phys.* **4**, 1887.

Sodha, M. S., Ghatak, A. K., and Tripathi, V. K. (1974), Self-focusing of laser beams in dielectrics, plasmas and semiconductors, Tata McGraw–Hill, New Delhi.

Sodha, M. S., Ghatak, A. K., and Tripathi, V. K. (1976), Self-focusing of laser beams in plasmas and semiconductors (a review), in: *Progress in Optics* (E. Wolf, ed.), Vol. XIII, p. 169, North-Holland Publ. Co., Amsterdam.

Stroke, G. W. (1965), Lensless Fourier-transform method for optical holography, *Appl. Phys. Lett.* **6**, 201.

Stroke, G. W. (1969), *An Introduction to Coherent Optics and Holography*, Academic Press, New York.

Svelto, O. (1974), Self-focusing, self-trapping and self-phase modulation of laser beams, in: *Progress in Optics* (E. Wolf, ed.), Vol. XII, p. 3, North-Holland Publ. Co., Amsterdam.

Thompson, B. J. (1958), Illustration of the phase change in two-beam interference with partially coherent light, *J. Opt. Soc. Am.* **48**, 95.

Thompson, B. J. (1971), Applications of holography, in: *Laser Applications* (M. Ross, ed.), Vol. I, Academic Press, New York.

Thompson, B. J., and Wolf, E. (1957), Two beam interference with partially coherent light, *J. Opt. Soc. Am.* **47**, 895.

Thompson, B. J., Ward, J. H., and Zinky, W. (1967), Application of hologram techniques for particle size analysis, *Appl. Opt.* **6**, 519.

Thyagarajan, K., and Ghatak, A. K. (1974), Perturbation theory for studying the effect of the $\nabla\varepsilon$ term in lens-like media, *Opt. Comm.* **11**, 417.

Thyagarajan, K., and Ghatak, A. K. (1976), Hamiltonian theory of third-order aberration of inhomogeneous lenses, *Optik* **44**, 329.

Thyagarajan, K., Rohra, A., and Ghatak, A. K. (1976), Third-order aberration of conical Selfoc fibres, *J. Opt. (India)* **5**, 27.

Tippet, J. T., Berkovitz, D. A., Clapp, L. C., Koester, C. J., and Vanderbugh, A., Jr. (eds.) (1965), *Optical and Electro-Optical Information Processing*, MIT Press, Cambridge.

Titchmarsh, E. C. (1948), *Introduction to the Theory of Fourier Integrals*, Clarendon Press, Oxford.

Tsujiuchi, J. (1971), Correction of optical images by compensation of aberrations and by spatial frequency filtering, in: *Progress in Optics* (E. Wolf, ed.), Vol. II, p. 133, North-Holland Publ. Co., Amsterdam.

Tsujiuchi, J., Matsuda, K., and Takeya, N. (1971), Correlation techniques by holography and its application to fingerprint identification, in: *Applications of Holography* (E. S. Barrekette, W. E. Kock, T. Ose, J. Tsujiuchi, and G. W. Stroke, eds.), p. 247, Plenum Press, New York.

Uchida, T., Furukawa, M., Kitano, I., Koizumi, K., and Matsumura, H. (1970), Optical characteristics of a light focusing fiber guide and its applications, *IEEE J. Quant. Electr.* **QE-6**, 606.

Van der Lugt, A. (1964), Signal detection by complex spatial filtering, *IEEE Trans. Inform. Theory* **10**, 139.

Van der Lugt, A. (1968), A review of optical data processing techniques, *Opt. Acta* **15**, 1.

Wagner, W. G., Haus, H. A., and Marburger, J. H. (1968), Large scale self-trapping of optical beams in the paraxial ray approximation, *Phys. Rev.* **175**, 256.

Washer, F. E., and Rosberry, F. W. (1951), New resolving power test chart, *J. Opt. Soc. Am.* **41**, 597.

Wolf, E. (1959), Coherence properties of partially polarized electromagnetic radiation, *Nuovo Cimento* **13**, 1165. (A reprint of this paper appears in *Polarized Light* (W. Swindell, ed.), Dowden, Hutchinson, and Ross, Stroudsburg, Pennsylvania.

Wuerker, R. F. (1971), Experimental aspects of holographic interferometry, in: *Applications of Holography* (E. S. Barrekette, W. E. Kock, T. Ose, J. Tsujiuchi, and G. W. Stroke, eds.), p. 127, Plenum Press, New York.

Index